# CATALYSIS AND
# SURFACE SCIENCE

# CHEMICAL INDUSTRIES

*A Series of Reference Books and Textbooks*

*Consulting Editor*
## HEINZ HEINEMANN
*Heinz Heinemann, Inc.,*
*Berkeley, California*

Additional Volumes in Preparation

# CATALYSIS AND SURFACE SCIENCE

## Developments in

- *Chemicals from Methanol*
- *Hydrotreating of Hydrocarbons*
- *Catalyst Preparation*
- *Monomers and Polymers*
- *Photocatalysis and Photovoltaics*

edited by

## Heinz Heinemann

*Lawrence Berkeley Laboratory*
*and Department of Chemical Engineering*
*University of California*
*Berkeley, California*

## Gabor A. Somorjai

*Lawrence Berkeley Laboratory*
*and Department of Chemistry*
*University of California*
*Berkeley, California*

CRC Press
Taylor & Francis Group
Boca Raton London New York

CRC Press is an imprint of the
Taylor & Francis Group, an **informa** business

# Preface

This book contains the papers presented at the Second Berkeley Catalysis and Surface Science Conference, held in January 1984 at the Lawrence Berkeley Laboratory, of the University of California. The Conference was supported by the U.S. Department of Energy and by a group of industrial organizations. It was cochaired by Gabor A. Somorjai and Heinz Heinemann, both of Lawrence Berkeley Laboratory and the University of California, Berkeley.

Like the First Berkeley Catalysis and Surface Science Conference in 1980, this meeting was designed to provide information on five major subjects in the field of catalysis, exposed from the viewpoints of basic research, exploratory and process research, and technology.

The papers presented were reviewed to assess the state of the art, rather than to provide new and unpublished information. The subjects treated and the topics published in this book are: (1) chemicals from methanol, (2) hydrotreating of hydrocarbons, (3) catalyst preparation, (4) monomers and polymers, and (5) photocatalysis and photovoltaics.

The first Conference in 1980 covered: (1) ammonia synthesis, (2) homogeneous catalysis by transition metals, (3) ethylene oxidation, (4) hydrogenation of carbon monoxide, and (5) zeolite catalysis. All of these papers were published in Catalysis Reviews — Science and Engineering, Vol. 23 (1 and 2) 1981. The two volumes that contain the papers from the First and Second Berkeley Catalysis and Surface Science Conferences thus constitute a review of perhaps the ten most important fields in the area of catalysis, which are of major industrial significance at the present time.

A principal aim of these continuing Conferences is to bring together workers who are engaged in research or development or who guide the technology in a given specific area of catalysis or surface science. As many of the chemical technologies become science driven high technology, a more intimate interaction between researchers and technologists becomes essential. Since most conferences address only one of these concerns, the editors felt the need for a gathering of researchers and technologists working in the same subfields for information exchange.

It is the editors' belief that the present book, along with the previously published proceedings of the First Conference, provide a significant overview of research and development trends in areas of large commercial importance. By tracing what is going on in fundamental work, as well as in process research and by combining this with the current state of commercial technology trends, the future directions of the catalytic industry can be deduced.

Heinz Heinemann
Gabor A. Somorjai

# Contents

# Contributors

P. ANDREU  Intevep S.A., Caracas, Venezuela

GEORGE C. BARILE  Mobil Chemical Company, Princeton, New Jersey

W. H. CALKINS  Central Research and Development Department, E. I. du Pont de Nemours & Co., Wilmington, Delaware

CLARENCE D. CHANG  Central Research Division, Mobil Research and Development Corporation, Princeton, New Jersey

R. R. CHIANELLI  Corporate Research Laboratories, Exxon Research and Engineering Co., Annandale, New Jersey

JAMES C. W. CHIEN  Department of Chemistry, Department of Polymer Science and Engineering, Materials Research Laboratories, University of Massachusetts, Amherst, Massachusetts

BJERNE S. CLAUSEN  Haldor Topsøe Research Laboratories, Lyngby, Denmark

F. M. DAUTZENBERG  Catalytica Associates, Inc., Mountain View, California

J. C. DE DEKEN  Catalytica Associates, Inc., Mountain View, California

E. K. DIENES  United Catalysts Inc., Louisville, Kentucky

R. GALIASSO  Intevep S.A., Caracas, Venezuela

W. GARCIA  Intevep S.A., Caracas, Venezuela

A. L. HAUSBERGER  United Catalysts Inc., Louisville, Kentucky

L. L. HEGEDUS   W.R. Grace & Co., Columbia, Maryland

ADAM HELLER   AT&T Bell Laboratories, Murray Hill, New Jersey

M. HENDEWERK   Materials and Molecular Research Division,
Lawrence Berkeley Laboratory; and Department of Chemistry,
University of California, Berkeley, Berkeley, California

H. L. HSIEH   Research and Development, Phillips Petroleum Company, Bartlesville, Oklahoma

P. JERUS   United Catalysts Inc., Louisville, Kentucky

WARREN W. KAEDING   Mobil Chemical Company, Princeton, New
Jersey

FREDERICK J. KAROL   UNIPOL Systems Department, Union Carbide Corporation, Bound Brook, New Jersey

W. M. KEELY   United Catalysts Inc., Louisville, Kentucky

G. KIM   W.R. Grace & Co., Columbia, Maryland

JAMES M. MASELLI   Davison Chemical Division, W.R. Grace & Co.,
Columbia, Maryland

C. J. PEREIRA   W.R. Grace & Co., Columbia, Maryland

ALAN W. PETERS   Davison Chemical Division, W.R. Grace & Co.,
Columbia, Maryland

M. M. RAMIREZ DE AGUDELO   Intevep S.A., Caracas, Venezuela

GABOR A. SOMORJAI   Materials and Molecular Research Division,
Lawrence Berkeley Laboratory; and Department of Chemistry,
University of California, Berkeley, Berkeley, California

HENRIK TOPSØE   Haldor Topsøe Research Laboratories, Lyngby,
Denmark

J. E. TURNER   Materials and Molecular Research Division, Lawrence Berkeley Laboratory; and Department of Chemistry, University of California, Berkeley, Berkeley, California

F. V. WALD   Mobil Solar Energy Corporation, Waltham, Massachusetts

IRVING WENDER   Department of Chemical and Petroleum Engineering, University of Pittsburgh, Pittsburgh, Pennsylvania

MARGARET M. WU   Mobil Chemical Company, Princeton, New Jersey

# CATALYSIS AND
# SURFACE SCIENCE

# Chemicals from Methanol

# Chemicals from Methanol

IRVING WENDER
Department of Chemical and Petroleum Engineering
University of Pittsburgh
Pittsburgh, Pennsylvania

For good reasons, there has been a spate of articles, much
thinking, and a fair amount of industrial action on making chemicals and fuels from methanol. Only methane and methanol are
made commercially in over 99% yields from synthesis gas.

Methanol, of course, is much more reactive than methane.
Methanol has special chemical properties, some due to the presence of the −OH group, some due to the absence of steric hin-
drance of the methyl group, and others a consequence of the fact
that, unlike any other alcohol, the −CH$_2$OH group is bound to a
hydrogen atom rather than to another carbon atom. Methanol
promises to be the raw material that may displace ethylene and
other petrochemical feedstocks from chemical syntheses. The
trend in processes for the synthesis of chemicals is shifting
away from high energy and usually expensive intermediates
toward lower energy, more available and secure materials such
as methanol and synthesis gas.

Some chemical reactions of methanol are outlined in Fig. 1; a
list of some of methanol's unusual properties follows:

It is commercially available at a reasonable price in high purity.
It is a stronger acid than any other alcohol but a somewhat weaker
    acid than water. The alkoxide ion, CH$_3$O$^-$, is thus a stronger
    base than the hydroxide (OH$^-$) ion
The carbon-oxygen bond in methanol is the strongest C−O bond
    of any alcohol, with only ethanol having a comparable bond
    energy
The presence of the electronegative oxygen atom makes methanol
    itself a weak base and a weak nucleophile
The hydroxide (OH) group is difficult to displace from methanol
    (OH is a poor leaving group in nucleophilic reactions). The
    CH$_3^+$ ion and CH$_3$· radical are the least stable of the corres-
    ponding alkyl species
The C−O bond in methanol, however, is easily broken in the
    presence of acids, especially hydrogen halides (HI > HBr >
    HCl). The reaction with HI proceeds rapidly because water
    is easily displaced from the protonated alcohol by the weakly
    basic halide ion which is both an excellent nucleophile and
    leaving group
Methanol is especially susceptible to nucleophilic attack in the
    presence of suitable acids because of the lack of steric hin-
    drance to such attack. Methanol is over 100 times more reac-
    tive than ethanol in this type of reaction
Methanol and transition metal complexes often react in the pres-
    ence of HI, I$_2$, or I$^-$ to form oxygenated chemicals. There is

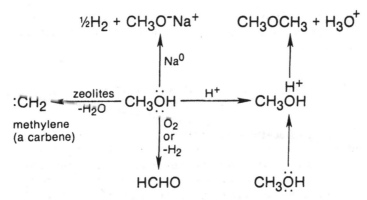

FIG. 1.  Some reactions of methanol.

much evidence that methanol reacts with HI to form $CH_3I$ as the
reactive intermediate in many syntheses

Methanol, especially in the presence of an iodide promoter, readily
forms $CH_3-M$ bonds with transition metal complexes.  The $CH_3$
group, when present in metal carbonyl complexes, readily
undergoes a migratory insertion reaction to form acyl-metal
$$(CH_3\overset{\text{O}}{\overset{\|}{C}}-M) \text{ bonds}$$
About half the methanol produced has historically been converted
to formaldehyde (HCHO), the simplest, most reactive aldehyde,
which has a diverse and fascinating chemistry of its own

Under certain conditions, usually in the presence of zeolites (i.e.,
ZSM-5), methanol may undergo reactions that correspond to
the loss of water to form a carbene (methylene in this case)

This article will be concerned chiefly with the synthesis of
oxygenated chemicals of mostly, but not exclusively, two carbon
oxygenated compounds. The only oxygenated chemicals now made
commercially from methanol are formaldehyde, acetic acid, methyl
acetate, and methyl formate.  The Tennessee Eastman plant for
the production of 500 million pounds per year of acetic anhydride
from 900 tons per day of high sulfur bituminous coal will be
in operation in 1984.  Methanol may be the source of emerging,
hopefully improved syntheses of acetic anhydride, acetaldehyde,
ethanol, ethyl acetate, vinyl acetate, and ethylene glycol.  The
routes to these chemicals usually involve either the catalyzed

carbonylation, reductive carbonylation, or oxidative carbonylation of methanol or formaldehyde by transition metal complexes or catalysis by bases.

## I. ACETIC ACID (AcOH) BY CARBONYLATION OF METHANOL

The history of acetic acid manufacture is a classic example of how raw material availability and changing technology have evolved over the years [1]. It also furnishes clues as to coming changes in the chemistry of the synthesis of other chemicals. Acetic acid and many other organic chemicals were originally produced by fermentation processes. The mercuric-ion-catalyzed hydrolysis of acetylene to acetaldehyde was the basis for the first major synthetic process for acetic acid production, lasting for some 40 years until the late 1950s. British Petroleum and Celanese then introduced the oxidation of short-chain paraffins using cobalt or manganese salts as free radical catalysts. Between 1957 and 1959, Wacker Chemie developed a process for HOAc synthesis by the Pd-Cu-catalyzed oxidative hydration of ethylene to acetaldehyde (AcH). (Ac denotes $CH_3\overset{\overset{\displaystyle O}{\|}}{C}-$.) The direct synthesis of acetic acid by the carbonylation of methanol using transition metal catalysts was first described by BASF in 1965; the carbonylation was carried out at high pressures using an iodide-promoted cobalt catalyst [2]. In 1968, Paulik and Roth of the Monsanto Co. announced the discovery of a low-pressure carbonylation of methanol using iodide-promoted rhodium or iridium catalysts [3]; a large plant based on this process was started up in 1970.

The iodide-promoted rhodium-catalyzed carbonylation of methanol to acetic acid,

$$CH_3OH + CO \xrightarrow[HI]{Rh\ complex} CH_3COOH$$

is a considerably more active system than either the cobalt or iridium catalytic system. A comparison of the cobalt- and rhodium-catalyzed reactions is given in Table 1 [4].

Essentially all new plants will use the rhodium process which, at present, has shutdown economics. But even this process is under attack, for research on a low-pressure methanol carbonylation based on cheaper nickel iodide and other catalysts is underway [5].

## TABLE 1

Comparison of Cobalt- and Rhodium-Catalyzed
Acetic Acid Synthesis[a]

|  | Cobalt | Rhodium |
|---|---|---|
| Metal concentration | $\sim 10^{-1}$ $\underline{M}$ | $\sim 10^{-3}$ $\underline{M}$ |
| Iodide concentration | $\sim 10^{-1}$ $\underline{M}$ | $\sim 10^{-1}$ $\underline{M}$ |
| Reaction temperature | 250°C | $\sim 180$°C |
| Reaction pressure | 715 atm | 30-40 atm |
| Selectivity (on $CH_3OH$) | 90% | >99% |
| Hydrogenation by-products | $CH_4$, $CH_3CHO$, $C_2H_5OH$ | None |
| Other by-products | $C_2H_5CO_2H$, alkyl acetates, 2-ethylbutanol | None |

[a]D. Forster, Adv. Organomet. Chem., 17, 256 (1979). Chemical Engineering, 148 (May 19, 1969). British Patent 1,450,993 (January 1974), assigned to BASF A.G. (Reproduced with permission from Hydrocarbon Processing, November 1982.)

The rhodium-catalyzed carbonylation of methanol to acetic acid is unusual in that the concentrations of reactants and products have no kinetic influences. The reaction is first order in iodide promoter (HI, $I_2$, or $CH_3I$) and in rhodium concentrations. The rate-determining step involves the oxidative addition of $CH_3I$ to an Rh(I) species. This anionic $[Rh(CO)_2I_2]^-$ species is of enhanced nucleophilicity toward $CH_3I$ (relative to a neutral Rh(I) complex) [6]. The role of iodide is clearly related to the formation of $CH_3I$ from methanol. The key metal-carbon bond is then formed by reacting $[Rh(CO_2)I_2]^-$ with $CH_3I$ to generate a methylrhodium(III) species. The $CH_3COI$ is then hydrolyzed to acetic acid. Forster has obtained evidence for the reaction pathway shown in Fig. 2 [6]. It is interesting that three elements, Co, Rh, and Ir, all in the same row of the periodic table, are capable of giving fast reaction rates for the carbonylation of methanol to acetic acid. Offhand, one might expect the syntheses to have similar mechanisms and sensitivities to reaction variables, but it is really no surprise that different sensitivities are observed with respect to variables such as iodide concentration, iodide form, and CO pressure [7].

One difference between the rhodium and cobalt systems is that the rhodium system is insensitive to hydrogen while the cobalt

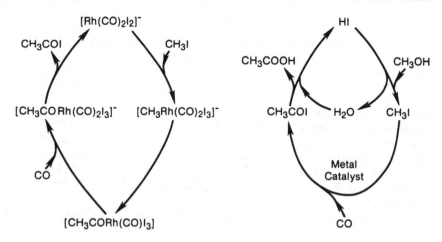

FIG. 2.    Cycles for catalytic action of rhodium and iodine in acetic acid synthesis [7].

system yields significant amounts of hydrogenated by-products. This is not unexpected in view of the well-known reaction of methanol with synthesis gas to give acetaldehyde and ethanol [8].

The cobalt-catalyzed reactions are complicated by reactions of key intermediates, $HCo(CO)_4$ and $Co_2(CO)_8$, with polar solvents and with acids:

$$Co_2(CO)_8 + 4HX \rightarrow 2CoX_2 + 2H_2 + 8CO, \quad X = I^- \text{ or } OAc^-$$

Mizoroki and Nakayama [9] found that Co(II) iodide complexes were the predominant cobalt species during the reactions. The high pressures of CO used in the cobalt-catalyzed reaction are almost certainly needed to generate significant amounts of the active cobalt carbonyl species rather than being involved in aiding the insertion of CO into the $CH_3-Co$ bond.

The iridium-catalyzed reaction is considerably more complicated mechanistically. Complex kinetic interactions were found among solvent, water, CO pressure, and the form of the iodide [10].

## II.   ACETIC ANHYDRIDE FROM METHANOL

Currently, acetic anhydride, $CH_3\overset{\text{O}}{\overset{\|}{C}}O\overset{\text{O}}{\overset{\|}{C}}CH_3$ or $Ac_2O$, is made mostly from ketene derived from acetic acid which, until a short

time ago, was made from ethylene via the Wacker process [11].
Eastman's upcoming process is outlined as follows:

$$CH_3OH + CH_3COOH \overset{H^+}{\rightleftharpoons} CH_3\overset{\overset{\displaystyle O}{\|}}{C}OCH_3 + H_2O$$

methyl
acetate

$$CH_3\overset{\overset{\displaystyle O}{\|}}{C}OCH_3 + CO \xrightarrow[80 \text{ atm}]{\underset{160°C}{Rh-I}} (CH_3\overset{\overset{\displaystyle O}{\|}}{C})_2O$$

$$(CH_3\overset{\overset{\displaystyle O}{\|}}{C})_2O + \text{cellulose} \rightarrow \text{cellulose acetate} + CH_3COOH$$

(recycled)

The only reaction in this sequence that is not already commercial is the reaction of methyl acetate with CO to form $Ac_2O$. Surprisingly, the carbonylation of methyl acetate to $Ac_2O$ was carried out by Reppe as far back as 1956; he used $CoI_2$ or $CoBr_2$ as
catalysts [12]. As in the carbonylation of methanol to acetic acid,
recent work on the conversion of methyl acetate to acetic acid has
used rhodium iodide-based catalysts. A large number of patents
have been issued to different companies for the conversion of the
methyl acetate to $Ac_2O$ using a variety of catalysts, including the
use of cheaper catalysts such as promoted nickel iodide. It is
probable that Eastman will use a catalyst based on $RhCl_3$ with
$CH_3I$ and β-picoline as promoters.

In 1980, Halcon SD announced a process for the manufacture of
acetyl compounds which replaced ethylene with CO and made unnecessary the separate synthesis of acetic acid [13]. This is accomplished by reacting 2 moles of methyl acetate with 2 moles of CO to
give 2 moles of $Ac_2O$. Half the anhydride is recycled and reacted
with 2 moles of methanol to give 2 moles of methyl acetate and 1
mole of water:

$$2CH_3\overset{\overset{\displaystyle O}{\|}}{C}OCH_3 + 2CO \rightarrow 2CH_3\overset{\overset{\displaystyle O}{\|}}{C}-O-\overset{\overset{\displaystyle O}{\|}}{C}CH_3$$

$$CH_3\overset{\overset{\displaystyle O}{\|}}{C}-O-\overset{\overset{\displaystyle O}{\|}}{C}CH_3 + 2CH_3OH \rightarrow 2CH_3\overset{\overset{\displaystyle O}{\|}}{C}OCH_3 + H_2O$$

$$2CH_3OH + 2CO \rightarrow CH_3\overset{\overset{\displaystyle O}{\|}}{C}O\overset{\overset{\displaystyle O}{\|}}{C}CH_3 + H_2O \quad \text{(net reaction)}$$

Clarification is needed for the mechanism of the conversion of methyl acetate to acetic anhydride with a rhodium complex catalyst in the presence of $CH_3I$. It is possible that $Rh(I)CO$ complexes, in the presence of promoters such as $\beta$-picoline, react with $CH_3I$ in an oxidative addition reaction to give a $CH_3Rh$-(III)CO(I) complex. The acetyl complex then formed undergoes a reductive elimination to release $CH_3COI$ and regenerate an $Rh(I)CO$ complex. The $CH_3COI$ may then react with methyl acetate to give acetic anhydride and regenerate $CH_3I$.

## III. VINYL ACETATE

The process now used commercially for the synthesis of vinyl acetate is based on the vapor-phase acetoxylation of ethylene:

$$CH_3COOH + H_2C{=}CH_2 + \tfrac{1}{2}O_2 \rightarrow CH_3\overset{\displaystyle O}{\overset{\displaystyle \|}{C}}OCH{=}CH_2 + H_2O$$
$$\text{vinyl acetate}$$

Since acetic acid is synthesized from methanol and CO, the process is already based on synthesis gas for 70% of its weight and has high yields and moderate processing costs. Halcon has developed the reductive carbonylation of methyl acetate or dimethyl ether to ethylidene diacetate (EDA) [4]:

$$CH_3COOCH_3 + 3CO + 3H_2 \rightarrow CH_3CH(OCOCH_3)_2 + H_2O$$
$$\text{EDA}$$

The EDA can then be pyrolyzed to give vinyl acetate and acetic acid:

$$CH_3CH(O\overset{\displaystyle O}{\overset{\displaystyle \|}{C}}CH_3)_2 \rightarrow CH_3\overset{\displaystyle O}{\overset{\displaystyle \|}{C}}OCH{=}CH_2 + CH_3COOH$$

The acetic acid can be treated with methanol to yield more methyl acetate.

It is of interest to note that the synthesis of EDA was carried out much earlier by the reaction of acetylene with acetic acid [14]. Its pyrolysis to a mixture of vinyl acetate and acetic anhydride was achieved.

The synthesis of EDA from methyl acetate involves a reductive carbonylation, so that the catalyst must have both carbonylation and reduction capabilities. The catalysts employed resemble those used in the synthesis of $Ac_2O$ from methyl acetate, but the addition of hydrogen (from synthesis gas) shifts the selectivity to EDA. Since the rhodium iodide catalyst system is insensitive to hydrogen, the rhodium catalytic system involved in the EDA synthesis must involve different rhodium species.

Halcon [15] claims the use of rhodium or palladium complexes; a source of iodide ion and a promoter are necessary. The rhodium catalyst is used with methyl iodide as a promoter and in the presence of β-picoline. The addition of CO and $H_2$ is evidently sequential, not simultaneous. The chemical reactions, starting from methyl acetate, may be written

$$CH_3\overset{O}{\overset{\|}{C}}OCH_3 \xrightarrow[\text{(2) } H_2]{\text{(1) CO}} \underset{79\%}{EDA} + \underset{17\%}{Ac_2O} + \underset{4\%}{CH_3CHO}$$

$$EDA \xrightarrow{\Delta} CH_2{=}CHOAc + CH_3COOH$$

Maximization of either $Ac_2O$ or vinyl acetate (via EDA) seems to be a function of the $H_2/CO$ molar ratio in the feed. Higher $H_2$ to CO ratios increase the vinyl acetate yield [16]. The fate of this route to vinyl acetate as a commercial process will not be apparent for some time.

## IV. THE HOMOLOGATION (REDUCTIVE CARBONYLATION) OF METHANOL

The homologation of methanol, consisting of its reaction with synthesis gas usually in the presence of cobalt carbonyl complexes, was looked upon as a possible route to ethanol from methanol [8]:

$$CH_3OH + 2H_2 + CO \xrightarrow{Co_2(CO)_8} CH_3CH_2OH + H_2O$$

But Wender et al. found that many products besides ethanol were produced in this reaction; these included not only alcohols but aldehydes, ethers, acetates, formates, methane, and minor amounts of acetic acid.

Since homologation is defined as a reaction by which a member of a chemical class is converted, by introduction of an additional carbon atom, to the next higher homolog, i.e., $R{-}X \rightarrow R{-}CH_2{-}X \rightarrow R{-}CH_2{-}CH_2{-}X$, etc., the cobalt-catalyzed reaction of methanol with synthesis gas is better termed a reductive carbonylation or a hydrocarbonylation reaction. In the two decades following the

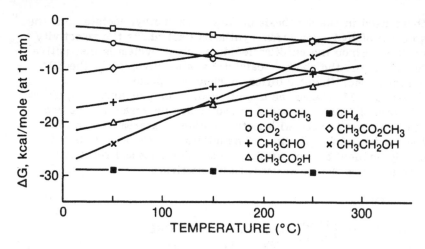

FIG. 3. Change in free energy with temperature for some products obtained in the reductive carbonylation of methanol [18].

discovery of the reductive carbonylation of methanol, it received little attention. In the 1970s, interest was revived as a result of the oil crisis. Several excellent reviews of this work have appeared [17]. A recent comprehensive review has been published by Fakley and Head [18].

The variation with temperature in free energy of some of the principal products from the cobalt-catalyzed reductive carbonylation of methanol are given in Fig. 3 [18]. Although the chief problem is kinetic control over the number of possible reactions, there is significant kinetic control away from hydrocarbons to oxygenated products. Selectivity to ethanol increases at the expense of $CH_3CHO$ with an increase in temperature. It is claimed that lower temperatures (<160°C) give $CH_3CHO$ almost exclusively. Acetaldehyde is obviously the primary product. An increase in total pressure increases methanol conversion while product selectivity does not change much, but high conversions result in more water being produced and the catalyst appears to be poisoned by the water that accumulates.

At high CO partial pressures, competing carbonylation reactions can give greater than 70% selectivity to methyl acetate. With high $H_2$ partial pressures, the catalyst is reduced to metal, and dimethyl ether and methane grow in abundance. Highest ethanol selectivities are obtained at $H_2/CO$ ratios of 1:1 although the stoichiometry to ethanol requires $2H_2:1CO$.

The most significant discovery in the reductive carbonylation of methanol has been the rate increase upon addition of $I_2$ or iodide. This may result, at least in part, from the conversion of methanol to $CH_3I$ which reacts much faster than methanol. Adding $I_2$ to a level of $I/Co = 0.5$ results in a fourfold increase in rate of methanol conversion and about a fourfold increase in catalyst turnover frequency (gram moles of carbonylated product per gram mole of Co per hour) [19].

Homogeneous hydrogenation cocatalysts are usually added to obtain higher yields of ethanol. Ruthenium is the best cocatalyst, converting essentially all $CH_3CHO$ to ethanol. Addition of a tertiary phosphine to the cobalt complex has three important desirable effects. First, the catalyst has greater thermal stability; second, ethanol selectivity is increased although methanol conversion is lowered somewhat; third, more acetaldehyde is hydrogenated to ethanol.

When using a cobalt carbonyl complex in the absence of iodine, the reaction may proceed as follows:

$$HCo(CO)_4 + CH_3OH \rightarrow [CH_3OH_2{}^+Co(CO)_4{}^-] \rightarrow CH_3Co(CO)_4 + H_2O$$

$$CH_3Co(CO)_4 \rightarrow CH_3COCo(CO)_3 \xrightarrow{CO} CH_3COCo(CO)_4$$

The $CH_3COCo(CO)_4$ is hydrogenated to acetaldehyde which is further reduced to ethanol with $HCo(CO)_4$. Attack of methanol on the acyl intermediate yields methyl acetate. Acetic acid is found in very small amounts—it is possibly esterified or decarboxylated. A possible pathway in the presence of $CH_3I$ involves a nucleophilic displacement of $I^-$ by $Co(CO)_4{}^-$, an $S_N2$ reaction [20].

Another plausible route to the acyl complex is oxidative addition of $CH_3I$ to $Co(CO)_4{}^-$ followed by rapid addition of iodide:

$$Co(CO)_4{}^- + CH_3I \xrightarrow{-CO} [CH_3CoI(CO)_3]^- \xrightarrow{CO} CH_3Co(CO)_4 + I^-$$

The cobalt-catalyzed reductive carbonylation of methanol is a complex reaction which achieves fairly good kinetic control over the thermodynamically favored mixture of products. However, selectivities to ethanol of about 75% are rare and ethanol selectivity is usually well below this value. The large mixture of products [21], coupled with low activity, makes the cobalt iodide reductive carbonylation of methanol an unlikely source of chemicals in the near future. A possible application of the product mixture is as a fuel blend [22].

Chen, Feder, and Rathke [23] have discovered a novel method to catalyze the selective conversion of methanol to ethanol at synthesis gas pressures near 300 atm and temperatures about 200°C. The overall reaction is

$$CH_3OH + 2CO + H_2 \rightarrow CH_3CH_2OH + CO_2$$

The reaction is efficient in the use of $H_2$ because $CO_2$ rather than $H_2O$ is produced; the main side reaction ($\sim 30\%$) is the formation of methane. Alcohols higher than ethanol are not formed, nor are acetates or acetals, which is in contrast to the cobalt-catalyzed reaction of methanol with synthesis gas. The reaction proceeds mainly by an $S_N2$ attack of a highly nucleophilic iron carbonyl anion ($HFe(CO)_4^-$) on the methyl group of a quaternary ammonium ion. Both of these reactants are formed catalytically under reaction conditions. $RhI_3$, $Ru_3(CO)_{12}$, $Mn_2(CO)_{10}$, and mixtures of $Mn_2(CO)_{10}$ and $Fe(CO)_5$ have also been found to be active catalysts for this reaction in amine-methanol solutions. This reaction bears further investigation although rates appear somewhat low.

## V. ETHYLENE GLYCOL ($HOCH_2CH_2OH$ OR EG)

The present commercial process for the synthesis of ethylene glycol (EG) involves the reaction of ethylene with oxygen over a silver catalyst. The ethylene oxide so formed is then hydrolyzed to EG. This process is vulnerable to replacement by processes based on synthesis gas or raw materials based on synthesis gas. Some of the routes to EG based on synthesis gas and/or methanol and/or formaldehyde are given in the following paragraphs [4].

The reaction for the direct production of EG from synthesis gas,

$$3H_2 + 2CO \xrightarrow[240°C]{\substack{Rh \text{ or } Ru \\ complexes}} HOCH_2CH_2OH$$
$$\sim 78\%$$

has already been realized, although at pressures too high to be of commercial use. But the raw material usage (pounds of feed per pound of product) is unity and the $H_2/CO$ ratio in the syngas is 1.5, making this route to EG very attractive. The synthesis of EG from syngas would present a case of high added value in the final product [24]. However, the reaction rate for the direct synthesis is too low, pressures are too high, and the production of methanol as a significant (20-40% carbon efficiency with rhodium) by-product have combined to make this route uneconomic at present. But the driving force for the development of a route from syngas to EG is high, and there are a number of new processes under development to achieve this end. At the current stage of development, however, only routes from methanol or from HCHO appear to be viable.

It is probable that, at least for the next decade or so, sequential process technology is better than trying to carry out several steps at once directly from synthesis gas. Methanol is efficiently made from synthesis gas and formaldehyde is easily made from methanol. Indeed, there is evidence that the mechanism for the production of EG directly from synthesis gas involves the synthesis of formaldehyde as a rate-limiting intermediate step [25]. Splitting the synthesis of EG into at least two separate reactions based on formaldehyde or methanol removes many kinetic and thermodynamic constraints to the synthesis of EG. A large research and development program is underway to synthesize EG from formaldehyde (or methanol), and some of the systems are described below.

DuPont operated a commercial process catalyzed by $H_2SO_4$ until 1968 but it required high pressure and temperature and neutralization of the $H_2SO_4$ to recover the product. Chevron has improved this route to EG by using HF as both catalyst and solvent, which results in much lower pressure and ease of catalyst separation and recycle. The acid is esterified and then hydrogenated to EG.

$$HCHO + CO + H_2O \xrightarrow[\text{100 psi}]{\text{HF}} HOCH_2COOH \text{ (hydroxyacetic acid)}$$

Celanese [26] and Redox Technologies [27] have developed an economically attractive process to EG from methanol and HCHO. It has significant raw material advantages over the ethylene oxide process. Methanol and HCHO react in the liquid phase at 125-200°C and 300-600 psig using a free radical initiator; selectivity to EG is high.

$$CH_3OH + HCHO \xrightarrow[\text{125-200°C}]{\text{peroxide}} HOCH_2CH_2OH$$
$$EG$$

## VI.  EG FROM OXALATE ESTERS (OXIDATIVE CARBONYLATION)

In this route to EG, oxidative coupling of methanol and CO (Pd-based) is separated from a water generation reaction (Cu-based) by using nitrous esters, which represents a preoxidation of the alcohol.  The dimethyl oxalate is hydrogenated to EG and methanol.  Ube and Union Carbide have announced plans to co-pilot this process using heterogeneous catalysts [28].  The equation may be written

$$2CH_3OH + \frac{1}{2}O_2 + 2CO \rightarrow \text{dimethyl oxalate} + H_2O$$

The dimethyl oxalate is hydrogenated to ethylene glycol and methyl alcohol.

## VII.  EG BY REACTION OF HCHO WITH SYNGAS

A number of companies have obtained patents using the route

$$HCHO + CO + 2H_2 \rightarrow HOCH_2CH_2OH$$

This reaction proceeds under mild conditions, but potentially serious problems may arise by side reactions of HCHO, i.e., reduction to methanol and the formation of formose (sugarlike) products.

## VIII.  OTHER ROUTES TO EG STARTING WITH FORMALDEHYDE

A newer route, studied by Monsanto [28], is shown by

$$HCHO + CO + H_2 \xrightarrow[\text{95\% conversion}]{\text{Rh-phosphine complex}} HOCH_2CHO \xrightarrow{H_2} EG$$

Another by Exxon [28]:

$$HCHO \xrightarrow[\text{CO} + H_2O,\ 150°C]{\text{solid acid resin}} HOCH_2COOH\ (70\%) \xrightarrow[\text{CH}_3OH]{H_2} EG$$

This last reaction may proceed as follows:

$$HCHO + H^+ \xrightarrow{\text{Nafion B resin}} H_2\overset{+}{C}=OH \xrightarrow{CO} HOCH_2\overset{+}{C}=O$$

$$HOCH_2\overset{+}{C}=O \xrightarrow{H_2O} HOCH_2CHO + H^+$$

## IX. SOME BASE-CATALYZED REACTIONS
## OF METHANOL

Methanol is converted to an alkali methoxide, say $NaOCH_3$, by addition of an alkali metal or its hydride to the alcohol. The methoxide ion is a strong base and nucleophile and it attacks, in this case, the electrophilic CO molecule:

$$CH_3O^- + {}^+C \equiv O^- \rightarrow [CH_3OCO]^-$$

$$[CH_3OCO]^- + CH_3OH \rightarrow H\overset{O}{\overset{\|}{C}}OCH_3 + CH_3O^-$$

$$CH_3OH + CO \xrightarrow{NaOCH_3} H\overset{O}{\overset{\|}{C}}OCH_3$$
$$\text{methyl}$$
$$\text{formate}$$

Formic acid, HCOOH, can be obtained by hydrolysis of methyl formate but, as retroesterification can readily occur, HCOOH is generally made by first synthesizing formamide:

$$H\overset{O}{\overset{\|}{C}}OCH_3 + NH_3 \xrightarrow[\text{5 atm}]{80\text{-}100°C} H\overset{O}{\overset{\|}{C}}NH_2 \text{ (formamide)}$$

Formamide is then hydrolyzed continuously to HCOOH:

$$H\overset{O}{\overset{\|}{C}}NH_2 \xrightarrow[85°C]{70\% \ H_2SO_4} HCOOH + (NH_4)_2SO_4$$

The synthesis of formates, run in excess of methanol, is first order with respect to CO. As expected, methanol reacts more slowly than any other alcohol [29].

It is of more than passing interest that methanol may be produced by a two-stage process involving carbonylation of an alcohol to the formate, followed by hydrogenation to give methanol and the original alcohol. The process was originally suggested by Christiansen [30]. While this process is more complex than the direct combination of CO and $H_2$, it has the advantage that reaction conditions may be less energy intensive. The reaction of the alcohol with CO proceeds at about 88°C and 50 atm while the hydrogenation with a copper chromite catalyst takes place at 160°C and about 30 atm.

Methyl formate has the same empirical formula as acetic acid and is converted to acetic acid under CO pressure in the presence of soluble rhodium complexes and methyl iodide at 200°C. It has been shown that this conversion involves the decarbonylation of methyl formate to produce methanol and CO. The methanol, in equilibrium with other methyl species in the system, reacts with CO in the presence of the rhodium catalyst in the expected manner to produce acetic acid [31].

## X. STYRENE FROM METHANOL AND TOLUENE

Traditionally, styrene is made by a two-step process: Benzene is alkylated to ethylbenzene which is then dehydrogenated to styrene. A one-step route to styrene via alkylation of toluene with methanol would offer advantages of both lower raw material costs and smaller energy costs. Unland and Barker [32] at Monsanto developed a zeolite catalyst and an alkylation process which can accomplish these objectives. The reaction may be written

$$C_6H_5CH_3 + CH_3OH \xrightarrow[\substack{4A\ pore \\ 400°C}]{CsBX\ zeolite} C_6H_5CH=CH_2 + H_2O$$
$$\text{styrene}$$

The geometry and acidity of the cesium-boron type X zeolite both activate the toluene molecule and protect it from attack at the undesirable para position; only the methyl group in toluene is exposed for alkylation. This route to styrene is unattractive at present; high toluene recycle is needed, the catalyst deactivates easily, and yields are low.

## XI. TEREPHTHALIC ACID FROM TOLUENE
## AND METHANOL

Terephthalic acid is usually made by the oxidation of p-xylene obtained from petroleum reformate. But toluene can be used to synthesize this important chemical:

$$C_6H_5CH_3 + CH_3OH \xrightarrow[300-500°C]{P\text{-modified H-ZMS-5}} p\text{-xylene}$$
$$\sim 80\%$$

$$p\text{-xylene} \xrightarrow{\text{oxidation}} \text{terephthalic acid}$$

The greater concentration of p-xylene, because of the restricted pore size of the ZSM-5 shape-selective catalyst, should greatly lower the costs of the separation process ordinarily used for its recovery from reformate.

## XII. SOME CONCLUDING REMARKS

This paper has not covered all the ways in which methanol can be used as a source of chemicals. There is a large patent literature on the subject, much of which has not yet reached the open literature.

Synthesis gas chemistry, with its ubiquitous and secure resource base, presents great opportunities and challenges for homogeneous and heterogeneous catalysis. But many present and emerging processes for oxygenated chemicals are best carried out sequentially, starting from methanol or formaldehyde using homogeneous catalysts. Homogeneously catalyzed reactions are carried out under mild conditions, are more selective, and can be fine-tuned by judicious choice of ligands, temperature, pressure, solvents, etc. Mechanistic studies of homogeneous reactions are easier, allowing better understanding and hence greater control of process reactions.

However, the future of methanol and synthesis gas may rest primarily on their uses as fuels. The fuel industry deals in units of barrels per day rather than pounds per year [33]. As with petroleum, fuel uses may govern and chemicals will be made from excess synthesis gas or from certain slipstreams or fractions. However, a major incentive for chemicals from methanol will be the high-value-added aspects of these chemicals compared with fuel-value methanol.

## REFERENCES

[1] D. Forster, Adv. Organomet. Chem., 17, 255 (1979).
[2] N. von Kutepow, W. Himmele, and H. Hohenschutz, Chem.-Ing.-Tech., 37, 383 (1965).
[3] F. E. Paulik and J. F. Roth, Chem. Commun., p. 1578 (1968).
[4] D. L. King, K. K. Ushiba, and T. E. Whyte, Jr., Hydrocarbon Process., p. 131 (November 1982).
[5] A. N. Naglieri and N. Rizkalla, U.S. Patent 4,134,912 (1979), to Halcon Internationa; Y. T. Isshiki and Y. M. Kizima, British Patent 2,007,212 (1979), to Mitsubishi Gas Chemical Co.

[6]   D. Forster, J. Am. Chem. Soc., 97, 951 (1975).

[7]   D. Forster, J. Mol. Catal., 17, 299 (1982).

[8]   I. Wender, R. A. Friedel, and M. Orchin, Science, 113, 206 (1951).

[9]   T. Mizoroki and M. Nakayama, Bull. Chem. Soc. Jpn., 38, 2876 (1965).

[10]  D. Forster, J. Chem. Soc., Dalton Trans., p. 1639 (1979).

[11]  J. Schmidt, W. Hafner, R. Jira, R. Sieber, J. Sedlmeir, and A. Sabel, Angew. Chem., Int. Ed. Engl., 1, 80 (1962).

[12]  W. Reppe, U.S. Patent 2,730,546 (1956).

[13]  J. L. Ehrler and B. Juran, Hydrocarbon Process., p. 109 (February 1982).

[14]  F. S. Wagner, Jr., in Kirk-Othmer Encyclopedia of Chemical Technology, Vol. 1, Wiley, New York, 1978, p. 154.

[15]  West German Offen. 2,610,035 (1976), to Halcon International.

[16]  A. M. Brownstein, in Catalysis of Organic Reactions (W. R. Moser, ed.), Dekker, New York, 1981, p. 3.

[17]  F. Piacenti and M. Bianchi, in Organic Syntheses via Metal Carbonyls, Vol. 2 (I. Wender and P. Pino, eds.), Wiley, New York, 1977, p. 13; D. W. Slocum, in Catalysis in Organic Chemistry (W. H. Jones, ed.), Academic, New York, 1980, p. 245; H. Bahrmann and B. Cornils, in New Synthesis with Carbon Monoxide (J. Falbe, ed.), Springer, New York, 1980, p. 226.

[18]  M. E. Fakley and R. A. Head, Appl. Catal., 5, 3 (1983).

[19]  W. R. Pretzer and M. M. Habib, Symposium on Catalytic Conversion of Synthesis Gas and Alcohols to Chemicals, 17th MARM Meeting, Hershey-Pocono, Pennsylvania, Plenum, In Press.

[20]  M. Y. Darensbourg, P. Jimenez, J. R. Sackett, J. H. Hanckel, and R. L. Kump, J. Am. Chem. Soc., 104, 1521 (1982).

[21]  G. S. Koermer and W. E. Slinkard, Ind. Eng. Chem., Prod. Res. Dev., 17, 231 (1978).

[22]  Chemical and Engineering News, p. 37 (April 7, 1980).

[23]  M. J. Chen, H. M. Feder, and J. W. Rathke, J. Am. Chem. Soc., 104, 7346 (1982).

[24]  A. Aquilio, J. S. Alder, D. N. Freeman, and R. J. H. Voorhoeve, Hydrocarbon Process., p. 57 (March 1983).

[25]  D. R. Fahey, J. Am. Chem. Soc., 103, 136 (1981); B. D. Dombeck, Ibid., 102, 6855 (1980); L. C. Costa, Catal. Rev.—Sci. Eng., 25, 325 (1983).

[26]  J. Kollar, U.S. Patent 4,337,371 (1982), to Celanese Corp.

[27]   J. Kollar, British Patent Appl. 2,083,037 (1982), to Redox
       Technologies.
[28]   Chemical and Engineering News, p. 41 (April 11, 1983).
[29]   S. P. Tonner, D. L. Trimm, and M. S. Wainwright, J. Mol.
       Catal., 18, 215 (1983).
[30]   J. E. Christiansen, U. S. Patent 1,302,011 (1919).
[31]   F. J. Bryant, W. R. Johnson and T. C. Singleton, Prepr.,
       Div. Pet. Chem., Am. Chem. Soc., Dallas, Texas, April,
       1973, p. 107.
[32]   M. L. Unland and G. E. Barker, in Catalysis of Organic
       Reactions (W. R. Moser, ed.), Dekker, New York, 1981,
       p. 51.
[33]   R. L. Pruett, Science, 211, 11 (1981).

# Methanol Conversion to Light Olefins

CLARENCE D. CHANG
Central Research Division
Mobil Research and Development Corporation
Princeton, New Jersey

I. INTRODUCTION

II. METHODOLOGY
    A. Shape-Selective Catalysis
    B. Partial Conversion with Recycle
    C. Reaction at Subatmospheric Partial Pressure
    D. High Temperature Conversion
    E. Reaction over Modified Zeolites and Other Catalysts

III. OLEFIN DISTRIBUTION

IV. CONCLUDING REMARKS

    REFERENCES

## I. INTRODUCTION

Light olefins will play a dominant role in any future methanol-based chemicals economy. Olefins are initial products in the conversion of methanol to hydrocarbons over zeolite catalysts [1]. The overall reaction path may be represented by

$$\frac{n}{2}[\,2CH_3OH \rightleftharpoons CH_3OCH_3 + H_2O\,] \xrightarrow{-nH_2O} C_nH_{2n} \longrightarrow n[CH_2]$$

where $[CH_2]$ = average formula of an aromatic-paraffin mixture. In general, the steps in this sequence are kinetically coupled. Selective olefin production hinges on finding means to decouple the

23

aromatization step. Several approaches to this problem have been
reported. These may be categorized as follows:

Shape-selective catalysis
Partial conversion with recycle
Reaction at subatmospheric partial pressure
High temperature conversion
Reaction over modified zeolites and other catalysts

This article presents a review of the above methods, as well as
an analysis of olefin distributions. Since the preponderance of information reported in the literature has concerned zeolite catalysts,
this review will concentrate mainly on these, with only brief men-
tion of nonzeolite catalysts.

## II. METHODOLOGY

### A. Shape-Selective Catalysis

Light olefins are the major products of methanol conversion in
the presence of small pore zeolites, i.e., those with pore openings
defined by 8-rings of oxygen atoms. Such small pore zeolites sorb
linear hydrocarbons but exclude branched and aromatic hydrocarbons. Because of these spatial restrictions, only molecules with
sufficiently small critical dimensions can diffuse out of the zeolite
interior.

Representative data from methanol conversion over small pore
zeolites, erionite, zeolite T, chabazite, and ZK-5 are shown in
Table 1 [2]. At 339-538°C the products are mainly $C_2$-$C_4$ olefins.
The presence of lower paraffins may be related to coke deposition.
Various workers have reported that cation-exchanged chabazites
are selective for light olefin synthesis [3-7].

Zeolite ZSM-34, a synthetic zeolite of the erionite-offretite family, has been found effective for converting methanol to hydrocarbons rich in ethylene and propylene [8]. Selectivity and activity are claimed to be improved upon in the exchange of ZSM-
34 with various cations [9]. At 400°C and 800 $h^{-1}$ GHSV, methanol
(12 vol% in $N_2$) was completely converted into hydrocarbons containing (mol%): $CH_4$ (13.8), $C_2H_4$ (42.9), $C_3H_6$ (33.4), $C_3H_8$ (5.4),
$C_4H_8$ (3.3), and $C_4H_{10}$ (1.2) over a Th-modified ZSM-34.

The conversion of methanol over zeolite T, an erionite-offretite
intergrowth, was studied by Ceckiewicz [25, 26]. At 450°C the
hydrocarbon product contained 38 wt% $C_2H_4$ and 39 wt% $C_3H_6$.

The shape-selectivity of zeolites may be increased by controlled
deposition of various substances in the zeolite channels and cavities

TABLE 1 [2]

Methanol Conversion over Small Pore Zeolites

| | Erionite[a] | Zeolite T | Chabazite | ZK-5 |
|---|---|---|---|---|
| Reaction conditions: | | | | |
| Temperature, °C | 370 | 341-378 | 538 | 538 |
| Pressure, atm | 1 | 1 | 1 | 1 |
| LHSV (WHSV), $h^{-1}$ | 1 | (3.8) | b | b |
| Conversion, % | 9.6 | 11.1 | 100 | 100 |
| Hydrocarbons, wt%: | | | | |
| $CH_4$ | 5.5 | 3.6 | 3.3 | 3.2 |
| $C_2H_6$ | 0.4 | 0.7 | 4.4 | 0 |
| $C_2H_4$ | 36.3 | 45.7 | 25.4 | 21.4 |
| $C_3H_8$ | 1.8 | 0 | 33.3 | 31.8 |
| $C_3H_6$ | 39.1 | 30.0 | 21.2 | 13.5 |
| $C_4H_{10}$ | 5.7 | 6.5 | 10.4 | 22.6 |
| $C_4H_8$ | 9.0 | 10.0 | | |
| $C_5^+$ | 2.2 | 3.1 | 2.0 | 7.5 |

[a]De-aluminized; $SiO_2/Al_2O_3$ = 16.
[b]Pulse microreactor, 1 μL MeOH in He, 500 $h^{-1}$ GHSV.

[10]. These substances increase the diffusion path of guest molecules and introduce new steric constraints. This technique was used by Rodewald [11] to increase the selectivity of ZSM-5, an intermediate pore (10-ring) zeolite, for light olefins from methanol. Silica (7.4%) was deposited in the pores of ZSM-5 by impregnation with a solution of methylhydrogensilicone in n-hexane and heating at 75°C followed by contacting with methanol at 320-370°C. At 370°C and 1 $h^{-1}$ LHSV, methanol was converted to the hydrocarbon mixture shown in Table 2. Included for comparison are data obtained with untreated ZSM-5 under comparable conditions [1]. The data demonstrate the suppression of the aromatization reaction upon silica modification. That the deposited silica was intracrystalline was confirmed by a reduction in the n-hexane sorption capacity of the silanated sample [12].

TABLE 2 [11]

Methanol Conversion to Hydrocarbons over ZSM-5.
Effect of Zeolite Silanation
(370°C, 1 LHSV)

| Hydrocarbon product (wt%) | Untreated ZSM-5 [1] | Silanated ZSM-5 [11] |
|---|---|---|
| Methane | 1.0 | 1.3 |
| Ethane | 0.6 | 0.0 |
| Ethene | 0.5 | 26.2 |
| Propane | 16.2 | 2.4 |
| Propene | 1.0 | 18.7 |
| Butanes | 24.2 | 5.4 |
| Butenes | 1.3 | 10.8 |
| $C_5^+$ Aliphatic | 14.0 | 16.0 |
| Aromatic | 41.2 | 19.2 |
| | 100.0 | 100.0 |

## B. Partial Conversion with Recycle

A straightforward approach toward maximizing olefins is to backtrack along the reaction path, recover the intermediate olefins, and recycle unreacted feed [13]. The reaction trajectory for conversion over ZSM-5 at 370°C is shown in Fig. 1. It is clear that olefin selectivity may be increased by decreasing contact time. However, some amount of aromatics will normally accompany the olefins save at very low conversion. Partial conversion can also be achieved by decreasing temperature.

This concept has been tested in experiments utilizing a fluid-bed reactor [14]. Methanol containing various amounts of water was partially converted at 299-343°C over ZSM-5 to give 39.9-52.5% $C_2$-$C_5$ olefin selectivity. Representative data are shown in Table 3 [14].

An alternative to reducing contact time is to decrease catalyst Bronsted acidity. This can be achieved by either partial

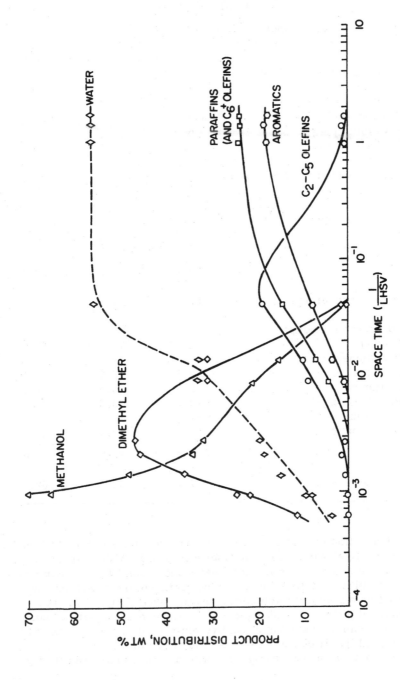

FIG. 1.  Methanol to hydrocarbons reaction path (371°C) [2].

TABLE 3 [14]

Conversion of Methanol to Light Olefins

| Reactant composition | 83% CH$_3$OH/ 17% H$_2$O | | 16% CH$_3$OH/ 84% H$_2$O |
|---|---|---|---|
| Conditions: | 1 | 2 | 3 |
| Temperature, °F | 570 | 650 | 615 |
| Pressure, psig | 6 | 6 | 27 |
| Methanol WHSV | 0.4 | 0.4 | 0.2 |
| Product yield, wt% of methanol: | | | |
| Hydrocarbons | 22.7 | 36.6 | 27.2 |
| Methanol | 15.9 | 8.9 | 26.8 |
| Dimethylether | 23.0 | 5.1 | 7.9 |
| Water | 38.3 | 49.2 | 38.0 |
| Coke, CO, CO$_2$, other | 0.1 | 0.2 | 0.1 |
| | 100.0 | 100.0 | 100.0 |
| Per pass conversion, % | 52 | 84 | 62 |
| Hydrocarbon product, wt%: | | | |
| Ethylene | 21.3 | 18.8 | 27.6 |
| Propylene | 17.2 | 11.7 | 17.5 |
| Butenes | 7.1 | 6.5 | 6.1 |
| Pentenes | 2.1 | 2.9 | 1.3 |
| Paraffins (C$_1$-C$_5$) | 18.3 | 20.8 | 24.7 |
| Aromatics | 14.5 | 16.3 | 6.6 |
| Nonaromatics | 19.5 | 23.0 | 16.2 |
| | 100.0 | 100.0 | 100.0 |
| C$_2$-C$_5$ Olefins | 47.7 | 39.9 | 52.5 |

neutralization of acid sites, or for some zeolites such as ZSM-5, by decreasing aluminum content (increasing SiO$_2$/Al$_2$O$_3$) during crystallization. These two methods have been found to give equivalent selectivities in olefin formation from methanol over zeolites compared on an equal proton concentration basis [15]. This is illustrated in Fig. 2, which plots hydrocarbon selectivity at constant T and SV vs catalyst acidity expressed as "equivalent SiO$_2$/Al$_2$O$_3$" [15]. It can be seen that the trajectory is similar to the reaction path shown in Fig. 1, with maximum olefin selectivity

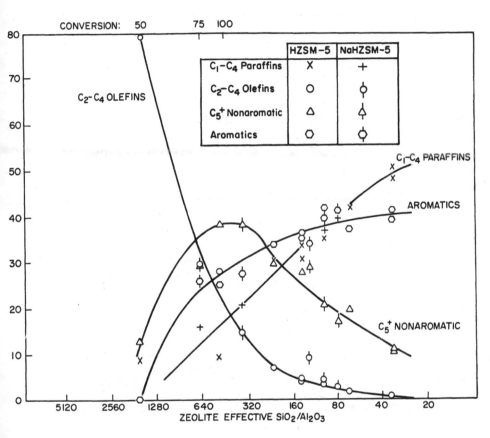

FIG. 2.  Methanol conversion over HZSM-5.  Hydrocarbon selectivity [15].   371 ± 1°C, 1 atm (101.3 kPa), 1 LHSV.

occurring at partial conversion.  When the space velocity was decreased to achieve complete conversion in the high $SiO_2/Al_2O_3$ regime, significant aromatization occurred, indicating a trade-off
between contact time and catalyst activity [15].  Thus, at this
temperature the relative rates of olefin and aromatics formation
are not altered by changing catalyst Bronsted acidity.

## C.  Reaction at Subatmospheric Partial Pressure

Reactant partial pressure has a profound effect on selectivity
in hydrocarbon formation from methanol [16].  The main effect
of varying partial pressure is to change the relative rates of

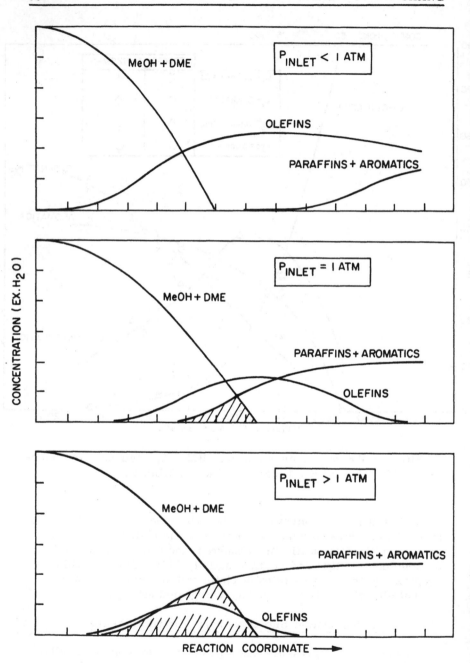

FIGURE 3.

### TABLE 4 [16]

Methanol Conversion at Low Pressures
(427°C, 1 LHSV, 1 atm total pressure[a])

| Methanol partial pressure, atm: | 1.00 | 0.25 | 0.17 | 0.07 |
|---|---|---|---|---|
| Conversion, %: | 99+ | 99+ | 99+ | 99+ |
| Hydrocarbon distribution, wt%: | | | | |
| Ethylene | 3.2 | 12.4 | 17.4 | 21.0 |
| Propylene | 4.8 | 18.2 | 26.5 | 38.7 |
| Butenes | 2.2 | 6.3 | 7.6 | 18.5 |
| Pentenes | 0.4 | 0.3 | 0.7 | 2.4 |
| (Total $C_2$-$C_5$ olefins) | (10.6) | (37.2) | (52.2) | (80.6) |
| Methane | 1.5 | 0.8 | 0.6 | 0.5 |
| $C_2$-$C_5$ Paraffins | 43.0 | 39.4 | 24.2 | 15.6 |
| $C_6^+$ Nonaromatic | 3.9 | 2.3 | 2.7 | 1.3 |
| Aromatics | 41.0 | 20.3 | 20.3 | 2.0 |

[a]Helium diluent at subatmospheric MeOH partial pressures.

the dehydration and aromatization steps in the reaction sequence
[16]. By lowering partial pressure, aromatization may be decoupled
from olefin formation without sacrificing feed conversion. This is
depicted in Fig. 3, where the shaded regions represent the degree
of overlap (coupling) between olefin and aromatics formation. Typical selectivities as a function of partial pressure at complete conversions are shown in Table 4 [16]. Upon lowering $P_{MeOH}$ from
atmospheric to 0.07 atmosphere pressure (427°C, 1 LHSV, He diluent), the $C_2$-$C_5$ olefin selectivity is seen to increase from 10.6
to 80.6%. Among the drawbacks to this approach, however, is the
necessity for product recovery from dilute streams.

### D. High Temperature Conversion

Olefin selectivity is increased at high temperature [1]. This
effect is further enhanced by the use of high $SiO_2/Al_2O_3$ (low
Bronsted acidity) zeolites [15, 17]. A kinetic analysis of the
interrelation between temperature and catalyst activity has been
made with ZSM-5 [15] and is summarized here.

MeOH CONVERSION OVER HZSM-5 (SiO$_2$/Al$_2$O$_3$ = 70)
500°C (932°F) ATM P

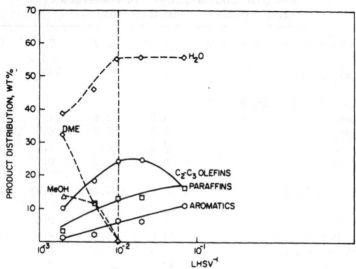

MeOH CONVERSION OVER HZSM-5 (SiO$_2$/Al$_2$O$_3$) = 142)
500°C (932°F) ATM P

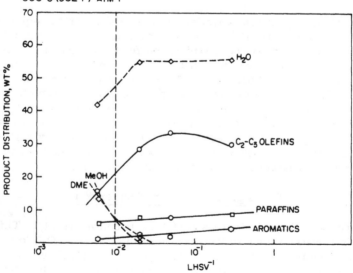

FIGURE 4 [15].

MeOH CONVERSION OVER HZSM-5 ($SiO_2/Al_2O_3$=500)
500°C (932°F) ATM P

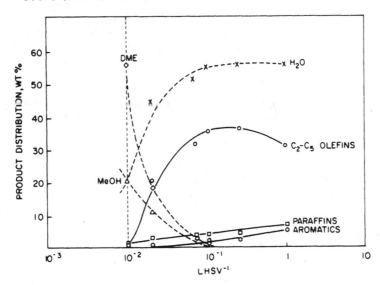

MeOH CONVERSION OVER HZSM-5 ($SiO_2/Al_2O_3$=1670)
500°C (932°F) ATM P

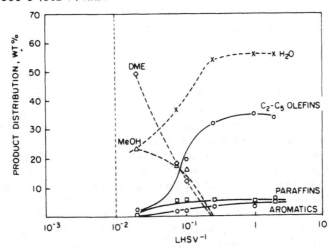

FIGURE 4 [15].  (continued)

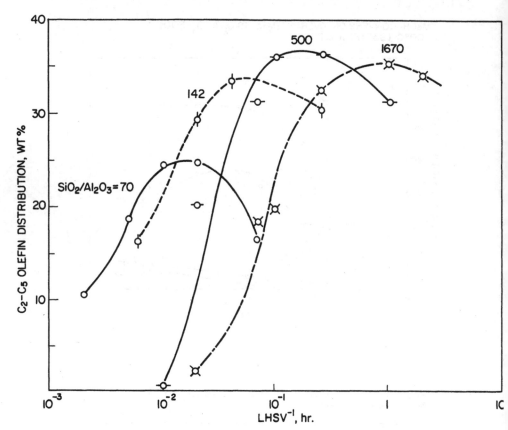

FIG. 5. MeOH conversion over HZSM-5 $SiO_2/Al_2O_3$ effect [15]. 500°C (932°F), 1 atm.

The effect of varying ZSM-5 catalyst acidity on selectivity at 370°C has already been discussed (Fig. 2). At this temperature significant improvement in olefin selectivity is obtained only at the expense of feed conversion. A different picture emerges upon raising the reaction temperature to 500°C. This may be seen in the series of plots in Fig. 4, which show changes in the reaction trajectory with increasing $SiO_2/Al_2O_3$ from 70 to 1670. The dashed lines in the plot are arbitrarily positioned at $LHSV^{-1} = 10^{-2}$ h to provide a common reference. The plots show that decoupling of aromatization is enhanced by increasing $SiO_2/Al_2O_3$. This effect is emphasized in Fig. 5, which plots $C_2$-$C_5$ olefin selectivity as a function of space-time. It is evident from Fig. 5 that little additional improvement is achievable at $SiO_2/Al_2O_3$ higher than about 500.

The high temperature synthesis of olefins from methanol has been represented by a simple kinetic model [15]:

$$A \xrightarrow{k_1} B \xrightarrow{k_2} C$$

where  A = oxygenates (as $CH_2$)
       B = olefins
       C = aromatics + paraffins

Both steps in this global model are assumed first-order. In Fig. 6 typical experimental data are plotted, with the fitted curves superimposed, for 400, 450, and 500°C ($SiO_2/Al_2O_3$ = 500). In Fig. 7 similar plots are indicated for $SiO_2/Al_2O_3$ = 400 and 1670 (T = 500°C). The Arrhenius plot of Fig. 8 indicates that the temperature coefficient for olefin formation is much larger that aromatization, while Fig. 9 shows that the ratio $k_1/k_2$ (olefin/aromatic) increases with $SiO_2/Al_2O_3$.

### E.  Reaction over Modified Zeolites and Other Catalysts

Kaeding and Butter [18] reported the conversion of methanol to $C_2$-$C_4$ olefins in 70% selectivity, at 100% conversion, over phosphorus-modified ZSM-5. The zeolite was reacted with trimethyl phosphite, and upon calcination in air, P was bonded to the framework. The following was proposed as the mechanism of P modification.

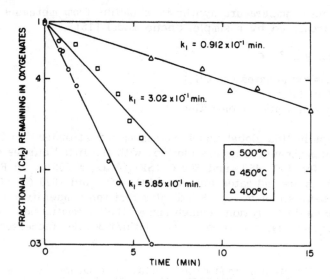

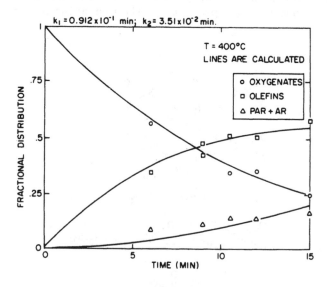

FIGURE 6 [15].

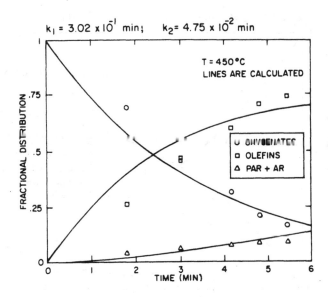

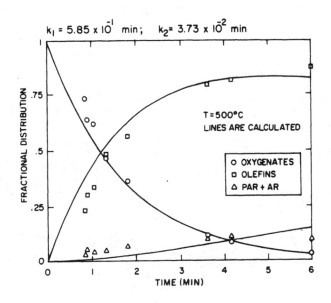

FIGURE 6 [15]. (continued)

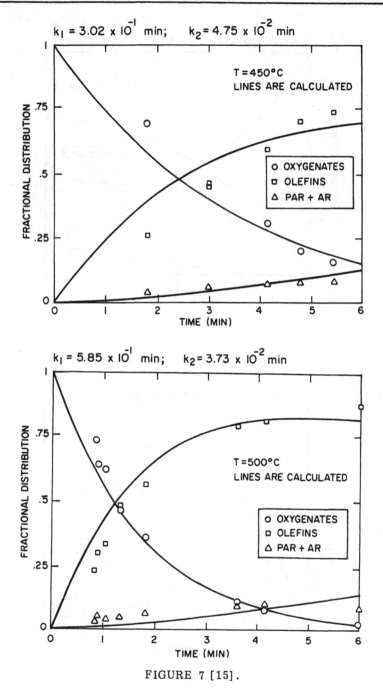

FIGURE 7 [15].

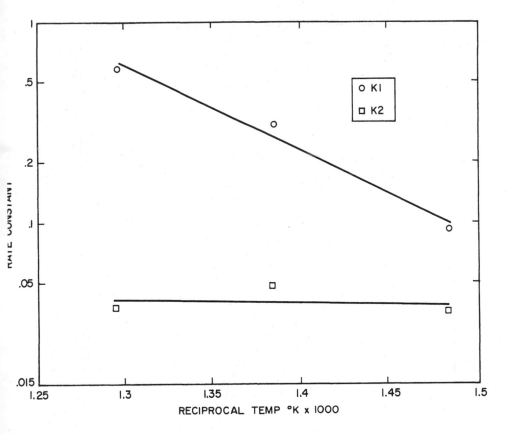

FIGURE 8 [15].

Table 5 contains detailed selectivity data for conversion of dimethyl ether over P-modified ZSM-5 at various temperatures. Total $C_2$-$C_4$ olefins increased from 56.5 to 74.6% with a temperature increase from 300 to 600°C. At 700°C the major products were $CH_4$, CO, and $H_2$ from ether dissociation and coking reactions.

Ion exchange of $Mn^{2+}$ into ZSM-5 has been reported to increase $C_2$-$C_4$ olefin selectivity [28], and $C_2$-$C_4$ olefin selectivities in excess of 90% were achieved by Chen and Liang [19] with a Mg-modified ZSM-5.

Zeolite-metal composites prepared by crystallization in the presence of metal-loaded alumina are reported effective for olefin synthesis [20]. The Ru- or Rh-loaded catalysts had improved life due to reduced coking.

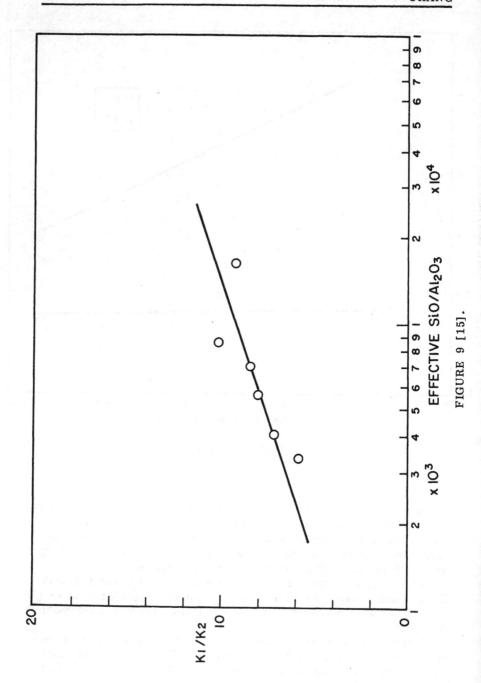

EFFECTIVE SiO/Al₂O₃

FIGURE 9 [15].

TABLE 5 [18]

Conversion of Methyl Ether to Hydrocarbons
over Phosphorus-Modified Zeolites[a]

| Reaction conditions:[b] | | | | | | |
|---|---|---|---|---|---|---|
| Temperature (°C) | 300 | 350 | 400 | 500 | 600 | 700 |
| Conversion (wt%)[c] | 1.8 | 22.2 | 80.9 | 100 | 100 | 100 |
| Hydrocarbon | Weight percentage | | | | | |
| Methane, CO, $H_2$ | 0.0 | 0.4 | 0.9 | 2.7 | 3.9 | 76.6 |
| Ethane | 11.4 | 1.4 | 0.2 | 0.4 | 0.6 | 1.0 |
| Propane | 8.2 | 1.8 | 0.2 | 0.7 | 1.3 | 0.2 |
| Butanes | 6.9 | 3.9 | 1.0 | 1.6 | 1.2 | 0.0 |
| Total $C_2$-$C_4$ Paraffins | 26.5 | 7.1 | 1.4 | 2.7 | 3.1 | 1.2 |
| Ethylene | 11.2 | 8.0 | 2.5 | 2.8 | 16.2 | 5.2 |
| Propylene | 32.8 | 30.9 | 29.4 | 34.6 | 38.7 | 7.3 |
| Butylenes | 12.5 | 16.2 | 36.4 | 21.6 | 19.7 | 3.6 |
| Total $C_2$-$C_4$ Olefins | 56.5 | 55.1 | 68.3 | 59.0 | 74.6 | 16.1 |
| $C_5$-$C_{10}$ Aliphatics | 12.0 | 30.4 | 24.1 | 33.2 | 13.1 | 2.5 |
| Aromatics | 5.0 | 7.0 | 4.3 | 2.4 | 5.3 | 3.6 |
| Total | 100.0 | 100.0 | 100.0 | 100.0 | 100.0 | 100.0 |
| $C_2$-$C_4$ Olefin/paraffin ratio | 2.1 | 7.8 | 48.8 | 21.9 | 24.1 | 13.4 |

[a]P = 3.48 wt%.
[b]Atmospheric pressure: weight hourly space velocity = 2.3.
[c]To hydrocarbons and water.

Mordenite modified by ion exchange with various cations or
mixed with $CrO_3$, $MoO_3$, $WoO_3$, or $Sb_2O_3$ showed improved selectivity to $C_2$-$C_4$ olefins [21, 22, 27].
Various layered silicate minerals, ion exchanged with $Ti^{3+}$,
have been found to convert methanol selectively to olefins with
chain growth limited at $C_6$ [23].

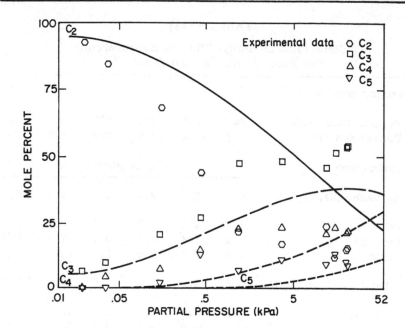

FIG. 10.   Olefin distribution at 500°C (lines are calculated)
[24].

## III.   OLEFIN DISTRIBUTION

Detailed distribution of $C_2$-$C_5$ olefins from methanol over ZSM-5 in the early stages of the reaction was reported by Chang et al. [16] and further analyzed by Chu and Chang [24].

Garwood [29] found that olefins ($C_2$-$C_{10}$) equilibrate rapidly in the presence of ZSM-5, yielding mixtures whose composition is largely governed by thermodynamics. To determine whether similar constraints apply to olefins from methanol conversion, Chu and Chang studied the reaction at low (<2) conversion at 300 and 500°C. Experimental data obtained at 500°C are shown in Fig. 10, where $C_2$-$C_5$ olefin selectivity is plotted against olefin partial pressure (at the reactor exit). Superimposed on the experimental points are curves representing the thermodynamic equilibrium distribution between the different olefin species.

It is clear from Fig. 10 that partial pressure exerts a profound influence on olefin distribution. Further, it is seen that the experimental distributions approach thermodynamic equilibrium only at low partial pressure (low conversion). As conversion approaches zero, $C_2H_4$ selectivity approaches 100%. This is a key observation which resolves a long-standing dispute as to the identity of the "first" olefin from methanol conversion to hydrocarbons [30].

With increasing conversion, olefin distribution deviates from thermodynamic equilibrium and becomes subject to kinetic control. This is particularly evident for $C_3H_6$, which undergoes a marked increase after an induction period. This was attributed to autocatalysis [31], whereby olefins are alkylated with reactive $C_1$ intermediates generated from methanol/dimethyl ether decomposition, resulting in homologation.

The methanol reaction differs, therefore, from Garwood's reaction [29], which is conducted in the absence of water. Furthermore, the mechanism of the latter reaction is believed to involve mainly oligomerization and $\beta$-scission, without a $C_1$ "chain propagation" reaction to compete against equilibration.

## IV. CONCLUDING REMARKS

This survey shows that a wide variety of methods is available for increasing total olefin selectivity from methanol conversion. However, the problem of limiting selectivity to an individual olefin, $C_2H_4$ in particular, has no satisfactory solution as yet at practical levels of conversion. No doubt the difficulty is rooted in the autocatalytic nature of the reaction. Nevertheless, there are indications that the selectivity of $C_2H_4$ relative to higher olefins can be altered under certain conditions, e.g., via catalyst modification [11, 19] or by reaction in the presence of excess steam [14]. It may be conjectured that these techniques suppress the autocatalytic step in some manner. At any rate, the problem remains a fascinating challenge to the catalytic chemist.

## REFERENCES

[1]  C. D. Chang and A. J. Silvestri, J. Catal., 47(2), 249 (1977).

[2]  C. D. Chang, W. H. Lang, and A. J. Silvestri, U.S. Patent 4,062,905 (1977).

[3]  J. T. Cobb, Jr., V. T. Coon, and P. Tipnis, Final Report, September 1978, Contract No. EW-78-S-02-4691, U.S. Department of Energy.

[4]  B. B. Singh and R. G. Anthony, Prepr. Can. Symp. Catal., 6, 113 (1979).

[5]   B. B. Singh, F. N. Lin, and R. G. Anthony, Chem. Eng.
      Commun., 4, 749 (1980).
[6]   F. A. Wunder and E. I. Leupold, Angew. Chem., Int. Ed.
      Engl., 19(2), 126 (1980).
[7]   W. Dettmeier, H. Litterer, H. Baltes, W. Herzog, E. I.
      Leupold, and F. A. Wunder, Chem.-Ing.-Tech., 54(6),
      590 (1982).
[8]   E. N. Givens, C. J. Plank, and E. J. Rosinski, U.S.
      Patent 4,074,095 (1978); U.S. Patent 4,079,096 (1978).
[9]   T. Inui, E. Araki, T. Sezume, T. Ishihara, and Y.
      Takegami, React. Kinet. Catal. Lett., 18, 1 (1981).
[10]  P. B. Weisz, Pure Appl. Chem., 52, 2091 (1980).
[11]  P. G. Rodewald, U.S. Patent 4,100,219 (1978).
[12]  P. G. Rodewald, U.S. Patent 4,145,315 (1979).
[13]  C. D. Chang, W. H. Lang, and A. J. Silvestri, U.S.
      Patent 4,052,479 (1977).
[14]  J. A. Brennan, W. E. Garwood, S. Yurchak, and W. Lee,
      Proceedings of the International Seminar on Alternate
      Fuels (A. Germain, ed.), Liege, Belgium, 1981, p. 19.
[15]  C. D. Chang, C. T-W. Chu, and R. F. Socha, J. Catal.,
      In Press.
[16]  C. D. Chang, W. H. Lang, and R. L. Smith, Ibid., 56,
      169 (1979).
[17]  W. W. Kaeding, German Offen. 2,935,863.
[18]  W. W. Kaeding and S. A. Butter, J. Catal., 61, 155 (1980).
[19]  G. Chen and J. Liang, China-Japan-U.S. Symposium on
      Heterogeneous Catalysis, Dalian, China, 1982, Paper A01C.
[20]  T. Inui, G. Takeuchi, and Y. Takegami, Appl. Catal., 4,
      211 (1982).
[21]  G. Pop, G. Musca, E. Pop, D. Herghelegiu, and P. Tomi,
      Rev. Chim. (Bucharest), 34(4), 293 (1983).
[22]  H. Itoh, T. Hattori and Y. Murakami, J. Chem. Soc.,
      Chem. Commun., p. 1092 (1982).
[23]  Y. Morikawa, F.-L. Wang, Y. Moro-oka, and T. Ikawa,
      Chem. Lett., p. 965 (1983).
[24]  C. T-W. Chu and C. D. Chang, J. Catal., In Press.
[25]  S. Ceckiewicz, Bull. Pol. Acad. Sci., 27(7-8), 624 (1979).
[26]  S. Ceckiewicz, J. Chem. Soc., Faraday Trans. 1, 77, 269
      (1980).
[27]  W. Zatorski and S. Krzyzanowski, Acta Phys. Chem., 24,
      347 (1978).
[28]  T. Fleckenstein, H. Litterer, and F. Fetting, Chem.-Ing.-
      Tech., 52(10), 816 (1980).
[29]  W. E. Garwood, Prepr. Am. Chem. Soc., Div. Pet. Chem.,
      27(2), 563 (1982).

[30]  For an account of this controversy, see C. D. Chang, Catal.
      Rev.–Sci. Eng., 25(1), 1 (1983).
[31]  N. Y. Chen and W. J. Reagan, J. Catal., 59, 123 (1979).

# Chemicals from Methanol

W. H. CALKINS
Central Research and Development Department
E. I. du Pont de Nemours & Co.
Wilmington, Delaware

## I. INTRODUCTION

Over the past 35 years, the chemical industry has largely changed its raw material base from coal to petroleum and natural gas. This has basically meant a shift from acetylene chemistry and coal tar chemistry to olefin chemistry (ethylene, propylene, isobutylene and butadiene) and petroleum-derived benzene, toluene, and xylenes. Methanol, ammonia, and hydrogen are still made from synthesis gas. However, synthesis gas is now made

47

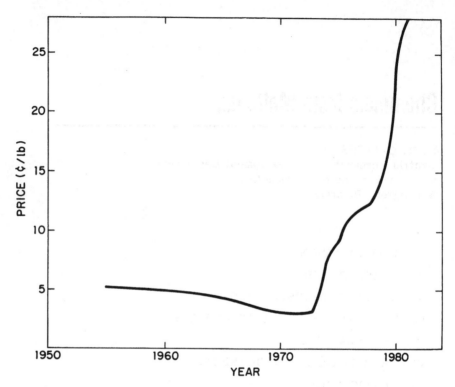

FIG. 1.  Price of ethylene (¢/lb) vs years (1955-1983) in the
United States.  Source:  U.S. International Trade Commission.

by methane reforming instead of coal gasification.  These conversions have involved a vast expenditure of money, time, and effort,
but the financial incentives have been huge and a great deal of
new chemistry has been uncovered.

     The 1973 Arab oil embargo and the ensuing petroleum and nat-
ural gas shortages, followed by the OPEC escalation of crude oil
prices with the attendant increases of natural gas prices, have
forcefully reminded the chemical industry that about 86% of our
domestic carbonaceous fossil fuel resources are coal and only
about 2% each are petroleum and natural gas.  The price of ethylene, the major petrochemical intermediate, which has increased
almost 10-fold since the embargo, dramatically illustrates (Fig. 1)
the problem.  A return to a coal base for organic chemical products at some point in time is clearly indicated and some companies
have already done so.

But the conversion back to coal at this time does not represent such an obviously attractive investment opportunity as the conversion away from coal did in the late 40s, 50s, and 60s. Many of the incentives for departing from coal are still valid. The investment in coal-based plants of economical size is usually very large, the operating costs high, and the current business climate is not particularly favorable and rather uncertain. At the present time, too, there is a surplus of natural gas in the United States [1, 2], and although most agree this situation is temporary, there is not general agreement as to how long it will last. Obviously, the timing of conversion back to coal is crucial.

The coal-based acetylene processes to organic chemical products used 20 and 35 years ago were both expensive and more hazardous than the present-day processes. We need new, better, and lower cost coal-based chemistry. Many people think this lies in new synthesis gas- and methanol-based processes, the so called $C_1$ chemistry.

The manufacture of methanol from synthesis gas has been steadily improved over the years. It is about the only route from synthesis gas to a single organic compound which can be made in high yield at relatively low cost with the current state-of-the-art. This makes methanol an attractive possibility as an intermediate to other chemicals. In the last 30 years a great deal of new chemistry has been developed, particularly in the field of homogeneous catalysis, using soluble complexes of transition metals. Much of it is applicable to syntheses starting with methanol.

In 1982, approximately 1096 MM gallons or 3.30 MM tons of methanol were produced in the United States, practically all from natural gas. They were consumed roughly as shown in Table 1.

TABLE 1

1982 Consumption Distribution of Methanol in the United States[a] (%)

| | |
|---|---:|
| Formaldehyde manufacture | 31 |
| Methylamines manufacture | 4 |
| Chloromethanes manufacture | 9 |
| Acetic acid manufacture | 12 |
| Methyl esters (phthalates, methacrylates, etc.) | 8 |
| Solvent uses | 11 |
| Miscellaneous (fuel, antiknock, antifreeze, etc.) | 25 |
| | 100 |

[a]Chemical Economics Handbook, SRI, October 1983.

In the first four of these uses, methanol is a chemical intermediate to other chemicals. The first three of these are old established processes growing with the economy at 4-5% or less per
year. Acetic acid production, however, represents a relatively
new use for methanol and one based on some of the new chemistry referred to earlier. It is growing more rapidly than the
economy as it displaces older processes based on butane and
other raw materials.

Even more recently, still another use of methanol as an intermediate to acetic anhydride has been commercialized by Eastman
Kodak Company with a plant in Kingsport, Tennessee. Dimethyl
ether, a dehydration product of methanol, is another methanol
derivative beginning to be used commercially as an aerosol propellant. This use may become of more importance in the future.

Each of these commercial products and processes, most of
which are catalytic, will be discussed briefly.


## II.  FORMALDEHYDE

Formaldehyde production is the largest single application of
methanol, with at least 16 manufacturers and accounting for over
30% of the methanol consumed. It, in turn, goes into a variety of
resin and chemicals manufactured as shown in Table 2.


TABLE 2

1981 Consumption Distribution of Formaldehyde
in the United States[a] (%)

| | |
|---|---|
| Urea-formaldehyde resins | 27 |
| Phenol-formaldehyde resins | 20 |
| Melamine-formaldehyde resins | 4 |
| Acetal resins | 8 |
| Pentaerythritol | 7 |
| Hexamethylenetetramine | 6 |
| Acetylenic chemicals | 8 |
| Other | 20 |

[a]Chemical Economics Handbook, SRI, January 1983.

Practically all the formaldehyde produced today (5729 MM lb of 37% solution in 1981) is made using variants of either of two processes, an oxidation-dehydrogenation (silver catalyst) process or a purely oxidation (metal oxide catalyst) process [3, 4]. Both of these types use air to oxidize methanol to formaldehyde or to supply in situ the endothermic heat of dehydrogenation (29 kcal) or both. Since air and methanol form explosive mixtures at concentrations of 6 to 37% by volume of methanol in air, the oxidation-dehydrogenation process operates on the methanol rich side and the oxidation process operates on the methanol lean side, so that the explosive range is avoided.

The reactions involved are:

$$CH_3OH + \frac{1}{2}O_2 \rightarrow HCHO + H_2O, \quad \Delta H = -37 \text{ kcal, 900 K} \quad (1)$$

$$CH_3OH \rightarrow HCHO + H_2, \quad \Delta H = +29 \text{ kcal, 900 K} \quad (2)$$

## III. SILVER CATALYST PROCESS

There are a number of variants of this process. The oxidation-dehydrogenation process as developed by Imperial Chemical Industries Ltd. [3] runs at 600-700°C and uses a shallow bed of silver crystals. Other variants use layers of silver gauze or combinations of silver crystals and gauze.

The gases exit the reactor, are cooled in a waste heat boiler, and then enter an absorption tower where they are absorbed in water. The aqueous solution is then usually fed to a distillation column where most of the excess methanol is removed by distillation to give a formaldehyde solution (up to 55% formaldehyde) in water containing about 1 to 1.5% methanol. The typical reactor operates at approximately 80% conversion. This results in a reaction product which must go to a distillation column to strip out and recycle the unconverted methanol. BASF has a modification of the process operating at somewhat higher temperatures and recycling some dilute formaldehyde solution which is reported to result in 98.6% conversion of the methanol feed, eliminating the need for the distillation step.

## IV. METAL OXIDE-CATALYST PROCESS

In contrast to the silver process, all of the formaldehyde produced by the oxide process is made by Eq. (1) only. Also,

unlike the silver process, the reaction temperature is generally in
the range of 300-400°C at atmospheric pressure [4]. Methanol
conversion is about 99%, with yields based on methanol in the range
of 88-91%. Vaporized methanol is mixed with an excess of air and
passed through catalyst-filled tubes in a heat exchanger with heat
removed by vaporization of heat transfer liquid on the outside of
the tubes. Product exiting the reactor is passed into an absorber
where the vapors are scrubbed with water. Solutions of up to 55%
formaldehyde can be obtained, and the methanol concentration is
low because of the high conversion in the reactor. No distillation
column is required. The small amount of formic acid formed is removed by an ion-exchange process.

The metal oxide catalyst commonly used today is the iron ox-
ide-molybdenum oxide type [5]. It can be a physical mixture of
ferric molybdate and molybdenum trioxide and may have small
amounts of other oxides as promoters. The oxide process has
the advantage of higher conversion, which makes distillation for
methanol removal and recovery unnecessary. Catalyst life is also
longer because the oxide catalyst is less sensitive to impurities
[6]. However, much greater volumes of air must be used, nec-
essitating larger equipment for the oxide than for the silver process. For that reason only about a fifth of the formaldehyde
produced in the United States if produced by the oxide process.


## V. METHYLAMINES

The production of mono-, di-, and trimethylamines consumes
approximately 4% of the methanol produced in the United States.
Approximately 270 MM lb of these compounds were produced in
1979 in the United States by four companies. They were used in
a wide variety of applications. Dimethylamine is produced to the
greatest extent followed by monomethylamine and then trimethylamine (Table 3).

The methylamines are formed by reaction of methanol and ammonia in the vapor phase over a dehydration catalyst at a temperature around 450°C and from atmospheric to approximately 300
psi pressure. The reactions involved are

$$CH_3OH + NH_3 \rightarrow CH_3NH_2 + H_2O, \quad \Delta H = -4.95 \text{ kcal/mol}, \quad (3)$$
$$300 \text{ K}$$

$$2CH_3OH + NH_3 \rightarrow (CH_3)_2NH + 2H_2O, \quad \Delta H = -14.59 \text{ kcal/mol}, 300 \text{ K} \quad (4)$$

$$3CH_3OH + NH_3 \rightarrow (CH_3)_3N + 3H_2O, \quad \Delta H = -27.37 \text{ kcal/mol}, 300 \text{ K} \quad (5)$$

## TABLE 3

Application Distribution for Methylamines (millions of pounds)
in the United States in 1979[a]

| | |
|---|---|
| Monomethylamine (insecticides, surfactants, etc.) | 76 |
| Dimethylamine (dimethylformamide and dimethylacetamide, alkyldimethylamine oxides, rubber chemicals, etc.) | 164 |
| Trimethylamine (oholine chloride, biocides, etc.) | 30 |

[a]Chemical Economics Handbook, SRI.

An excess of ammonia over methanol (2 to 6/1 depending on the
amine product ratio desired) is used, and the reaction comes to approximate thermodynamic equilibrium which is a function of the am-
monia/methanol ratio [7, 8]. Under production conditions the more
desired dimethylamine is obtained in only 30-40% of the total methylamines produced. The other amines in excess of market requirements, together with the additional ammonia, must be recycled to
the reactor. (Except for a small amount of breakdown of methanol
to CO and $H_2$, there is little yield loss and yield based on methanol
can be in the range of 99%.) Excess ammonia with some trimethylamine is stripped off the reaction mixture and recycled. Simple
distillation is not suitable for separating the remaining product
mixture which contains the three methylamines along with small
amounts of impurities. Extractive distillation with water as the
extractive solvent is used to remove the remaining trimethylamine,
and the mono- and dimethylamines are separated from water by distillation [9, 10].
    Preferred catalysts generally are silica/alumina compositions
which may be impregnated with promoters, such as silver phosphate, rhenium heptasulfide, molybdenum sulfide, or cobalt sulfide [8].
    It has been reported that the reaction may be made more se-
lective for dimethylamine by use of zeolite catalysts [11]. This
process does not appear to be in commercial use, however.

## VI.  CHLORINATED HYDROCARBONS

The chloromethanes are produced either from methane or methanol [12]. Over 2.1 billion lb of these compounds were produced
in the United States in 1979, distributed as shown in Table 4.

## TABLE 4

### 1979 Production of Chloromethanes (millions of pounds) in the United States[a]

| | |
|---|---|
| Methyl chloride | 463 |
| Methylene chloride | 633 |
| Chloroform | 356 |
| Carbon tetrachloride | 715 |
| | 2167 |

[a]Chemical Economics Handbook, SRI.

Methyl chloride is mainly used in the manufacture of silicones, tetramethyl lead, and as a solvent in butyl rubber manufacture. Methylene chloride is a multipurpose solvent, a degreasing agent, and a blowing agent. Chloroform is an intermediate in the manufacture of refrigerants, aerosol propellants, and fluorocarbon resins, and carbon tetrachloride is an intermediate for aerosol propellants, refrigerants, and insecticides. There are at least 10 United States companies which produce some or all of these compounds, and there are many foreign manufacturers.

The reactions involved, starting from methane, are

$$CH_4 + Cl_2 \rightarrow CH_3Cl + HCl, \quad \Delta H = -24.7 \text{ kcal/mol, 600 K} \tag{6}$$

$$CH_4 + 2 Cl_2 \rightarrow CH_2Cl_2 + 2HCl, \quad \Delta H = -48.5 \text{ kcal/mol, 600 K} \tag{7}$$

$$CH_4 + 3Cl_2 \rightarrow CHCl_3 + 3HCl, \quad \Delta H = -71.5 \text{ kcal/mol, 600 K} \tag{8}$$

$$CH_4 + 4Cl_2 \rightarrow CCl_4 + 4HCl, \quad \Delta H = -92.6 \text{ kcal/mol, 600°C} \tag{9}$$

If methanol is used as the starting material, the reactions are

$$CH_3OH + HCl \rightarrow CH_3Cl + H_2O, \quad \Delta H = 8.17 \text{ kcal/mol, 600 K} \tag{10}$$

$$CH_3Cl + Cl_2 \rightarrow CH_2Cl_2 + HCl, \quad \Delta H = -23.84 \text{ kcal/mol, 600 K} \tag{11}$$

$$CH_3Cl + 2Cl_2 \rightarrow CHCl_3 + 2 HCl, \quad \Delta H = -46.76 \text{ kcal/mol, 600 K} \tag{12}$$

$$CH_3Cl + 3Cl_2 \rightarrow CCl_4 + 3HCl, \quad \Delta H = -67.86 \text{ kcal/mol, 600°C} \tag{13}$$

For every mole of chlorine introduced in the methane process, 1 mole of HCl must be disposed of, usually as 31 wt% aqueous HCl. In the methanol-based processes, much of the HCl is recycled and utilized. For that reason the predominant raw material for methyl chloride and methylene chloride production in the United States is methanol which accounts for 95% of the methyl chloride and approximately 70% of the methylene chloride produced.

The methanol-hydrogen chloride process to produce methyl chloride is typically carried out by mixing equimolar amounts of gaseous methanol and HCl and passing the mixed gases over an alumina gel catalyst at about 350°C [12]. Other acid catalysts such as cuprous chloride or zinc chloride are also used. The reaction can also be run in the liquid phase. Subsequent chlorinations to produce the other chloromethanes can be run either thermally or photochemically. Chlorine and the chloromethanes are fed to a tubular reactor at 490-530°C. The product distribution can be controlled by changing feed ratios.


## VII.  ACETIC ACID

The production of acetic acid consumed about 12% of the methanol produced in 1982. This is the most rapidly growing chemical application for methanol. It is also the first commercial application to methanol of the new homogeneous catalytic processes uncovered in the last 10 years or so. This process for acetic acid was developed by Monsanto [13-15], and it is the process being used in practically all new acetic acid facilities. There are approximately eight producers of acetic acid in the United States and at least four of these now use the methanol-based process.

The overall reaction is as follows:

$$CH_3OH + CO \rightarrow CH_3CO_2H, \quad \Delta H = -29.5 \text{ kcal}, 400 \text{ K} \tag{14}$$

The reaction runs at 150-200°C at about 450 psi. The catalyst is rhodium salts (e.g., RhCl) with certain ligands and in the presence of an iodine compound such as $CH_3I$. The reaction usually takes place in a stirred reactor and the product then flows into a flash tank and then to a refining train.

Because of the high cost and scarcity of rhodium, much effort has been devoted to searching for a nonnoble metal catalyst that will do as well. To date, no commercially practical alternative to rhodium appears to have been found.

## VIII.  ACETIC ANHYDRIDE

The most recent commercial development of a process based on methanol is the Tennessee Eastman process for acetic anhydride [16].  Eastman has a large demand for acetic anhydride for its cellulose acetate-based film, cigarette filters, and plastics. Until recently, their acetic anhydride production has been based on ethylene, a petrochemical as starting material via acetaldehyde, acetic acid, and ketene.

Eastman's new process is based on gasification of coal in a Texaco gasifier to make synthesis gas which they convert into methanol.  The methanol is converted into methyl acetate by esterification with acetic acid and then carbonylated using chemistry related to that of the Monsanto acetic acid process [17].

$$CH_3OH + CH_3COOH \rightarrow CH_3\overset{\overset{\displaystyle O}{\|}}{C}OCH_3 + H_2O, \tag{15}$$
$$\Delta H = -3.5 \text{ kcal/mol, } 400 \text{ K}$$

$$CH_3\overset{\overset{\displaystyle O}{\|}}{C}OCH_3 + CO \rightarrow CH_3\overset{\overset{\displaystyle O\ \ O}{\|\ \ \|}}{C}OCCH_3, \quad \Delta H = -13.5 \text{ kcal/mol} \tag{16}$$

The carbonylation step was developed by Eastman but is similar to the Halcon processes for vinyl acetate and acetic anhydride and uses rhodium salt catalysts with ligands and an iodide promoter [18].  Despite the recent drop in petroleum prices, Eastman still believes the process is economically attractive [19].

## IX.  DIMETHYL ETHER

A new commercial use for methanol is the production of dimethyl ether (DME).  A compound which can function as an aerosol propellant, DME can be made from methanol by passing the vapors over a dehydration catalyst such as silica/alumina at temperatures of 300°C or higher [20].

$$CH_3OH \rightarrow CH_3OCH_3 + H_2O, \quad \Delta H = -5.3 \text{ kcal/mol, } 600 \text{ K} \tag{17}$$

Still relatively low in volume, this could become much larger if it becomes widely accepted for aerosol use.

## X.  THE FUTURE

The so-called "energy crisis" impacts on methanol growth potential in other ways than its use as a chemical feedstock.  Methanol is already being added to gasoline at about 5% concentration as an antiknock additive.  It is also physically mixed with tert-butanol and added to gasoline and also reacted with tert-butanol to form methyl tert-butyl ether for the same purpose.  Extensive tests are also underway to use methanol alone as a fuel for internal combustion engines.

Methanol has great potential as a turbine fuel for peaking loads in electrical utilities.  These and other fuel uses may eventually result in an enormous increase in methanol demand and consequently production far beyond the needs, however important, for chemical manufacture.  Large increases in methanol manufacture, which inevitably must come from coal, will mean lower unit costs.  This will make methanol even more attractive as a chemical feedstock and lead to conversion of other processes to methanol.

The potential of methanol for chemical feedstock and other uses has, of course, been recognized in other countries.  Saudi Arabia has a 600,000 ton a year methanol plant on stream as part of a huge petrochemicals complex and has another large plant under construction [21].  Saudi Arabia, Mexico, and other underdeveloped countries with petroleum and natural gas resources can see the opportunity to convert gas (much of it now being flared) to methanol, which can be more easily transported than gas, as a way to enter the world chemical markets.  These and other similar developments will inevitably tend to keep methanol prices down.

While it is difficult to predict what the future will bring, methanol appears destined to play an increasingly important role in the future of the chemical industry.

## REFERENCES

[1]  Oil Daily, p. 10 (November 8, 1983).
[2]  Oil Daily, p. A9 (November 15, 1983).
[3]  Kirk-Othmer, Encyclopedia of Chemical Technology, 3rd ed.,
     Vol. 11, p. 237.
[4]  R. N. Hader, "Formaldehyde from Methanol," Ind. Eng.
     Chem., 44, 1508 (1952).
[5]  V. E. Meharg and H. Adkins, U.S. Patent 1,913,405 (June 13,
     1933), to Bakelite Corp.
[6]  U. Tsao, "Formaldehyde Flowscheme Features Tough Catalyst,"
     Chem. Eng., p. 118 (May 18, 1970).

[7]  Kirk-Othmer, Encyclopedia of Chemical Technology, 3rd
     ed., Vol. 2, p. 286.
[8]  J. O. Leonard, U.S. Patent 3,387,032 (June 8, 1968).
[9]  E. I. du Pont de Nemours & Co., U.S. Patent 2,999,053
     (February 18, 1959).
[10] Allied Chemical Corp., U.S. Patent 2,998,355 (July 1,
     1959).
[11] I. Mochida, A. Yasutaki, H. Fujitsu, and K. Takeshita,
     J. Catal., 82, 313-321 (1983).
[12] Kirk-Othmer, Encyclopedia of Chemical Technology, 3rd
     ed., Vol. 5, pp. 668—713.
[13] Kirk-Othmer, Ibid., 3rd ed., Vol. 1, p. 135.
[14] F. E. Paulik, Catal. Rev., 6, 49 (1972).
[15] R. P. Lowry and A. Aquilo, Hydrocarbon Process., 53(11),
     103 (1974).
[16] Chemical and Engineering News, p. 6 (January 14, 1980).
[17] Hydrocarbon Processing, p. 131 (November 1982).
[18] Hydrocarbon Processing, p. 109 (February 1982).
[19] Chemical Week, p. 19 (June 1, 1983).
[20] British Patent 278,353 (March 25, 1929), to Delco Light Co.
[21] Oil Daily, pp. 8—9 (November 14, 1983).

# Hydrotreating of Hydrocarbons

# Fundamental Studies of Transition Metal Sulfide Hydrodesulfurization Catalysts

R. R. CHIANELLI
Corporate Research Laboratories
Exxon Research and Engineering Co.
Annandale, New Jersey

## I. INTRODUCTION

Hydroprocessing catalysts based upon the transition metal sulfides have been widely used for over 60 years and catalysts such as $Co/Mo/Al_2O_3$ remain the industry "workhorses" in hydroprocessing of petroleum-based feedstocks [1]. Such applications include sulfur removal (hydrodesulfurization), nitrogen removal (hydrogenitrogenation), and product quality improvement (hydrotreating, hydroconversion). Original interest (prior to World War II) in these catalysts centered on their activity in the hydrogenation of coal liquids which contain considerable amounts of sulfur, thus maintaining the transition metal in the sulfided state.

It was quickly discovered that Co, Ni, Mo, and W sulfides and their mixtures were the most active and least expensive of the transition metal sulfides [2]. Later (post-World War II) their major uses shifted to hydroprocessing of sulfur- and nitrogen-containing petroleum-based feedstocks with Co- and Ni-promoted Mo and W catalysts usually supported on $Al_2O_3$. However, as petroleum feedstock supplies dwindle, we are required to process larger quantities of "dirtier" feeds containing larger amounts of sulfur, nitrogen, and metals. In order to meet these requirements in the future, a new generation of transition metal sulfide-based catalysts will be needed which have higher activities, greater selectivity to desired products, and greater resistance to poisons.

In spite of the importance of these catalysts, little is understood regarding the general fundamental basis for and origin of their catalytic activity, although specific knowledge regarding Co/Mo catalysts is progressing rapidly. This paper reviews some of our recent studies of the fundamental properties of transition metal sulfide catalysts, which govern their ability to catalyze a given reaction. It is convenient to divide the discussion of properties into a discussion of three effects which become apparent during experimentation: electronic (effect of a particular transition metal in a sulfur environment), geometric (effect of structure for a given transition metal sulfide), and chemical (effect of local active site configuration). This division is done for simplicity even though we know that all effects are operating simultaneously in a catalytic reaction. Finally, we discuss the promotional effect which requires an understanding of the first three effects for explanation.

## II.  ACTIVITY MEASUREMENTS AND CATALYST PREPARATION

The hydrodesulfurization (HDS) of dibenzothiophene (DBT) was chosen as a model reaction in these studies because it is a compound representative of the organic compounds in real feeds which are the most difficult to desulfurize:

| DBT | BP | CHB |

DBT = dibenzothiophene      BP = biphenyl      CHB = cyclohexyl benzene

A typical test feed was prepared by dissolving 4.4 g of DBT in 100 cc of hot decalin containing 5 wt% DBT or about 0.89 wt% sulfur. The disappearance of the reactant and the appearance of the products as depicted above were followed by gas chromatography. A stirred autoclave reactor was used to evaluate HDS activity of the catalyst at 350°C and/or 400°C under conditions described in Ref. 3. At 400°C the partial pressure of hydrogen is low, and thus the product distribution consists primarily of BP. At 350°C hydrogenation is sufficiently favored to yield a variety of hydrogenation products and the reaction is kinetically controlled with minimal diffusional perturbations. The Carberry batch autoclave (100 cc) was modified to allow constant flow of hydrogen through the feed, removal of reaction inhibiting $H_2S$, and liquid sampling from the activity evaluation.

All rate constants reported are zero-order as determined from the best fit to a straight line (linear regression analysis) obtained from a plot of total conversion of DBT vs time for a given temperature. All measurements were made up to 50% total conversion of DBT, and linear plots resulted for all catalysts reported [3].

Of prime concern in this study was the uniform preparation of the transition metal sulfides across the Periodic Table so that the activity measurements would be as preparation independent as possible. The technique developed to achieve this involves the precipitation from nonaqueous solution of the amorphous transition metal sulfide starting with the corresponding transition metal halide [4, 5]. By using this procedure we were able to obtain transition metal sulfides from Groups IV-VIII which were oxide free and of moderate surface area ($\sim$10-60 $M^2/g$). Prior to activity testing, the catalysts were pretreated in $H_2S$ for 1 h at 400°C, washed with 12% acetic acid to remove the LiCl produced in the reaction, then treated again at 400°C in 15% $H_2S/H_2$. This procedure converts the amorphous sulfides to the poorly crystalline sulfide phases shown in Table 1. Also indicated in Table 1 are the phases which were present after the HDS reaction. These phases were determined by x-ray diffraction where possible but in many cases the resulting diffraction patterns were either too broad for unambiguous interpretation or completely amorphous. Thus, Table 1 contains approximate stoichiometries for V, Fe, Os, and Ir, and further work is required to define some of these phases. Most catalysts undergo loss of sulfur during reaction, with this loss being most pronounced for the case of Ru and Rh. Nevertheless, the sulfides listed in Table 1 are the stable phases which come out of the reactor after activity testing. These sulfide phases form the basis for the discussions below.

TABLE 1

Stable Binary Sulfides

| $H_2/15\% H_2S$ (400°C) | HDS reactor (400°C) |
|---|---|
| $TiS_2$ | $TiS_2$ |
| $VS_x$ | $VS_x$ |
| $Cr_2S_3$ | $Cr_2S_3$ |
| MnS | MnS |
| $FeS_x$ | $FeS_x$ |
| $Co_9S_8$ | $Co_9S_8$ |
| $NiS_x$ | $NiS_x$ |
| | |
| $ZrS_2$ | $ZrS_2$ |
| $NbS_2$ | $NbS_2$ |
| $MoS_2$ | $MoS_2$ |
| $RuS_2$ | $RuS_{2-x}$ |
| $Rh_2S_3$ | $Rh_2S_{3-x}$ |
| PdS | PdS |
| $SnS_2$ | - |
| | |
| $HfS_2$ | $HfS_2$ |
| $TaS_2$ | $TaS_2$ |
| $WS_2$ | $WS_2$ |
| $ReS_2$ | $ReS_2$ |
| $OsS_2$ | $OsS_x$ |
| $IrS_x$ | $IrS_x$ |
| PtS | PtS |
| Au° | Au° |

The preparation of $RuS_2$ may serve as an example of a typical catalyst preparation, and the reader is referred to Refs. 3-5 for further information. $RuS_2$ may be prepared from $RuCl_4$ or $RuCl_3$. In a typical preparation, 7.4 g of $RuCl_4$ was dissolved in 100 mL of ethyl acetate and 2.80 g $Li_2S$ was added with stirring. After 4 h of stirring, the solution was filtered, yielding a black powder which was still wet with ethyl acetate. The filtrate was partially green, indicating suspended particles of $RuS_2$. The sample was then heat treated in pure $H_2S$ at 400°C for 1.5 h, cooled to room temperature, washed with 12% acetic acid, filtered, and heated again in 15% $H_2S/H_2$ for 1.5 h. This procedure yielded pure $RuS_2$ as determined by x-ray diffraction and chemical analysis.

## III.  THE ELECTRONIC EFFECT

The general periodic variation of the ability of the transition
metal sulfides to catalyze the HDS of sulfur-bearing molecules
was recently reported [3].  This ability varies smoothly over
three orders of magnitude of catalytic activity from Group IVB
to Group VIIB, yielding "volcano" curves for the second and
third row elements with maxima occurring in Group VIIIB with
the first row elements being relatively inactive.  Such curves
had not been previously reported for transition metal sulfides,
although Wakabagashi et al. [6] recently discussed the HDS of
thiophene over alumina-supported metals but found no smooth
variation of activity, probably because of superimposed effects
of the $Al_2O_3$ support.

Initially the rates of desulfurization of DBT were measured at
400°C under conditions described above.  A comparison of the
transition metal sulfides as hydrodesulfurization catalysts indicates that carbon-sulfur hydrogenolysis activity varies with the
position that the metal occupies in the Periodic Table (Fig. 1).
The hydrodesulfurization activity varies by about 3 orders of
magnitude across a given period and down a given group.  The
maximum activity occurs in the second and third transition series
with a peak occurring near $RuS_2$ in the second row and Os in the
third row.  The first row transition metal sulfides are relatively
inactive compared to the second and third row transition series.

When normalized to surface area (Fig. 2), only slight changes
occur in the curves.  For example, the most active catalysts at
the peak of the curves change position; Rh becomes slightly more
active than Ru, Os becomes more active relative to Ir, etc.  Also,
the trends in the first row activities become smoother.  These
changes are not considered significant because it was found that
the hydrodesulfurization activities of the sulfides do not in general correlate to BET surface areas due to specific morphological
effects of structural and geometric origin [7].  Thus, at this
writing the normalization of the activity to a per metal basis
(Fig. 1) best reflects the intrinsic activity of the transition metal
sulfides.  These results indicate that the nature of the transition
metal in the sulfide (primary effect) dominates the role of anisotropy as discussed below.  The shape of the curves at 350°C remains essentially the same.

The catalysis literature contains numerous examples of model
reactions which display periodic maxima or "volcano" relationships.  Sinfelt [8] has reviewed broad relationships between
catalytic activity of various metals in hydrogenation, hydrogenolysis, isomerization, hydrocarbon oxidation, and ammonia

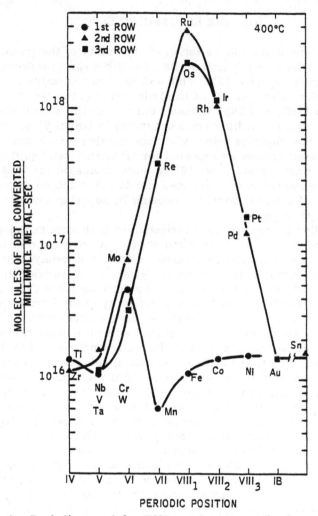

FIG. 1. Periodic trend for TMS catalyst normalized per milli-mole of catalyst.

synthesis-decomposition reactions and in the positions of the metals in the Periodic Table. The Group VIII transition metals display maximum activity when compared to Groups IV-VII and Groups I and IIB. Within Group VIII, the position of the maxima fluctuates depending on the reaction or upon the particular transition series under study. In general, the catalytic activity of these studies can be correlated with the electronic configuration of the

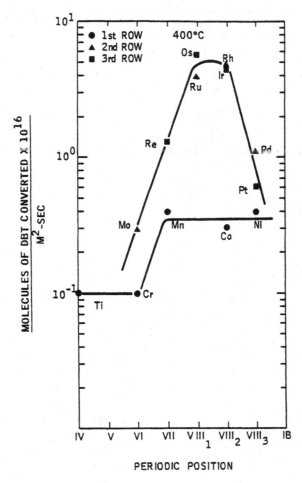

FIG. 2.  Periodic trend for TMS catalyst normalized per m$^2$.

d-orbitals as "percentage d character" (of the metallic bond based
upon Paulings valence bond theory) or with the strength of the
metal adsorbate bond.

The periodic trends for HDS activity in the case of the transition metal sulfides follow the correlations described above for
other catalytic systems.  A relation exists among catalytic activity, the heat of adsorption of a reacting molecule, and the heat
of formation of the corresponding sulfide [3].  This relation,
which is the well-known principle of Sabatier [9], states that compounds exhibiting maximum activity for a given reaction will have

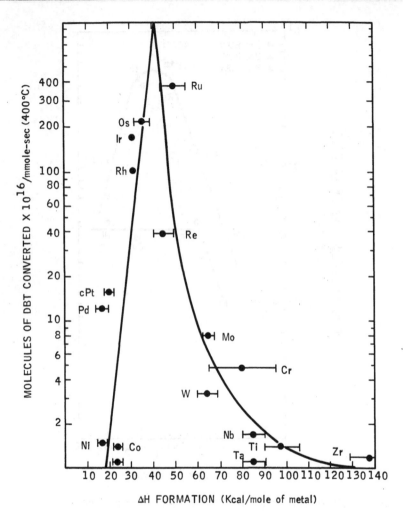

FIG. 3.  Heat of formation of TMS vs HDS activity.

intermediate heats of formation.  For sulfides catalyzing the HDS
reaction, the compounds exhibiting maximum activity will have intermediate heats of formation, presumably because the stability of
the surface complex formed by the sulfur-bearing molecule will be
intermediate.  The metal sulfur bond strengths of the transition
metal sulfides decrease continuously across the Periodic Table.
For the second and third transition series the most active catalysts
have intermediate values of the heat of formation (30-55 kcal/mol)
as seen in Fig. 3.  This suggests that the bond strength of the

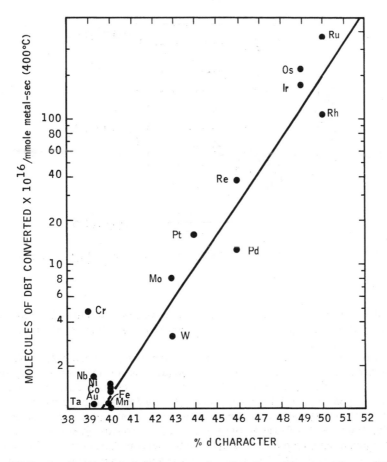

FIG. 4.  Pauling % d character for the transition metals vs HDS
activity.

metal-sulfur bond at the surface of the catalyst must not be too
strong or too weak to obtain the maximum desulfurization rate.
This appears to be consistent with the commonly accepted idea
that sulfur vacancies on the surface of the catalyst are the active HDS sites [1].  However, this cannot be the entire picture
since MnS in the first transition series falls within the required
range (51 kcal/mol) but shows very low activity.  Additionally,
Pauling percentage d character for the transition metals correlates very well to HDS activity although we do not understand
why this should be so (Fig. 4).  The most that can be stated
regarding these correlations is that the strength of the metal

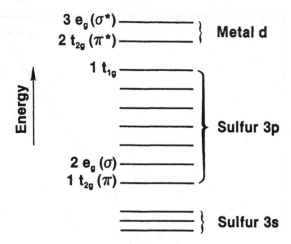

FIG. 5. Valence energy levels for octahedial $MS_6^{n-}$ cluster.

d-sulfur bond at the surface of the catalyst is important in determining HDS activity, as well as the presence of 4d and 5d electrons
in the catalyst.

   In order to provide a basis for further insight into this problem,
a theoretical study was undertaken to determine how the electronic
structure of the transition metal sulfides varies as a function of
periodic position and to determine if a relation exists between the
calculated electronic structure of the transition metal sulfides and
their observed catalytic behavior [10-12]. Scattered wave Xα calculations were employed to study the electronic structure of a series of octahedral $MS_6^{n-}$ clusters. In these clusters the transition
metal M was varied systematically across the first transition series from Ti to Ni and across the second transition series from Zr
to Pd. In a discussion of these cluster calculations and the resulting energy levels and charge distributions [10], several interesting trends in the electronic structure of the sulfides were
observed. By considering the results of the cluster calculations
in more detail, it was possible to identify electronic factors which
do correlate with the catalytic activity of the metal sulfides [11,
12]. A schematic diagram of the valence energy levels calculated
for the clusters is shown in Fig. 5. It was found that the calculated quantities relevant to a discussion of catalysis are all related to the orbitals having $t_{2g}$ and $e_g$ symmetry. These are the
orbitals through which the metal-sulfur d-p covalent interactions
occur. The molecular orbitals of interest are the $2e_g$ and $1t_{2g}$

levels, which are the sigma and pi bonding orbitals between the
metal d and sulfur 3p orbitals and the $3e_g$ and $2t_{ag}$ levels, which
are the higher energy antibonding counterparts of the bonding
$2e_g$ and $1t_{2g}$ orbitals. The bonding $2e_g$ and $1t_{2g}$ orbitals have
high sulfur 3p character, lie low in energy within the occupied
valance levels, and are fully occupied in all the clusters. The
quantity of interest for these bonding orbitals is the metal contribution to each orbital. This contribution provides a good measure of the relative covalency of the metal-sulfur bonding, a
larger metal contribution corresponding to greater metal-sulfur
covalency. In a comparison of the metal contribution to the bonding orbitals, it was found that in the 4d series this contribution
is not only larger but also has both a sigma and pi component,
whereas for the 3d metals only the sigma component is important.
The antibonding $3e_g$ and $2t_{2g}$ orbitals, on the other hand, have
high metal d orbital character and are the highest energy occupied valence levels. The occupation of these orbitals was found
to vary as the transition metal is varied, and it is these orbital
occupations as well as the covalency of the bonding of the orbitals which prove to be of interest from the catalytic point of
view.

It is possible to identify factors which are related to these
metal-sulfur bonding and antibonding molecular orbitals and which
correlate directly with catalytic activity. The first such factor is
the number of electrons in the highest occupied molecular orbital
(HOMO) of each cluster. The HOMO is either the $3e_g$ or $2t_{2g}$ orbital, so that this factor, which we call n, can be thought of as
the number of metal d electrons in the HOMO. The trends in n
follow very closely the trends in catalytic activity of the sulfides,
i.e., the sulfides which have a large value of n also have high
HDS activity (Fig. 6). The other factor which correlates with
catalytic activity is the relative metal-sulfur d-p covalent bond
strengths. Although this quantity could not be obtained directly
from the calculations, a parameter (B) whose relative size provides
a measure of the relative metal-sulfur d-p covalent bond strength
was defined as

$$B = n_\sigma D_\sigma + n_\pi D_\pi$$

where $D_\sigma$ and $D_\pi$ are the metal d orbital contributions to the $2e_g$
(sigma bonding) and $1t_{2g}$ (pi bonding) orbitals, respectively, and
$n_\sigma$ and $D_\pi$ are the net number of sigma and pi bonding electrons.
Since $D_\sigma$ and $D_\pi$ provide a measure of the sigma and pi covalency
and $n_\sigma$ and $n_\pi$ measure the net number of electrons which take

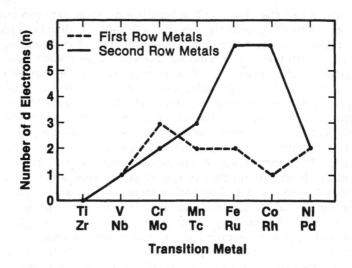

FIG. 6. Number of d electrons in HOMO (n) vs periodic position.

part in the d-p sigma and pi bonds, the products $n_\sigma D_\sigma$ and $n_\pi D_\pi$ provide a measure of the relative metal-sulfur d-p sigma and pi bond strengths. The sum of these quantities, B, gives a relative measure of the overall metal-sulfur d-p covalent bond strength. In general, throughout either the 3d or 4d transition series, $D_\sigma$ and $D_\pi$ increase while $n_\sigma$ and $n_\pi$ correlate between relative metal-sulfur covalency and high HDS activity. That is, a larger value of B is associated with higher HDS activity (Fig. 7). Comparison of the two factors n and B with the catalytic activity of the metal sulfides suggests that the better catalysts have larger values of n and/or B. It was assumed that both n and B are directly related to activity, and a correlation between the calculated electronic structure and the experimental catalytic activity of the sulfides was obtained. By defining an activity parameter A, which is the product of n and B:

$$A = nB$$

A plot of A calculated for each cluster versus periodic position is shown in Fig. 8. Also shown in this figure, plotted against the scale on the right, are the experimentally-measured HDS activities of the corresponding sulfides. Although the agreement is not exact, the overall behavior of the activity parameter A follows

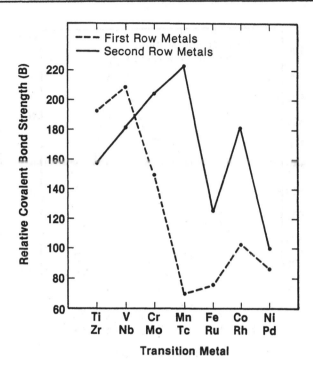

FIG. 7.  Relative covalent bond strength (b) vs periodic position.

very closely the trends in measured activities.  Particularly noticeable is the strong differentiation between the 3d and 4d transition metals.  The activity parameter A thus appears to be a good predictor of the HDS activity of the transition metal sulfides, and this behavior indicates that electronic factors are related to catalytic activity in the case of HDS.  From the simplest point of view, these factors suggest that for an optimum HDS catalyst:  (1) there should be a high d electron density available on the transition metal, and (2) the transition metal should be capable of forming a strong covalent bond with sulfur 3p orbitals.  The calculated results indicate that this covalent interaction should have both a sigma and pi component, and suggest that both the formation of a vacancy and the bonding capability of the transition metal at the vacancy are important for high HDS activity.  An HDS mechanism where (1) the reacting heterocyclic molecule binds to the transition metal at vacancy through the ring sulfur atom and (2) the transition metal donates electron

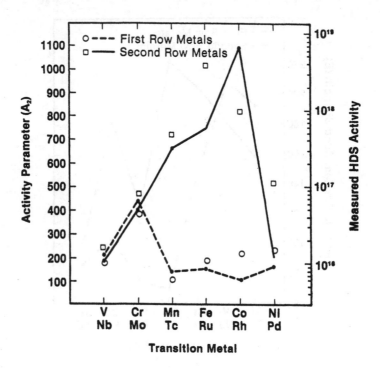

FIG. 8.  Activity parameter (A) vs periodic position.

density into a ring pi* orbital, thus facilitating the breaking of
C−S bonds, would be consistent with the calculated trends and
the electronic properties of the more active HDS catalysts [11,
12].

Although electronic factors which appear to provide a basis
for the observed trends in catalytic activity of a wide variety of
transition metal sulfides have been identified, it is still necessary
to define the relation between these calculated quantities which
are based on the bulk electronic structure of the metal sulfides
and the electronic structure of active sites on the catalyst surface.  In this case, the strong correlation between these factors
and activity on the surface suggests that the surface interactions
which are important catalytically are related to the interactions
measured by these factors.  Presumably, because the underlying
bulk electronic structure provides the "base" electrons which determine the energetic and symmetry properties of the active surface electrons.  While more work needs to be done, we need to

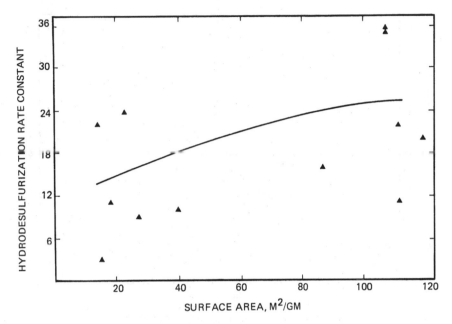

FIG. 9.  HDS activity vs surface area for $MoS_2$ catalysts.

define the relation between bulk and surface electronic structures.  In order to do this we need to understand the effect of
crystal structure on catalytic activity.

### IV.  THE GEOMETRIC EFFECT

As described above, crystal structure is of secondary importance in determining the activity of a TMS catalyst.  However,
structure becomes quite important when considering the activity
optimization of a particular TMS.  We define this secondary effect
in TMS catalysts as the "geometric effect."  The geometric effect
is clearly manifested by the heat treating of a particular TMS catalyst under different conditions (changing properties such as surface area, pore size, and crystallite size) and observing the effect
on the catalytic activity.  It is suggested that in any catalyst there
is an ideal set of heat treatment conditions where the HDS activity
is at a maximum.  Previously, this effect has been studied for $MoS_2$,
where due to its highly anisotropic layered structure, HDS activity
does not correlate well with BET surface area [7].  This can be
seen upon inspection of Fig. 9 which shows this lack of correlation

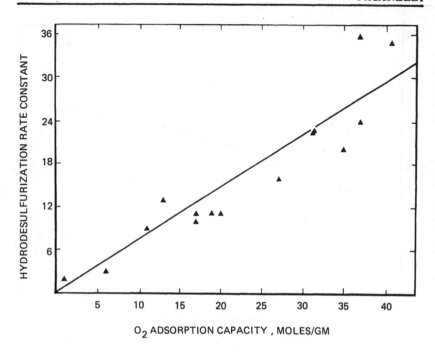

FIG. 10. HDS activity vs $O_2$ adsorption capacity for $MoS_2$ catalysts.

for 13 catalysts. In addition to the absence of linearity, the scatter of the data around any curve is very large. A vastly improved correlation emerges if catalytic activity is plotted versus $O_2$ chemisorption capacity as shown in Fig. 10. The relationship, despite some scatter, is seen to be linear and furthermore appears to pass through the origin.

Voorhoeve and Stuiver [13] have proposed that the edge plane in $MoS_2$ and $WS_2$ is the site of their hydrodesulfurization activity, based on ESR studies of $NiS/WS_2$ catalysts. Their model of pseudo-intercalation by $Ni^{2+}$ at the $WS_2$ edge plane was supported by electron micrographs obtained by Farragher and Cossee [14]. The present results support this hypothesis and demonstrate that the chemisorption of oxygen, when applied in a dynamic mode, is effective in determining the edge plane area of $MoS_2$. The poor correlation of activity with total surface area apparently reflects the fact that the edge plane/basal plane ratio is capable of wide variations among preparations of $MoS_2$, sometimes for reasons that are difficult to identify or control.

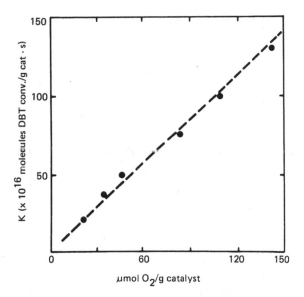

FIG. 11. $O_2$ chemisorption vs surface area for $RuS_2$ catalysts.

$RuS_2$, on the other hand, has a completely isotropic cubic structure identical to pyrite ($FeS_2$). The isotropic nature of $RuS_2$ is clearly demonstrated in linear correlations between HDS activity, surface area, and $O_2$ chemisorption [15]. These relations are indicated in Figs. 11 and 12 for $RuS_2$ catalysts prepared in $H_2/H_2S$ from amorphous $RuS_2$ [16]. $RuS_2$ is an example of a well-behaved catalyst system unlike $MoS_2$ where anisotropy plays a large role. $RuS_2$ and $MoS_2$ exhibit examples of the effect of anisotropy in HDS catalyst. Preliminary results indicate that all other TMS catalysts described in the previous section fall somewhere in between these examples. Further work is in progress to determine the "geometric effect" in the most active TMS. When complete, a picture will emerge of the periodic trends for HDS including both electronic and geometric factors.

The chemisorption studies cited above give strong evidence for the activity of the edge planes of $MoS_2$. Tanaka and Okuhara showed that the edge planes of $MoS_2$ were reactive for certain types of reactions by cutting single crystals into pieces and comparing the rates of cut and uncut crystals [17]. However, confirming studies on single crystals using modern surface science techniques are still rare. One such study was the adsorption and binding of thiophene and other molecules on the basal plane of

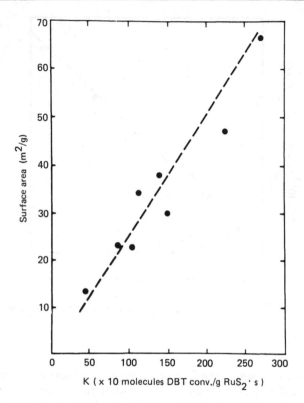

FIG. 12.  Surface area vs HDS activity for $RuS_2$ catalysts.

$MoS_2$ single crystals [18]. The investigations were carried out in
an ultrahigh vacuum chamber that was equipped with several tools
for surface analysis. These included low energy electron diffraction (LEED) for surface structure determination and Auger electron spectroscopy (AES) for analysis of the surface composition.
The sample holder permitted cooling of the specimen to 77 K or
heating to over 1500 K, and a chromel-alumel thermocouple was
attached firmly to the sample through a hole pierced near one
edge. The crystals were in the form of thin sheets with typical
dimensions of approximately 10 X 15 X 0.5 mm and were mounted
on a gold foil forming a frame around its edges. Prior to insertion in the ultrahigh vacuum chamber (base pressure 1 X $10^{-9}$
torr), both sides of the crystal were exfoliated in air. The chemical and structural inertness of the crystal was revealed by the
fact that a good quality LEED pattern was obtained readily just

after evacuation of the chamber to $10^{-7}$ torr, without any heat
treatment of the crystal. Also, the Auger spectrum showed that
only negligible amounts of carbon are present apart from the S
and Mo signals. In the adsorption experiments, the crystal was
cooled through the Au support to near liquid nitrogen temperatures ($\sim 77$ K). Only at these low temperatures could thiophene
be adsorbed when using exposures in the range of $10^{-6}$ torr·s.
The crystal was subsequently heated by radiation from a hot
tungsten filament located at approximately 1 cm from its back
face. A linear temperature ramp of approximately 10 K/s was
produced in all of the desorption experiments. The desorption
products were analyzed by means of a quadrupole mass-spectrom-
eter. Thiophene desorbed without decomposition as revealed by
the absence of $H_2$-evolution and of carbon residues left on the
surface as monitored by Auger spectroscopy. The desorption of
intact $C_4H_4S$ molecules followed a first-order process as shown in
Fig. 13 by the constancy of the desorption peak temperature, at
$165 \pm 5$ K, with changing surface coverage. The binding energy
of $C_4H_4S$ on the basal plane of $MoS_2$ was estimated to be E = 9.5
kcal/mol. The adsorption of 1,3-butadiene, an important product in HDS chemistry, yielded similar results. In order to adsorb this molecule, the crystal must be cooled to below 170 K.
Upon heating, $C_4H_6$ desorbs without decomposition with a maximum rate at $150 \pm 15$ K. From this value a binding energy of approximately 8.5 kcal/mol was estimated. Finally, $H_2S$ was also adsorbed on $MoS_2$ below 170 K. Intact molecules desorbed at a maximum rate at temperatures around 165 ($\pm$ 15) K. Some $H_2$ desorption was observed simultaneously at approximately the same temperature. The Auger spectrum after these experiments, however, showed no detectable decomposition of $H_2S$ on the $MoS_2$
basal plane. These experiments show that the basal planes of
$MoS_2$ exhibited very little chemical activity when exposed to
$C_4H_4S$, $C_4H_6$, or $H_2S$ at low pressures.

It was further shown that the basal plane of $MoS_2$ was inert to
$O_2$ exposure at 520 K [19]. Only sputtering with He ions, which
caused the destruction of the hexagonal LEED pattern, would induce reactivity toward $O_2$. The basal plane could be annealed at
100 K and its inertness recovered. From this study it was concluded that defects may be introduced into the surface by sputtering, and this produced a drastic increase in the rate of oxidation of the surface and in removal of sulfur. This indicates that
oxygen chemisorption (and thus HDS) is associated with defect
sites in $MoS_2$ and not with an ordered basal plane. It should also
be kept in mind that the concept of edge and basal planes is an
idealized one and that by "edge sites" is meant defects which

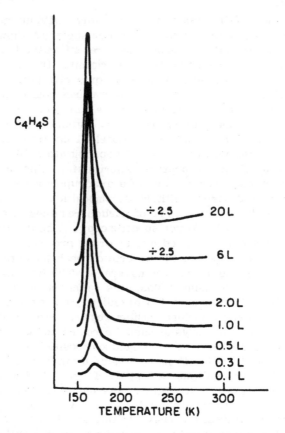

FIG. 13. Desorption of thiophene on $MoS_2$ basal plane as a function of temperature.

have an idealized edge-like local structure and that these defects can occur in disordered basal planes. Finally, we may note that from the above it is clear that we would like to study the edge planes of layered sulfides. However, here nature does not co-operate with us because $MoS_2$ by the nature of its anisotropic structure can only be grown in very thin crystals which do not permit easy study of edge planes. Some recent attempts have been made to grow edge-like surfaces on Mo single crystals, but the relation of these surfaces to real catalysts awaits demonstration [20].

TABLE 2

Characterization and Catalytic Data for Bulk Molybdenum
Sulfide Catalysts (tested)

| Starting material | Activity ($\mu$mol DBT / g·s) | BET surface area ($m^2/g$) | $O_2$ Chemisorption ($\mu$mol/g) | ESR intensity (X $10^{18}$ spins/g) |
|---|---|---|---|---|
| $MoS_3$ | 0.37 | 114 | 31 | 2.30 |
| $MoS_3$ | 0.33 | 120 | 35 | 1.85 |
| $MoS_3$ | 0.27 | 88 | 27 | 1.23 |
| $MoS_3$ | 0.18 | 112 | 19 | 0.83 |
| $MoS_2$ | 0.18 | 35 | - | 0.74 |
| $MoS_2$ | 0.15 | 27 | 11 | 0.51 |
| $MoS_2$ | 0.13 | 43 | - | 0.84 |

## V. THE CHEMICAL EFFECT

In the previous section evidence for associating the active site
for HDS catalysis on the edge planes of $MoS_2$ crystallites was presented. What is the structural nature of this site and what is the
concentration of these sites on the edge plane? Such questions
are lumped in the term "chemical effect" in TMS catalysts. A recent paper describes a correlation between electron spin resonance (ESR) and catalytic activity for the HDS of DBT by TMS
catalysts [21]. ESR is a tool of great potential utility in catalyst studies. It is a highly sensitive probe of defect sites in a
material. Unpaired electrons and "dangling bonds" which are
likely associated with active sites are easily detected even at the
part per million level.

Seven $MoS_2$ catalysts were prepared (as described in Ref. 13)
and activity tested as previously described. The characterization and catalytic data for the activity tested catalysts are shown
in Table 2. Also shown are the BET surface area, the $O_2$ chemisorption, and the ESR intensity. The ESR intensities for these
catalyst defects have been determined with respect to a Varian
weak pitch carbon standard and are presented for a series of sulfide samples. The intensities, $\sim 10^{18}$ spins/g-sample, correspond

to $\sim 2 \times 10^{-4}$ spins/atom of molybdenum. These ESR intensities are compared with BET surface areas, oxygen chemisorption levels, and the catalytic activity level for the HDS of DBT in Fig. 14. Figure 14(A) shows the almost linear variation of ESR intensity with DBT activity, which suggests a connection between the molybdenum defects and the catalytic process. A similar linear variation is seen for the level of oxygen chemisorbed (Fig. 14B). By contrast, the correlation between BET surface area and HDS activity is weak (Fig. 14C). This suggests that not only the $MoS_2$ surface area but also the types of surfaces are significant. The surfaces most likely to adsorb oxygen and to contain paramagnetic molybdenum atoms are the defects (edges, corners, etc.) in the layered $MoS_2$ structure.

While both ESR and $O_2$ chemisorption reflect the HDS activity of these catalysts, a simple estimate of the magnitude of these two effects suggests that they are not necessarily associated with the same site. Taking the first catalyst of Table 2 as an example, 31 $\mu$mol of $O_2$ chemisorbed per gram of the final $MoS_2$ product implies that there are $9.9 \times 10^{-3}$ oxygen atoms adsorbed per molybdenum atom. Conversely, $2.3 \times 10^{18}$ thio-$Mo^{5+}$ spins/g-sample implies $6.1 \times 10^{-4}$ paramagnetic molybdenum species observed per molybdenum. This fifteenfold difference in number of sites observed can be explained by attributing oxygen chemisorption to all edge sites on the $MoS_2$ crystallites and asserting that only a subset of these edge sites, perhaps corner sites, are seen in ESR. Recent magnetic susceptibility measurements on $MoS_3$ materials suggest that all the edge sites are, in fact, paramagnetic but unobservable because of exchange interactions between neighboring sites [22].

The organometallic complex molybdenum trisdithiolene, which can be quantitatively reduced by reaction with $NaBH_4$ in diglyme, was chosen as a system to model the ESR signal of the $MoS_2$ catalysts. The resulting molybdenum species is formally pentavalent and coordinated with six sulfur atoms in a configuration similar to that encountered in $MoS_2$. The resulting solution ESR signal provides a measure of the isotropic g-value which might be anticipated for sulfur-coordinated, thio-$Mo^{5+}$, species. A narrow (splitting between derivative maxima, $\Delta H_{pp} \sim 7$ G), symmetric signal is observed with $g = 2.0091$, in good agreement with previous reports [23]. In the resulting ESR for $MoS_2$-prepared catalysts, shown in Table 2, absorption consists of a narrow signal with $g \simeq 2.004$ superimposed on a main broader ($\sim 120$ G) spectrum. The narrow signal was attributed to sulfur radical species which were eliminated during the HDS activity test. The broad component is asymmetric and typical of an axially symmetric g-value tensor, with $g = 2.0380$ and $g = 2.0038$. The mean g-value,

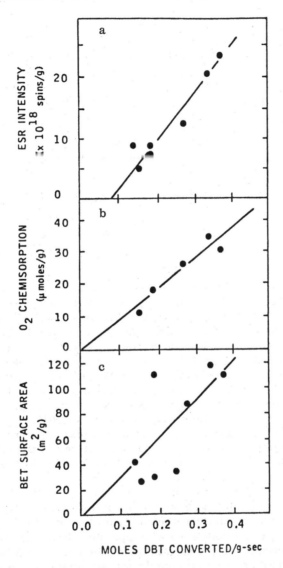

FIG. 14. Physical parameters vs HDS activity for $MoS_2$ catalysts: (a) ESR vs activity, (b) $O_2$ vs activity, (c) surface area
vs activity.

obtained by finding the center of gravity of the absorption, is
2.0096, quite similar to that seen for Mo-tris dithiolene as shown
in Table 3.

TABLE 3

Comparison of ESR g-values

|                     | Mo-Trisdithiolene | $MoS_2$ | $WS_2$[a] |
|---------------------|-------------------|---------|-----------|
| Isotropic g-value   | 2.0091            | 2.0096  | -         |
| gz                  | -                 | 2.0380  | 2.060     |
| gy                  | -                 | 2.0038  | 2.030     |
| gx                  | -                 | -       | 2.010     |

[a]From Ref. 25.

Samples with compositions of approximately $MoS_3$ are readily obtained by thermal decomposition of $(NH_4)_2MoS_4$ at temperatures of
300°C. The ESR absorption is different from the $MoS_2$ case. The
line is broader ($\sim$150 G) and the derivative signal is significantly
different. The narrow component seen for $MoS_2$ is not of the same
form. This ESR "signature" is reproducibly obtained and similar
in general form to the absorption for $MoS_3$ recently reported by
Johnston et al. [22]. The absorption has an obviously lower symmetry than the $MoS_2$ case. Assuming a single paramagnetic species
and a nonaxial g-value tensor, values of $g_1 = 2.0492$, $g_2 = 2.0180$,
and $g = 1.9736$ are obtained. In spite of these apparent differences, the mean g-value, 2.0162, is also similar to that of the
other thio-$Mo^{5+}$ defects. $MoS_3$ is converted to $MoS_2$ in the reactor.

The ESR results clarify the character of the defect sites. The
assignment of the observed defect site is primarily based upon the
closeness of the g-values in the catalysts to that of the trigonal
prismatic Mo-trisdithiolene model compound. However, the catalyst sites, because of the asymmetry of the observed ESR spectra,
are not isotropic and would contain the asymmetry to be expected.
Because the g-values are frequency independent, the fact that
the ESR absorption does not change in character at temperatures as low as 5 K also suggests that the molybdenum defects occur with a nominal $Mo^{5+}$ valence. By contrast, $Mo^{3+}$ species in low
symmetry crystalline field environments would be expected to have
highly anisotropic g-values and a finite crystal field splitting term,
producing frequency-dependent changes in the position and shape

of the ESR absorption [24]. $Mo^{3+}$ species in high symmetry sites would have rapid spin lattice relaxation rates and exhibit significant variations in lineshape with temperature. The g-value magnitudes are still the subject of calculation. As in the $Mo^{5+}$-trisdithiolene case, a g-value of $\sim 2.01$ is consistent with the electronic properties of the system. A detailed analysis of the g-values of model sulfide systems will appear elsewhere, and we must await these results for conclusive assignment of the active site symmetry and electronic structure. Finally, we may again note the work of Voorhoeve [25, 26] who correlated defect site density, as determined by integration of the ESR adsorption curve, and the activity for benzene hydrogenation over $WS_2$ catalysts. A linear relation was observed, similar to the one discussed here for the HDS reaction over $MoS_2$ catalysts. However, Voorhoeve assigned the observed ESR signal to $W^{+3}$ defects. The g-values for the $WS_2$ catalysts may be compared to those for $MoS_2$ catalysts in Table 3. The similarity is quite close, and it is probable that whatever the correct assignment for these defects is, both $MoS_2$ and $WS_2$, which are isostructural, contain the same type of ESR active site.

## VI. THE PROMOTIONAL EFFECT

It is well known that the presence of a second metal can in some cases lead to catalysts which have activities greater than the simple sum of the activities of catalysts based on binary sulfides. Such "promotion" of $MoS_2$ or $WS_2$ by Co and Ni occurs in either supported or unsupported catalysts. Although the subject of promotion has been well studied, no consensus exists as to the origin of this effect. It has been suggested that a relation exists between the activity of the well-known Co- and Ni-promoted $MoS_2$ and $WS_2$ catalysts and the periodic trends for the binary sulfides [3]. Studies of promotion have led to the idea of "contact synergy" for sulfided catalysts containing Co or Ni together with Mo or W [27]. Even though this specific idea appears to be incorrect, Ni/Mo, Co/Mo, Co/W, and Ni/W can be said to behave as "synergic pairs" which retains the idea that the members of these pairs "work together or cooperate." Although the bulk binary phases for these elements $Ni_3S_2$, $Co_9S_8$, $MoS_2$, and $WS_2$ are present in the unsupported catalysts, recent work by Topsøe et al. [28] has shown the presence in both supported and unsupported CoMo catalysts of a unique form of sulfided Co (the "CoMoS" phase) which correlates with activity. Moreover, although $Co_9S_8$ or $Ni_3S_2$ phases have been described as "promoters" of $MoS_2$ or $WS_2$, it has been shown that these phases themselves have activities which are of the same order as $MoS_2$ [29]. The synergic pairs of active sulfides give enhancements of activity far greater than would be expected by the simultaneous presence of noninteracting phases. A relation between the activity of these synergic pairs and the periodic trends for the binary sulfides appears by examining the average heats of formation of the sulfides of the synergic pairs.

As previously discussed, the heats of formation of the most active binary sulfides fall into an "optimum" range as shown in Fig. 3. The maximum in activity for the second transition series occurs near $RuS_2$ (where $\Delta H_f = 49.2$ kcal/mol) and in the third transition series Os (where $\Delta H_f = 35.3$ kcal/mol). There is some uncertainty in the position of the exact maxima due to incomplete correlation of HDS activity to BET surface area as described above. The most active catalysts have heats of formation in the range of 30-50 kcal/mol. Recent work shows that heats of formation for the sulfides are linearly correlated with heats of adsorption of sulfur on transition metal surfaces [30]. Presumably this correlation of $\Delta H_{ads}$, $\Delta H_f$, and activity reflects the optimum metal-sulfur bond strength on the surface of the catalyst. Under catalytic conditions this quantity is related both to the ease of formation of sulfur vacancies and to the strength of binding of S-containing reactants to the surface. The elements to the left of the Periodic Table have high heats of formation, bind sulfur or sulfur-bearing molecules too strongly, and are in a sense "poisoned" by sulfur. Sulfides of elements to the right of the Periodic Table have low heats of formation and sulfur-bearing molecules are likely bound too weakly for reaction to occur. Those sulfides which have intermediate heats of formations, bind sulfur-bearing molecules neither too strongly nor too weakly, are effective catalysts.

If heats of formation of the individual sulfide components of the synergic pairs are averaged, these averages fall into the optimum range for the binary sulfides. A list of the average heats of formation of the first transition series sulfides and $MoS_2$ or $WS_2$ is shown in Table 4. The known synergic pairs all lie near the center of the range for the optimum value of the heat of formation ($\sim 40$ kcal/mol). This relation suggests that the synergic pairs behave catalytically as second or third row pseudobinary sulfides and implies a fundamental relation between the periodic trends for the binary sulfides and the activity of the synergic pairs (Fig. 15). Presumably, the sulfides of the synergic pairs may have surface properties which reflect the average surface

TABLE 4

Average Heats of Formation of Pairs of
Transition Metal Sulfides

|         |      | MoS$_2$ | WS$_2$ |
|---------|------|---------|--------|
|         |      | 65.8    | 62     |
| MnS     | 51.1 | 58.5    | 56.6   |
| FeS$_2$ | 42.6 | 54.2    | 52.3   |
| FeS     | 24   | 44.9    | 43     |
| Co$_9$S$_8$ | 19.8 | 42.8 | 40.9   |
| Ni$_3$S$_2$ | 17.2 | 41.5 | 39.6   |
| CuS     | 12.7 | 39.3    | 37.4   |
| ZnS     | 46   | 57.3    | 54     |

properties and therefore the average bulk properties of their individual components. Thus, the sulfided Co/Mo or Ni/Mo catalysts behave at the surface as sulfides of hypothetical elements of
periodic position between those of the members of the pair, hence
the term "pseudobinary sulfide." The average value of the heat of
formation of the synergic-pair sulfides and other pairs of sulfides
which are not promoters are shown in Table 4. While the average
heat of formation of the synergic pairs falls in the 35-50 kcal/mol
range, that for pairs (with the exception of the closest metals to
Co and Ni, namely Fe and Co) which are not known as promoters
falls outside this range. Solid curves in Fig. 15 illustrate qualitative trends in relative activity which may occur for the sulfide
pairs. Further measurements may more accurately determine the
optimum heat of formation for maximum activity. In the case of
iron there is ambiguity because the Fe/S phase present under reaction temperature is not unequivocally known due to the complexity of the Fe/S phase diagram. If we chose the heat of FeS$_2$, the
Fe/Mo average heat of formation falls outside the optimum range.
As iron sulfide does not promote molybdenum sulfide catalysts [1],
this choice is consistent with the literature and with our own measurements discussed below. A similar argument can be made for
Cu which is also not known as a promoter.

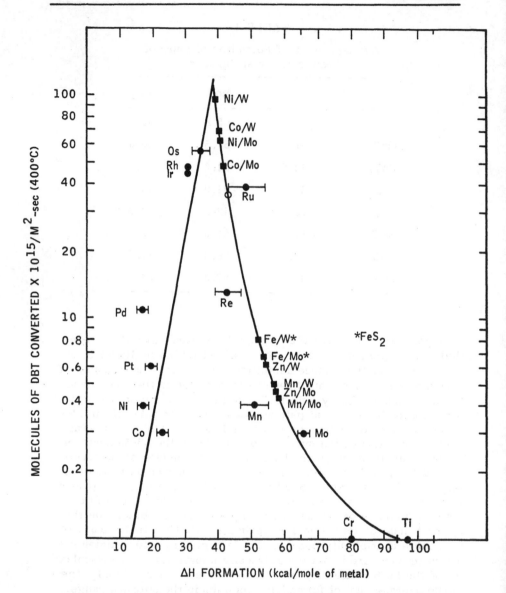

FIG. 15. HDS activity of the transition metal sulfides vs heat of formation of sulfides (circles are the binary sulfides, squares are the average heat of formation of the pairs of binary sulfides taken from Table 4). Open circle is a measured activity for Co/Mo catalyst at 400°C.

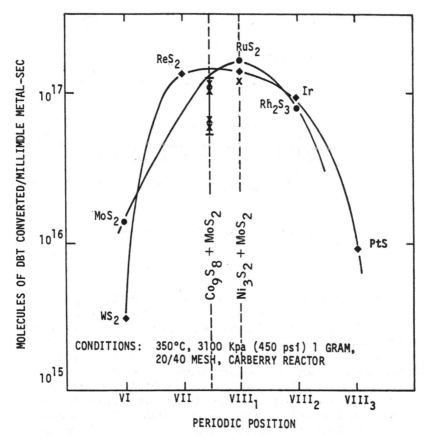

FIG. 16. Activity of the second and third row transition metal sulfides and the relation of the "synergic pairs" $Co_9S_8/MoS_2$ and $Ni_3S_2$ to the binary sulfides.

The activity of the synergic-pair Co/Mo and Ni/Mo system prepared in an analogous manner to the binary sulfides described above approximates that of a hypothetical pseudobinary having the average properties of $Co_9S_8$ and $MoS_2$ (Fig. 16). That some of the synergic catalysts fall below the trend curve can be attributed to differences in physical properties of the catalysts. Because of the difficulty in correlating BET surface area to HDS activity previously noted, we must await turnover number information for more precise quantitation of the suggested relation.

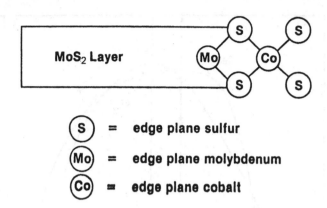

FIG. 17.  Hypothetical position for Co at MoS$_2$ edge plane.

This is particularly troublesome in the case of the synergic pairs,
and catalysts of the same nominal composition and surface area
have substantially different activities.  However, we note that in
general the surface areas of the synergic pairs are lower than
those of the pure phases (usually in the range of 15-25 m$^2$/g),
whereas MoS$_2$ prepared in a similar manner has surface areas in
the range of 50 m$^2$/g [1].  This suggests that in the synergic
pairs the number of active sites is not increasing but rather that
the quality of the active site may be enhanced.  In fact, a data
point for Co/Mo at 400°C normalized to surface area falls quite
close to the point (Fig. 15) for the average value of the heat of
formation of the synergic pair Co/Mo.  This idea is supported by
a recent paper on the role of Co in unsupported MoS$_2$ catalysts
which shows that the surface area for a series of catalysts (where
$0 \leq M/M + Mo \leq 1$) is equal to or lower than that of unpromoted
MoS$_2$ [31].  Furthermore, in the same paper the activation energy for the same series of catalysts was measured for dibenzothiophene conversion and found to be constant (20.7 ± 0.7 kcal/
mol) for the entire series.

Several microscopic theories for promotion have been presented in the literature and are consistent with the suggested
pseudobinary relation.  This relationship calls for a CoMoS unit
to be present at the surface, such as is shown in Fig. 17.  For
Co/Mo catalysts this suggestion is consistent with the Co being
located somewhere at the edge of the MoS$_2$ crystallites.  For example, the Co could be located at the edge of a single MoS$_2$ layer

as suggested by Ratnasamy and Sivasanker [32], or it could be
located between the layers near the edge (pseudointercalation) as
suggested by Voorhoeve and Stuiver [13]. Co located at the edge
of $MoS_2$ is also consistent with the "surface enrichment" model of
Phillips and Fote [33]. The essential point is that somewhere at
the edge, sulfur atoms (which upon leaving create vacancies) are
shared by Co and Mo, and behave in an average electronic fashion.
These special sulfur atoms are similar to sulfur atoms attached to
the surfaces of noble metal sulfides as in $RuS_2$ [34]. Presumably,
if such electronic interactions occur at the surface of promoted catalysts as experiment suggests, the strength of those metal-sulfur
bonds which lead to vacancies are adjusted to intermediate values
which, as the binary periodic trends suggest, are necessary for
high activity. Thus, in the case of Co and Ni promoters, such
surface states exist with sulfurs shared between the promoter
and Mo, and which have optimum bond strength for vacancy formation as indicated in Fig. 17.

## Acknowledgments

The author wishes to acknowledge the work of T. A. Pecoraro,
S. J. Tauster, S. Harris, J. D. Passeratti, G. A. Somorjai, A.
Wold, M. Salmerion, B. G. Silbernagel, D. C. Johnston, A. J.
Jacobson, E. I. Stiefel, W.-H. Pan, and T. R. Halbert, without
which this article would not have been possible.

## REFERENCES

[1]  O. Weisser and S. Landa, Sulfide Catalysts: Their Prop-
     erties and Applications, Pergamon, Oxford, 1973.
[2]  E. E. Donati, Adv. Catal., 8, 39 (1956).
[3]  T. A. Pecoraro and R. R. Chianelli, J. Catal., 67, 430-445
     (1981).
[4]  R. R. Chianelli and M. B. Dines, Inorg. Chem., 17, 2758
     (1978).
[5]  J. D. Passaretti, R. C. Collins, A. Wold, T. A. Pecoraro,
     and R. R. Chianelli, Mater. Res. Bull., 14, 1967 (1979).
[6]  K. Wakabagashi, H. Abe, and Y. Orito, Kogio Kagaku
     Zasshi, 74(7), 1317 (1971).
[7]  S. J. Tauster, T. A. Pecoraro, and R. R. Chianelli, J.
     Catal., 63(2), 515-519 (1980).
[8]  J. Sinfelt, Prog. Solid State Chem., 10(2), 55 (1975).
[9]  P. Sabatier, Ber. Deutsch. Chem. Ges., 44, 2001 (1911).
[10] S. Harris, Chem. Phys., 67, 229 (1982).

[11]   S. Harris and R. R. Chianelli, Chem. Phys. Lett., 101, 603-605 (1983).

[12]   S. Harris and R. R. Chianelli, J. Catal., 86, 400-412 (1984).

[13]   R. S. H. Voorhoeve and J. C. M. Stuiver, Ibid., 23, 228-241 (1971).

[14]   A. L. Farragher and P. Cossee, in Proceeding, 5th International Congress on Catalysis, Palm Beach, 1972 (J. W. Hightower, ed.), North-Holland, Amsterdam, 1973, p. 1301.

[15]   J. D. Passaretti, S. J. Tauster, and R. R. Chianelli, J. Catal., Submitted.

[16]   J. D. Passaretti, R. C. Collins, A. Wold, R. R. Chianelli, and T. A. Pecoraro, Mater. Res. Bull., 14, 1967 (1979).

[17]   K. Tanaka and T. Okuhara, in Proceedings of the Third International Conference on the Chemistry and Uses of Molybdenum, Climax Molybdenum, 1979, pp. 170-175.

[18]   M. Salmeron, G. A. Somorjai, A. Wold, R. R. Chianelli, and K. S. Liang, Chem. Phys. Lett., 90(2), 105-107 (1983).

[19]   M. H. Farias, A. J. Gelman, G. A. Somorjai, R. R. Chianelli, and K. S. Liang, Chem. Phys. Lett., 90, 105 (1982).

[20]   M. Salmeron, G. A. Somorjai, and R. R. Chianelli, Surf. Sci., 127, 526-540 (1983).

[21]   B. G. Silbernagel, T. A. Pecoraro, and R. R. Chianelli, J. Catal., 78, 380-388 (1982).

[22]   D. C. Johnston, A. J. Jacobson, B. G. Silbernagel, S. P. Frysinger, S. M. Rich, and L. A. Gebhard, Solid State Commun., In Press.

[23]   E. I. Stiefel, R. Eisenberger, R. C. Rosenberg, and H. B. Gray, J. Am. Chem. Soc., 88, 2956 (1966).

[24]   S. E. Radhakrisha, B. V. R. Chowdari, and A. K. Viswanath, J. Chem. Phys., 66, 2009 (1977).

[25]   R. J. H. Voorhoeve and Z. Wolters, Z. Anorg. Allg. Chem., 376, 165 (1970).

[26]   R. J. H. Voorhoeve, J. Catal., 23, 236 (1971).

[27]   B. Delmon, in Proceeding of the Third International Conference on the Chemistry and Uses of Molybdenum, Ann Arbor, Michigan, August 19-23, 1979, pp. 73-85.

[28]   H. Topsøe, B. S. Clausen, R. Candia, C. Wivel, and S. Morey, J. Catal., 68(2), 433 (1981).

[29]   V. H. J. De Beer, J. C. Duchet, and R. Prins, Ibid., 72, 369-372 (1981).

[30]   J. Bernard, J. Oudar, N. BarBouth, E. Margot, and Y. Berthier, Surf. Sci., 88, L35-L41 (1979).

[31]   M. L. Vrinat and L. DeMourgues, Appl. Catal., 5, 43 (1983).

[32]   P. Ratnasamy and S. Sivasanker, Catal. Rev.–Sci. Eng., 22(3), 401 (1980).

[33] R. W. Phillips and A. A. Fote, J. Catal., 41, 168 (1976).
[34] R. R. Chianelli, T. A. Pecoraro, T. R. Halbert, W.-H. Pan, and E. I. Stiefel, Ibid., 86, 226-230 (1984).

# Importance of Co-Mo-S Type Structures in Hydrodesulfurization

HENRIK TOPSØE AND BJERNE S. CLAUSEN
Haldor Topsøe Research Laboratories
Lyngby, Denmark

## I. INTRODUCTION

The great current interest in hydrodesulfurization (HDS) and other hydrotreating reactions is related to the need for efficient upgrading of crude oil fractions or coal-derived liquids. The catalysts used for such reactions generally consist of molybdenum (or tungsten) supported on high surface area aluminas with cobalt or nickel added as promoters. Great efforts have been devoted to the understanding of the structural and chemical form

in which the different elements are present in the active catalyst
and to the establishment of correlations between such information
and the various catalytic functions. This massive research effort
has given valuable information on many aspects of such catalyst
systems (for recent reviews of the extensive literature, see, e.g.,
Refs. 1-11). However, it has not been possible to reach general
agreement on the types of structures present in the active catalysts and the origin of promotion.

In view of recent results, the lack of a detailed structural description of such catalysts can be related to the observation that
many different types of structures may be present and that some
of these are x-ray amorphous. Moreover, a very strong dependence on preparation parameters has been observed which makes
direct comparison difficult between the different studies reported
in the literature. Furthermore, suitable in situ techniques have
not been available previously and the catalytic activities have
therefore usually been related to chemical and structural parameters determined after removal of the catalysts from the sulfiding environment. Consequently, there has been a great need
for new in situ techniques. Recently, we have found that many
of the difficulties may be overcome by the application of two novel
in situ techniques, Mössbauer emission spectroscopy (MES) and
extended x-ray absorption fine structure (EXAFS). By combining MES and EXAFS studies with studies using infrared (IR)
spectroscopy, x-ray photoelectron spectroscopy (XPS), high
resolution electron microscopy (HREM), and other techniques, it
has been possible to elucidate the different types of structures
present in HDS catalysts. Of particular importance is the observation of a mixed Co-Mo sulfide structure (the so-called Co-
Mo-S structure) since it has been found that the promotion of
the HDS activity can be related to the fraction of cobalt atoms
which is present as Co-Mo-S. This review will focus on these
developments and will emphasize the recent insight obtained on
the nature of Co-Mo-S and the role played by the promoter
atoms. The applicability of the "Co-Mo-S model" to other HDS
catalyst systems will also be dealt with.


## II.  NATURE OF PHASES PRESENT IN
Co-Mo/Al$_2$O$_3$ CATALYSTS

Many different structural models have been proposed to describe the active state of Co-Mo/Al$_2$O$_3$ and related catalysts. Some
of the models most often referred to are: the "monolayer model,"
where cobalt is proposed to be associated with the alumina; the

"contact synergy model," where the cobalt is supposed to be present as $Co_9S_8$; and the "intercalation models," where cobalt is supposed to intercalate into the layer structure of $MoS_2$. These and
other models have been extensively reviewed in the past [1-6, 8,
10].

In the absence of direct in situ structural information on the
working catalyst, it has been difficult to judge the validity of
the different models. The introduction of Mössbauer spectroscopy has changed this situation, and detailed information is now
available concerning the state of Co. Some of the important features of Mössbauer spectroscopy are outlined below. Mössbauer
spectroscopy can readily be applied to systems containing iron,
and this technique has provided important information on the
state of iron in various catalyst systems [12-14]. However,
Mössbauer absorption spectroscopy (MAS) measurements cannot
be used to study catalysts which do not contain suitable Mössbauer isotopes. This problem can partly be solved by doping
such catalysts with Mössbauer isotopes (see, e.g., Ref. 14).
Such an approach was used in the first Mössbauer spectroscopy
study of Co-Mo catalysts which were doped with iron [15]. Some
indications of the presence of an interaction between Co and Mo
were found from this study. However, more direct information
on the state of Co in such catalysts can be obtained by carrying out so-called Mössbauer emission spectroscopy (MES) studies of catalysts containing radioactive $^{57}Co$ [16]. Some of the
advantages which make MES studies particularly useful are the
following (for more details, see Refs. 8, 17-19). First of all,
the technique is very sensitive to small changes in the chemical and structural environments. This sensitivity is demonstrated in Fig. 1(a-c) which shows MES spectra of three of the
compounds proposed to be present in Co-Mo/$Al_2O_3$ catalysts.
Second, MES allows in situ studies of real catalysts in their
working state (even at high pressures). Finally, MES enables
quantitative information to be obtained about all the Co atoms
irrespective of their presence in microcrystalline or x-ray amorphous structures. This latter advantage is particularly important since it makes it possible (see Section III) to establish a
bridge between the structural data obtained by MES and the
catalytic properties.

The MES results have revealed the presence of several different types of Co configurations in sulfided Co-Mo/$Al_2O_3$ catalysts.
Some of the Co species have been easy to identify since they correspond to well-known structures. For example, it was observed that in general Co-Mo/$Al_2O_3$ catalysts have part of their
Co atoms located in the alumina lattice (Co:$Al_2O_3$) (see, e.g.,

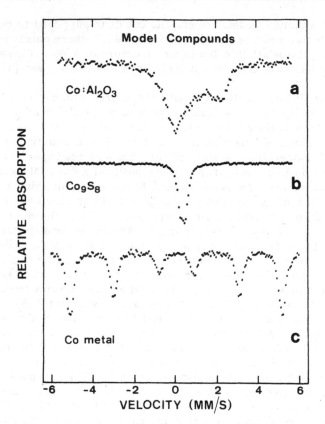

FIG. 1. Examples of Mössbauer emission spectra of model compounds. (a) Co:Al$_2$O$_3$; (b) Co$_9$S$_8$; (c) Metallic Co. (Figure adapted from Ref. 19.)

Refs. 8, 16, 17, 19-22). The spectral component arising from
Co:Al$_2$O$_3$ is indicated in Fig. 2. The fraction of Co atoms present in the alumina support is usually quite small and depends on
preparation conditions. For example, quite large amounts of Co:
Al$_2$O$_3$ may be present in catalysts calcined at high temperature
[22].

Another cobalt compound which has been observed in sulfided
Co-Mo/Al$_2$O$_3$ catalysts is Co$_9$S$_8$. This is perhaps not surprising
since Co$_9$S$_8$ represents the thermodynamically stable cobalt compound under reaction conditions. In general, there is an increased tendency toward formation of Co$_9$S$_8$ with increasing Co
loading (see Fig. 2). Moreover, the formation of Co$_9$S$_8$ is also

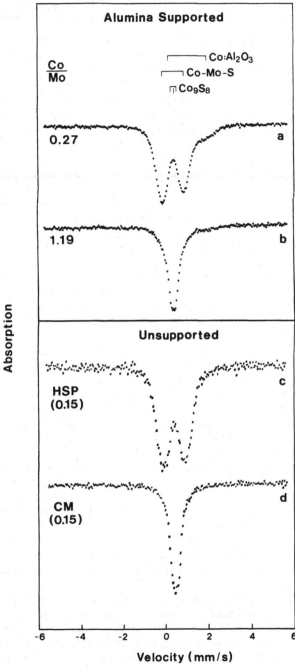

FIG. 2. Examples of MES spectra of alumina supported and un-supported Co-Mo catalysts. The unsupported catalysts were pre-pared using the homogeneous sulfide preparation (HSP) method and the comaceration (CM) method, respectively. (Figure adapted from Refs. 20 and 27.)

observed to depend strongly on preparation parameters, and it
may, for example, depend critically on the order of impregnation
[20-22] and the sulfiding conditions [23].

The presence of Co as $Co:Al_2O_3$ or $Co_9S_8$ has been considered
by many previous investigators, and in certain nontypical catalysts it has been possible, for example, to observe the presence
of bulk $Co_9S_8$ by x-ray diffraction [24, 25]. Nevertheless, the
above MES results are interesting since earlier it was difficult to
obtain definite proof for the presence of these phases in typical
$Co-Mo/Al_2O_3$ catalysts and little was known about their quantitative amounts.

Apart from $Co:Al_2O_3$ and $Co_9S_8$, another Co configuration was
also observed in the Mössbauer emission spectra of Co-Mo catalysts (see Fig. 2). This spectral component does not correspond
to any known Co bulk compounds [17]. Rather, the results lead
to the conclusion that it corresponds to Co present in a mixed Co-
Mo sulfide phase which was termed Co-Mo-S. The first MES study
showed that Co-Mo-S has a $MoS_2$-like structure since by doping
bulk $MoS_2$ crystals with Co this special Mössbauer spectrum could
be obtained [16]. This clearly shows that an alumina support is
not a necessity for forming this phase. In fact, Co-Mo-S has recently been observed to be a general feature of different types of
unsupported Co-Mo catalysts with greatly varying Co content [16,
17, 26-28]. An example of an unsupported catalyst with high Co
content and exhibiting Co-Mo-S is shown in Fig. 2(c).

It has been observed that the formation of Co-Mo-S is very
sensitive to preparation parameters [17, 26] such as the loading
(compare Figs. 2a and 2b) and the mode of preparation (compare
Figs. 2c and 2d).

As a consequence of the strong influence of preparation parameters on the type and amount of Co species present, it is not possible to give a generally valid description of the structure of sulfided $Co-Mo/Al_2O_3$ catalysts. These may contain one or more of
the above-mentioned Co phases simultaneously. In fact, in many
typical high activity catalysts we have observed all the above-
mentioned Co phases present in the same catalyst (see Section
III). A schematic representation of such a catalyst is shown in
Fig. 3.

Subsequent to the identification of the different Co phases in
Co-Mo catalysts, much of the work has been aimed at understanding the catalytic significance of the different phases. The results
have shown, as will be discussed in Section III, that the promotion
of the HDS activity is linked to the presence of the Co-Mo-S structures. In view of this, in Section IV we try to give a detailed
structural and physicochemical description of Co-Mo-S and in

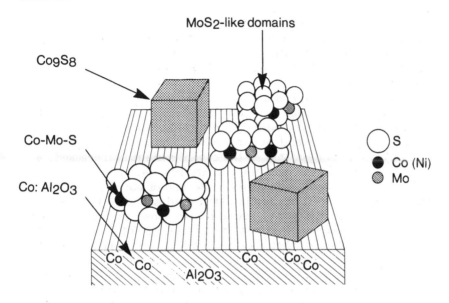

FIG. 3. Schematic representation of the different phases present in a typical alumina-supported catalyst.

Section V the nature of the active sites is discussed. Co-Mo-S
type structures have also been observed to be present in other
hydrodesulfurization catalysts than those based on Co and Mo.
These results are discussed in Section VI.

### III. CATALYTIC SIGNIFICANCE OF Co-Mo-S

The origin of the strong promotion of the HDS activity, which
is observed when Co or Ni is added to Mo- or W-based catalysts,
is one of the phenomena which has attracted the most attention.
Many different proposals have been put forward (for reviews,
see, e.g., Refs. 1, 2, 8) and in several of these proposals the
promotion has been suggested to be linked to the presence of
specific Co (or Ni) configurations. However, the lack of knowledge concerning the type of structures present in the working
catalysts has hindered significant progress in the understanding
of the origin of the promotion.

By use of Mössbauer emission spectroscopy it has recently
been possible both to elucidate the type of cobalt structures

present and to determine quantitatively their amount. The combination of such MES studies with measurements of the HDS activity has provided information on the origin of the promotion for many different catalyst systems. Some of these results are summarized in Fig. 4.

Figure 4(a) shows MES and activity results for a series of Co-Mo/$Al_2O_3$ catalysts which exhibit the typical synergistic promotion behavior of the catalytic activity [20]. All the catalysts have the same molybdenum content but the amount of the promoter atoms was varied. It is seen that as the cobalt content is increased, a drastic increase in the catalytic activity results. The catalytic activity reaches a maximum at a Co/Mo ratio of about 1. Further increase in the cobalt loading is found to give rise to a decrease in the catalytic activity. Figure 4(a) also shows the cobalt phase distribution obtained in situ by means of MES for the different catalysts. It is seen that as the Co/Mo atomic ratio is varied, the cobalt phase distribution also varies drastically. At low Co content (or Co/Mo ratio), Co is mainly present as Co-Mo-S. As the Co content increases, $Co_9S_8$ starts to form and this phase becomes the dominant Co phase at very high Co content. The amount of Co present in the alumina (Co:$Al_2O_3$) is low for all catalysts. It is evident from a comparison of the Co phase distribution with the observed HDS activity that the promotional effect is linked to the Co present as Co-Mo-S. The results shown in Fig. 4(a), as well as the other results given in Fig. 4, clearly show that $Co_9S_8$ does not have a significant promoting effect on the HDS activity. This in spite of the fact that $Co_9S_8$ has an activity by itself. However, recent results [20, 30] have shown that the intrinsic activity (per surface atom) of $Co_9S_8$ is much smaller than that of Co-Mo-S. Consequently, it is only in special cases that $Co_9S_8$ will contribute significantly to the overall activity.

By sulfiding at somewhat higher temperatures than those employed in most studies, it has recently [23] been possible to prepare catalysts which have a rather constant HDS activity over a wide range of Co/Mo ratios (Fig. 4b). The results confirm the above conclusions, and it is noteworthy that in spite of very large variations in the amount of $Co_9S_8$, the catalytic activity is observed to parallel the amount of Co present as Co-Mo-S.

An example of the effect of changing the calcination temperature on the activity and the Co phase distribution after sulfiding is shown in Fig. 4(c) [22]. With increasing calcination temperature, the amount of Co interacting with the alumina support increases and the amount which is present as Co-Mo-S after sulfiding decreases. Again we observe that from the in situ MES determinations of the amount of Co present as Co-Mo-S, one may understand the HDS activity behavior.

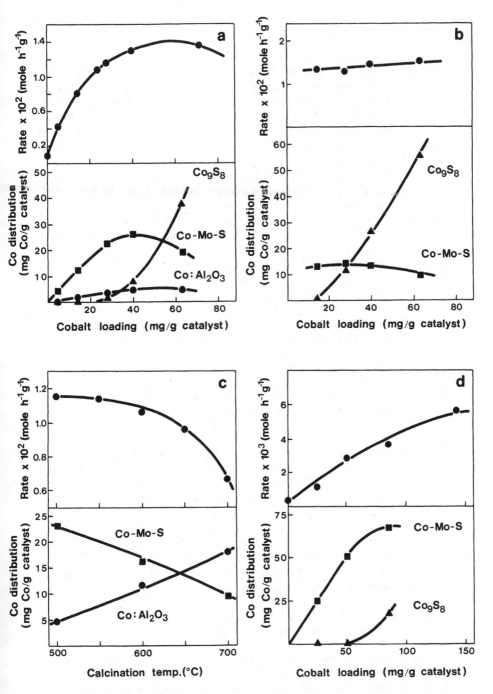

FIG. 4. Thiophene HDS rate parameters and Co phase distributions for different catalyst systems. For details see text. (Figure adapted from Refs. 20, 22, 23, and 27.)

The data discussed above (Figs. 4a-c) were obtained for alumina-supported Co-Mo catalysts which all have the same Mo content.  It is evident from an inspection of the figures that the catalytic activity of catalysts with the same metal loading may be very
different depending on the exact mode of preparation and sulfiding.
It is believed that such differences are the main reason why apparently similar catalysts have been reported to exhibit very different
catalytic activities.

It has been observed that unsupported Co-Mo catalysts may
also exhibit catalytic synergism similar to that shown in Fig. 4(a)
for alumina-supported catalysts (see, e.g., Refs. 2, 3).  $Co_9S_8$
has been observed to be present in such unsupported catalyst
systems, and the promotion of the catalytic activity has been proposed by some investigators [2, 5, 31, 32] to be linked to the
presence of this phase.  However, it is clear from the results
discussed in Section II (see also Fig. 2c) that such catalysts
may also have Co atoms present as Co-Mo-S, and recent studies [27] have in fact shown (see Fig. 4d) that the promotion
of the HDS activity for such catalysts is related to the Co atoms
present as Co-Mo-S and not to those present as $Co_9S_8$.  The origin of the promotion in the many different catalyst systems therefore appears to be the same.

In the above studies it was found that the promotion of the
HDS activity is roughly proportional to the amount of Co present as Co-Mo-S.  Thus, the Co atoms which are located in surface positions (see Section IV), create new and more active sites
(see Section V).  The activity per Co atom present as Co-Mo-S
is, however, different for different catalyst systems.  For example, for unsupported catalysts the activity per Co atom present as Co-Mo-S is quite low.  This is probably related to the
fact that in these catalysts, many of the Co atoms are not easily
accessible to the reactants.  Furthermore, the activity per Co
atom present as Co-Mo-S may not be the same for various alumina-supported catalysts.  For example, from a study of catalysts subjected to different sulfiding temperatures, it was observed [23] that two types of Co-Mo-S exist (Fig. 5) with a
different specific activity per Co atom.

## IV.  STATE OF Co AND Mo IN Co-Mo-S

Previously, it has been debated whether Mo in alumina-supported catalysts is present as $MoS_2$ or an oxysulfide (see, e.g.,
Refs. 33-36).  The MES results provide evidence for Co-Mo-S
having a $MoS_2$-like structure but these results do not allow one

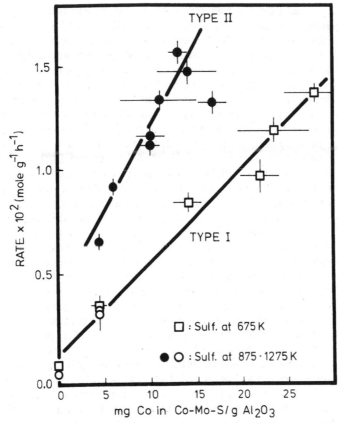

FIG. 5. Activity versus the amount of Co in Co-Mo-S for Co-Mo/Al₂O₃ catalysts sulfided at different temperatures. (Figure according to Ref. 23.)

to exclude that some Mo is present in other forms. The previous observation of $MoS_2$ in aged or high temperature sulfided catalysts [8, 23-25, 37-39] does not allow one to draw conclusions regarding the state of Mo in more typical catalysts.

Direct structural information regarding the state of Mo has been obtained by means of EXAFS. This technique has many of the same advantages as MES. The studies can be performed in situ, and information may be obtained on the type of x-ray amorphous or microcrustalline phases present in typical hydrodesulfurization catalysts. In the first EXAFS study of sulfided

Co-Mo/Al$_2$O$_3$ catalysts it was found that the Mo atoms are predominantly present in MoS$_2$-like structures [40]. EXAFS studies of other sulfided Co-Mo or Ni-Mo catalysts seem to be in qualitative agreement with this conclusion [41-43]. No evidence for Mo in oxygen surroundings was obtained after typical sulfiding conditions [40]. Thus, if such species are present, they must account for less than about 5% of the Mo atoms. Recent studies [29, 42, 44] have shown that even when the sulfiding temperature is lowered to around 300°C, the same picture seems to hold. However, at substantially lower sulfiding temperatures an appreciable amount of Mo atoms is coordinated to oxygen [42].

The EXAFS results [40] furthermore showed that the MoS$_2$ phase was present as very small ($\sim$10 Å) domains or crystallites. In this connection it is interesting that EXAFS is able to provide information on the dimension of the MoS$_2$ slabs parallel to the basal planes. In other words, the "edge dispersion" can be estimated. As will be discussed below, this is an important parameter. In a recent study [30], values for the Mo–S bond distances and the sulfur and molybdenum coordination numbers have been estimated (Fig. 6). It is seen that the least-squares fit of the contribution to the EXAFS from the second shell gives almost identical values of the bond distances and coordination numbers for Mo/Al$_2$O$_3$ and Co-Mo/Al$_2$O$_3$ catalysts. In the fits, parameters for well-crystallized MoS$_2$ had been used. This is seen (Fig. 6) to give an excellent fit to the data for the Mo/Al$_2$O$_3$ catalyst for all k values, whereas this is not the case for the Co-Mo/Al$_2$O$_3$ catalyst. This difference in the second coordination shell could be due to the presence of Co associated with the MoS$_2$-like phase (Co-Mo-S).

Unsupported Co-Mo catalysts have been helpful to the understanding of the nature of Co-Mo-S. The advantage of these systems is related to the observation that samples can be prepared where Co-Mo-S is the only Co phase even in cases of quite high Co concentrations [17, 19, 26-28, 45]. Studies of such catalysts by XRD, HREM, or EXAFS also confirm that Co-Mo-S is a MoS$_2$-like structure [26, 45].

MoS$_2$ crystallizes in a layer structure. Each layer consists of a S-Mo-S slab made up of a Mo layer having a close-packed sulfur layer on both sides. The Mo atoms are surrounded by six sulfur atoms in a trigonal prismatic coordination. The S-Mo-S layers are held together by relatively weak van der Waals forces. In the unsupported catalysts exhibiting Co-Mo-S, three-dimensional MoS$_2$ structures are present, i.e., several MoS$_2$ slabs are stacked on top of each other as in bulk MoS$_2$. However, in typical Mo/Al$_2$O$_3$ or Co-Mo/Al$_2$O$_3$ catalysts the Mo has recently been

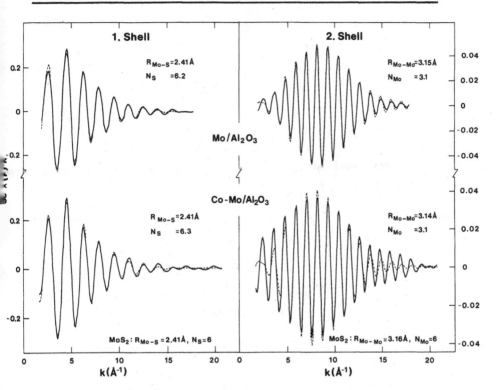

FIG. 6. Contributions to the EXAFS from the first and the second shells in sulfided Mo/Al₂O₃ and Co-Mo/Al₂O₃ catalysts. The experimental data are shown with a full line, whereas the least-squares fits are shown with a broken line. (Figure adapted from Ref. 30.)

found to be present mainly as single slab $MoS_2$ structures. Evidence for this has been obtained from IR studies of the surface hydroxyl groups in sulfided catalysts [46]. Figure 7(a) shows that the concentration of the OH bands decreases with increasing Mo loading with essentially no difference between the calcined and sulfided catalysts. This suggests that the Mo monolayer coverage present in the calcined state is maintained after sulfiding. This conclusion has been confirmed by XPS measurements [47] of the Mo/Al intensity ratio as a function of Mo loading (Fig. 7b). Single slab structures have also recently been observed [48] by use of HREM (Fig. 7c). The single slab Co-Mo-S structures are quite stable, but after extensive use or high temperature sulfiding the

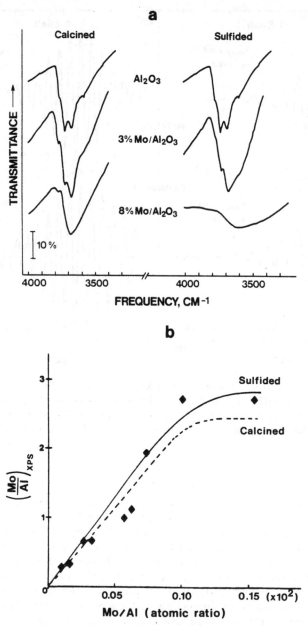

FIG. 7. (a) Infrared spectra of surface hydroxyl groups of alumina and $Mo/Al_2O_3$ catalysts in the calcined and sulfided state (adapted from Ref. 46). (b) XPS results of the Mo/Al intensity ratio as a function of Mo loading (adapted from Ref. 47).

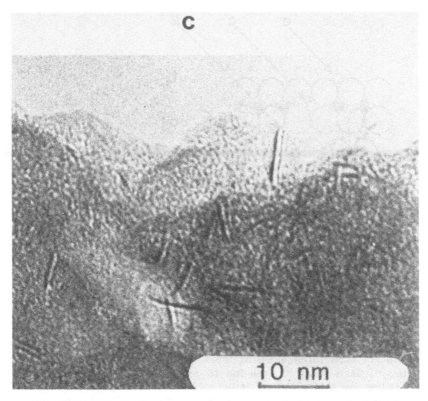

FIG. 7 (continued). (c) Electron micrograph of a used Ni-Mo/
Al$_2$O$_3$ catalyst (adapted from Ref. 48).


MoS$_2$ slabs will stack on top of each other and three-dimensional
MoS$_2$ may be observed by several methods [23, 25].

In a MoS$_2$ structure one can think of different possible locations of the Co atoms. These are schematically indicated in Fig.
8. In the past it has been proposed [34, 36, 37, 49] that Co may
enter different types of intercalation positions (i.e., positions in
the van der Waals gap). However, measurements on Co-Mo-S have
revealed that it is not an intercalation structure. For example,
Co EXAFS results [45, 50] have revealed that the Co--S bond distance in Co-Mo-S is significantly lower than that of Co in an intercalation site. Also, for Co-Mo-S samples exhibiting a three-dimensional MoS$_2$ structure, the lattice spacing observed by XRD
[26] was equal to that of bulk MoS$_2$, a result which is also in-
consistent with Co being in intercalation positions. Furthermore,
the results (discussed above), which show the presence of mainly

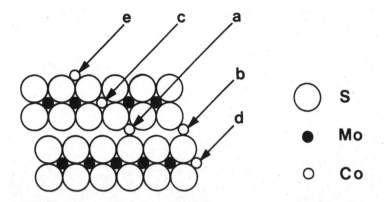

FIG. 8. Schematic picture of $MoS_2$ with different proposed Co locations: (a) bulk intercalation; (b) "pseudo-intercalation;" (c) bulk substitution; (d) edge substitution; (e) basal plane position. (Figure according to Ref. 19.)

single $MoS_2$ slabs, also allow one to exclude the presence of in-tercalation structures since at least two S-Mo-S slabs stacked on top of each other are needed to form the van der Waals gap necessary for intercalation.

Several results have shown that the Co atoms are present in surface positions of $MoS_2$. It has, for example, been found by MES that the Co atoms are bounded relatively weakly to the $MoS_2$ structure [17]. Furthermore, EXAFS data [45, 50] have shown that the Co atoms have a low sulfur coordination number and are affected by exposure to oxygen. Infrared studies [51] have shown that the Co atoms in Co-Mo-S are also accessible to NO. From these results it can be concluded that the Co present as Co-Mo-S must be located at the surface of $MoS_2$ in either basal or edge plane positions. Many of the early results provided in-direct evidence for the latter position (see, e.g., Refs. 4, 8, 19). Recently, it has been possible by the combined use of MES, HREM, and analytical electron microscopy (AEM) to con-firm the edge location of Co [52, 53]. Unsupported Co-Mo-S structures were studied. The crystals investigated were pre-pared such that they were much larger than the diameter of the electron beam used for the AEM measurements. In this way one could separately examine the Co concentration in the edge and basal plane regions. The results (see Table 1) clearly show that Co is present at the $MoS_2$ edges.

## TABLE 1

Co Concentration Measured by Analytical Electron Microscopy
(AEM) on Large Crystals Exhibiting Co-Mo-S [53]

| Basel plane region (Co atoms/cm$^2$) | Edge region (Co atoms/cm$^2$) |
|---|---|
| $(0.02 \pm 0.1) \times 10^{15}$ | $(0.58 \pm 0.1) \times 10^{15}$ |

Infrared studies [51] of NO adsorption on sulfided Co-Mo/Al$_2$O$_3$ catalysts also give evidence for the surface edge location of Co. From these studies information on both Mo and Co surface atoms is obtained simultaneously (Fig. 9a). Figure 9(b) shows the Co and Mo band intensities as a function of Co/Mo ratio (or Co content). The Mo atoms adsorbing NO are coordinately unsaturated sites located at the edges. EXAFS measurements indicated that the edge dispersion is approximately the same for all the catalysts [30, 40]. Nevertheless, the increasing amount of Co is seen to result in a decrease in the number of Mo atoms adsorbing NO. This decrease therefore indicates that Co is associated with the edges of the MoS$_2$, thereby "covering" some of the Mo sites. The EXAFS results [30, 40, 45, 50], which show that the Co—S distance ($\sim 2.27$ Å) is much shorter than the Mo-S distance in MoS$_2$ (2.41 Å), indicate that the Co sites do not occupy perfect edge substitutional sites. Also the MES results indicate that a distribution in surroundings may exist [17, 20, 54].

In view of the structure of Co-Mo-S it may be appropriate to regard the MoS$_2$ as the "primary support" for the Co atoms, whereas the alumina (or other supports) acts as a "secondary support" whose main role is to allow the preparation and stabilization of highly dispersed MoS$_2$ structures capable of accommodating high contents of Co as Co-Mo-S necessary for achieving a large promotion of the HDS activity.

In the above we have focused on a structural description of Co-Mo-S. Recently, it has also been possible to obtain information on the chemical properties of the Co atoms by means of XPS [28] and magnetic susceptibility [52].

## V.  NATURE OF ACTIVE SITES

The chemisorption of different probe molecules has recently been extensively used in order to get information about the

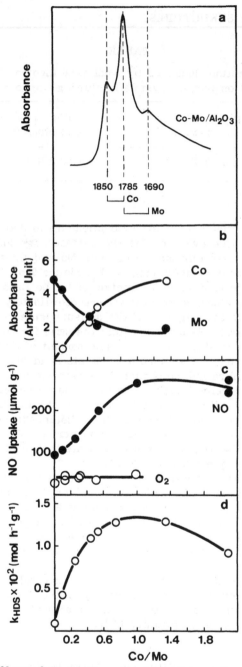

FIG. 9.  Effect of Co loading on adsorption and activity of a
series of sulfided Co-Mo/Al$_2$O$_3$ catalysts.  (a) IR spectrum of NO
adsorbed on Co-Mo/Al$_2$O$_3$ (Co/Mo = 0.44).  (b) Co and Mo band
intensities.  (c) Total volumetric NO and O$_2$ uptakes.  (d) The
thiophene HDS rate parameter.  (Figure adapted from Refs. 30
and 51.)

active sites in HDS catalysts. For unsupported as well as alumina-supported Mo catalysts, it has been found [55, 56] that the HDS activity correlates with the oxygen uptake. Oxygen has been observed [57] to adsorb on the edges of $MoS_2$ and it was therefore suggested [55] that the HDS reaction occurs on these planes and that anion vacancies are involved.

In contrast to the results for unpromoted catalysts, oxygen chemisorption results for promoted catalyst systems do not show similar correlations between the catalytic activity and the oxygen uptake [30, 58, 59] (Fig. 9c). This lack of correlation between the HDS activity and the oxygen uptake (Figs. 9c and 9d) is due to the fact that oxygen adsorb both on less active unpromoted Mo edge sites and—as evidenced by EXAFS [45, 50]—on the Co edge sites associated with Co-Mo-S.

It has recently been shown [51] that it may be advantageous to use NO chemisorption instead of oxygen chemisorption since by using NO one may simultaneously study the adsorption on promoted and unpromoted sites (Fig. 9a). In agreement with the MES results it is seen that for the promoted catalysts the activity is related to the Co sites adsorbing NO (Figs. 9b and 9d). These Co sites are in fact related to the Co atoms present as Co-Mo-S [51]. Furthermore, for the promoted catalysts, unpromoted Mo sites can still be observed. However, their intrinsic activity is much smaller than that of the promoted sites [29, 30], and therefore the unpromoted sites do not contribute significantly to the overall measured HDS activity for most catalysts. For reactions other than HDS, the unpromoted sites may play a more significant role [30].

The early MES results [16] of Co-Mo/$Al_2O_3$ catalysts indicated that reversible changes in the valence state of the promoter atoms in Co-Mo-S may take place upon changing the gaseous environment (e.g., the $H_2/H_2S$ ratio). These results therefore suggest (in agreement with the NO adsorption results discussed above) that the active sites are related to anion vacancies associated with the Co sites in Co-Mo-S. Recently, we have investigated in more detail the effect of hydrogen treatment on sulfided Co-Mo/$Al_2O_3$ catalysts [54]. Figure 10 shows MES spectra of two catalysts with different Co/Mo ratios. The spectra of the catalysts in the sulfided state show the presence of mainly Co in Co-Mo-S. The spectra after reduction in hydrogen show that a large fraction of the promoter atoms has undergone reduction. This effect is most pronounced for the catalysts with the high Co loading, i.e., the ease with which vacancies are formed seems to increase with the amount of Co in Co-Mo-S. As expected, it was also observed

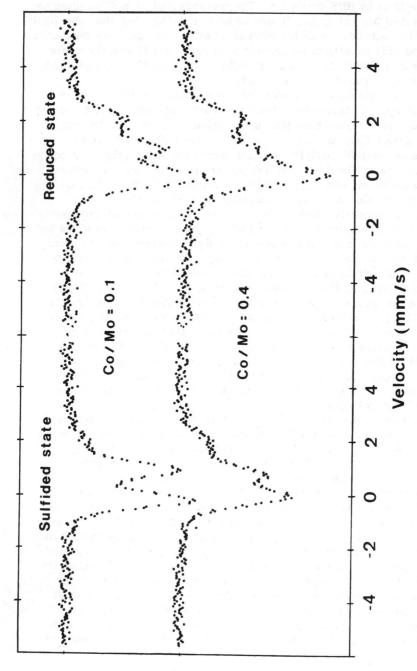

FIG. 10.  MES spectra of Co-Mo/Al₂O₃ catalysts in the sulfided and reduced state.  (Figure adapted from Ref. 54.)

that the vacancy concentration increases with increasing reduction temperature [54].

There may be several explanations for the fact that the sites associated with the Co atoms in Co-Mo-S have higher intrinsic activity for the HDS reaction than the unpromoted Mo sites (for a discussion, see Ref. 8). For example, the above-mentioned MES results, which reveal that a valence change of the Co atoms occurs, may be taken as an indication that Co promotes charge transfer or a redox mechanism. It has also been suggested [60] that the high activity of Co-Mo-S is related to a weakening of the Mo—S bond caused by the presence of the neighboring Co atoms.

## VI. PRESENCE OF Co-Mo-S TYPE STRUCTURES IN RELATED HDS CATALYST SYSTEMS

It has been mentioned above that unsupported catalysts behave quite similarly to alumina-supported catalysts. This can be ascribed to the presence of the catalytically active Co-Mo-S structure in both systems. Although the support is not necessary for the creation of Co-Mo-S, it may play an important role for the morphology, texture, stability, and dispersion of the structures. Co-Mo-S has also been observed in silica- [61] and carbon-supported [21, 61] Co-Mo catalysts.

It is noteworthy that Co-Mo-S type structures are not restricted to Co-Mo catalyst systems. Specifically, Mössbauer spectroscopy studies have shown the presence of Co-W-S in alumina-supported Co-W catalysts [19] and Fe-Mo-S in both unsupported and alumina-supported Fe-Mo catalysts [16, 18, 19]. Recent NO adsorption studies [51] of Ni-Mo/$Al_2O_3$ catalysts (Fig. 11) show that these behave quite analogously to Co-Mo/$Al_2O_3$ catalysts (Fig. 9). This strongly suggests that similar Ni-Mo-S structures are present in such catalysts. Also, the promotion of the HDS activity in Ni-Mo/$Al_2O_3$ catalysts is seen to be linked to the presence of Ni-Mo-S (Fig. 11). Adsorption of CO on the Ni-Mo/$Al_2O_3$ system has also been interpreted in terms of the presence of Ni-Mo-S [62-64]. Finally, evidence for similar structures has also recently been found for Ni-W/$SiO_2$ catalysts [65].

## VII. CONCLUSION

Previously, very little was known about the type of structures present in working HDS catalysts. Consequently, it has not been

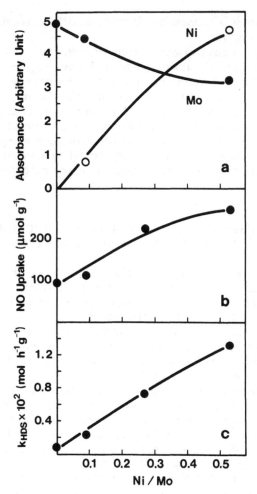

FIG. 11. Effect of Ni loading on the NO adsorption and HDS activity of sulfided Ni-Mo/Al$_2$O$_3$ catalysts. (a) The Mo and Co band intensities. (b) Total volumetric NO uptakes. (c) The thiophene HDS activities. (Figure according to Ref. 51.)

possible to understand the catalytic properties of such catalysts. However, as discussed in this review, the introduction of novel in situ techniques such as MES and EXAFS has given quite detailed structural insight, and based on this knowledge, it has been possible to understand the catalytic properties for a variety of HDS catalysts.

Specifically, it has been found that in spite of the fact that many different phases may be present, the promotion of the HDS activity is linked to the presence of a Co-Mo-S structure. This structure can be regarded as a $MoS_2$-like phase with the Co atoms located at the edges. Furthermore, it is likely that the active sites are anion vacancies associated with the Co atoms present as Co-Mo-S.

It is noteworthy that for a long time Mössbauer spectroscopy was the only technique that revealed the Co-Mo-S structure. However, recently a number of studies has given additional evidence for the presence of this phase, and the Co-Mo S model appears now to be gaining increasing acceptance in the literature. Finally, evidence for similar structures in other catalyst systems has been given. It is therefore likely that Co-Mo-S-type structures are general features of HDS catalysts.

## Acknowledgments

The authors are grateful to N.-Y. Topsøe and R. Candia for many valuable discussions.

## REFERENCES

[1]  F. E. Massoth, Adv. Catal., 27, 265 (1978).
[2]  B. Delmon, in Proceedings of the Climax Third International Conference on Chemistry and Uses of Molybdenum (H. F. Barry and P. C. H. Mitchell, eds.), Climax Molybdenum Co., Ann Arbor, Michigan, 1979, p. 73.
[3]  E. Furimsky, Catal. Rev.–Sci. Eng., 22, 371 (1980).
[4]  P. Ratnasamy and S. Sivasanker, Ibid., 22, 401 (1980).
[5]  P. Grange, Ibid., 21, 135 (1980).
[6]  B. C. Gates, J. R. Katzer and G. C. A. Schuit, in Chemistry of Catalytic Processes, McGraw-Hill, New York, 1979, Chap. 5.
[7]  P. C. H. Mitchell, in Catalysis (C. Kemball and D. A. Dowden, eds.), Specialist Periodical Reports, Royal Society of Chemistry, London, Vol. 4, 1980, p. 175.
[8]  H. Topsøe, in Surface Properties and Catalysis by Non-metals: Oxides, Sulfides, and other Transition Metal Compounds, Reidel, Dordrecht, 1983, p. 326.
[9]  W. K. Hall, in Proceedings of the Climax Fourth International Conference on Chemistry and Uses of Molybdenum (H. F. Barry and P. C. H. Mitchell, eds.), Climax Molybdenum Co., Ann Arbor, Michigan, 1982, p. 224.

[10]  F. E. Massoth and G. MuraliDhar, in Proceedings of the
      Climax Fourth International Conference on Chemistry and
      Uses of Molybdenum (H. F. Barry and P. C. H. Mitchell,
      eds.), Climax Molybdenum Co., Ann Arbor, Michigan,
      1982, p. 343.
[11]  M. L. Vrinat, Appl. Catal., 6, 137 (1983).
[12]  J. A. Dumesic and H. Topsøe, Adv. Catal., 26, 121 (1977).
[13]  W. N. Delgass, in Spectroscopy in Heterogenous Catalysis,
      Academic, New York, 1979.
[14]  H. Topsøe, J. A. Dumesic, and S. Mørup, in Applications
      of Mössbauer Spectroscopy, Vol. II (R. L. Cohen, ed.),
      Academic, New York, 1980, p. 55.
[15]  H. Topsøe and S. Mørup, in Proceedings, International
      Conference on Mössbauer Spectroscopy, Vol. 1 (A. Z.
      Hrynkiewicz and J. A. Sawicki, eds.), Cracow, Poland,
      1975, p. 305.
[16]  B. S. Clausen, S. Mørup, H. Topsøe, and R. Candia, J.
      Phys. Colloq., 37, C6-249 (1976).
[17]  H. Topsøe, B. S. Clausen, R. Candia, C. Wivel, and S.
      Mørup, J. Catal., 68, 433 (1981).
[18]  S. Mørup, B. S. Clausen, and H. Topsøe, J. Phys. Colloq.,
      40, C2-88 (1979).
[19]  H. Topsøe, B. S. Clausen, R. Candia, C. Wivel, and S.
      Mørup, Bull. Soc. Chim. Belg., 90, 1189 (1981).
[20]  C. Wivel, R. Candia, B. S. Clausen, S. Mørup, and H.
      Topsøe, J. Catal., 68, 453 (1981).
[21]  M. Breysse, B. A. Bennett, D. Chadwick, and M. Vrinat,
      Bull. Soc. Chim. Belg., 90, 1271 (1981).
[22]  R. Candia, N.-Y. Topsøe, B. S. Clausen, C. Wivel, R.
      Nevald, S. Mørup, and H. Topsøe, in Proceedings of the
      Climax Fourth International Conference on Chemistry and
      Uses of Molybdenum, Ann Arbor, Michigan, 1982, p. 374.
[23]  R. Candia, H. Topsøe, and B. S. Clausen, in Proceedings
      of the 9th Ibero-american Symposium on Catalysis, Lisbon,
      Portugal, 1984, p. 211.
[24]  S. S. Pollack, L. E. Makovsky, and F. R. Brown, J. Catal.,
      59, 452 (1979).
[25]  N.-Y. Topsøe and H. Topsøe, Bull. Soc. Chim. Belg., 90,
      1311 (1981).
[26]  R. Candia, B. S. Clausen, and H. Topsøe, Ibid., 90, 1225
      (1981).
[27]  R. Candia, B. S. Clausen, and H. Topsøe, J. Catal., 77,
      564 (1982).
[28]  I. Alstrup, I. Chorkendorff, R. Candia, B. S. Clausen,
      and H. Topsøe, Ibid., 77, 397 (1982).

[29]. H. Topsøe, Paper Presented at Advances in Catalytic Chemistry II Symposium, Utah, 1982, In Press.

[30] R. Candia, B. S. Clausen, J. Bartholdy, N.-Y. Topsøe, B. Lengeler and H. Topsøe, in Proceedings 8th International Congress on Catalysis, Verlag Chemie, Weinheim, 1984, p. II-375.

[31] B. Delmon, Prepr., Div. Pet. Chem., Am. Chem. Soc., 22, 503 (1977).

[32] G. Hagenbach, Ph. Courty, and B. Delmon, J. Catal., 23, 295 (1971).

[33] G. C. A. Schuit and B. C. Gates, AIChE J., 19, 417 (1973).

[34] V. H. J. De Beer, T. H. M. Van Sint Fiet, G. H. A. M. Van der Steen, A. C. Zwaga, and G. C. A. Schuit, J. Catal., 35, 297 (1974).

[35] F. E. Massoth, Ibid., 36, 164 (1975).

[36] A. L. Farragher and P. Cossee, in Proceedings, 5th International Congress on Catalysis, 1972 (J. W. Hightower, ed.), North-Holland, Amsterdam, 1973, p. 1301.

[37] A. L. Farragher, Symposium on the Role of Solid State Chemistry in Catalysis, ACS Meeting, New Orleans, 1977.

[38] J. V. Sanders, Phys. Scr., 14, 141 (1978/79).

[39] J. M. Thomas, G. R. Milward, and L. A. Bursill, Philos. Trans. R. Soc., A300, 43 (1981).

[40] B. S. Clausen, H. Topsøe, R. Candia, J. Villadsen, B. Lengeler, J. Als-Nielsen, and F. Christensen, J. Phys. Chem., 85, 3868 (1981).

[41] H. Harnsberger, Private Communications.

[42] T. G. Parham and R. P. Merrill, J. Catal., 85, 295 (1984).

[43] J. L. S. Arrieta, PhD Dissertation, Stanford University, 1983.

[44] B. S. Clausen, B. Lengeler, R. Candia, and H. Topsøe, Unpublished Results.

[45] B. S. Clausen, H. Topsøe, R. Candia, and B. Lengeler, in Catalytic Materials: Relationship between Structure and Reactivity, ACS Symposium Series, No. 248, San Francisco, California, 1984, p. 71.

[46] N.-Y. Topsøe, J. Catal., 64, 235 (1980).

[47] J. Grimblot, P. Dufresne, L. Gengembre, and J. P. Bonelle, Bull. Soc. Chim. Belg., 90, 1261 (1981).

[48] S. S. Pollack, J. V. Sanders, and R. E. Tischer, Appl. Catal., 8, 383 (1983).

[49] R. J. H. Voorhoeve and J. C. M. Stuiver, J. Catal., 23, 243 (1971).

[50] B. S. Clausen, B. Lengeler, R. Candia, J. Als-Nielsen, and H. Topsøe, Bull. Soc. Chim. Belg., 90, 1249 (1981).

[51]  N.-Y. Topsøe and H. Topsøe, J. Catal., 84, 386 (1983).
[52]  H. Topsøe, N.-Y. Topsøe, O. Sørensen, R. Candia, B. S.
      Clausen, S. Kallesøe, and E. Pedersen, in Symposium on
      Role of Solid State Chemistry in Catalysis, ACS Meeting,
      Washington, D.C., 1983, p. 1252.
[53]  O. Sørensen, B. S. Clausen, R. Candia, and H. Topsøe,
      Appl. Catal., In Press.
[54]  P. H. Christensen, S. Mørup, B. S. Clausen, and H. Topsøe,
      in Proceedings of International Conference of the Applica-
      tions of Mössbauer Effect, Alma-Ata, USSR, 1983, In Press.
[55]  S. J. Tauster, T. A. Pecoraro, and R. R. Chianelli, J.
      Catal., 63, 515 (1983).
[56]  J. Bachelier, J. C. Duchet, and D. Cornet, Bull. Soc.
      Chim. Belg., 90, 1301 (1981).
[57]  O. P. Bahl, E. L. Evans, and J. M. Thomas, Proc. R. Soc.
      London, Ser. A, 306, 53 (1968).
[58]  R. Burch and A. Collins, in Proceedings of the Climax
      Fourth International Conference on Chemistry and Uses
      of Molybdenum (H. F. Barry and P. C. H. Mitchell, eds.),
      Climax Molybdenum Co., Ann Arbor, Michigan, 1982, p. 379.
[59]  W. Zmierzcak, G. MuraliDhar, and F. E. Massoth, J. Catal.,
      77, 432 (1982).
[60]  R. R. Chianelli, T. A. Pecoraro, T. R. Halbert, W.-H. Pan,
      and E. I. Stiefel, Ibid., 86, 226 (1984).
[61]  H. Topsøe, B. S. Clausen, N. Burriesci, R. Candia, and
      S. Mørup, in Preparation of Catalyst II (B. Delmon, P.
      Grange, P. Jacobs, and G. Poncelet, eds.), Elsevier,
      Amsterdam/New York, 1979, p. 479.
[62]  J. Bachelier, Doctor Thesis, University of Caen, 1982.
[63]  J. Bachelier, J. C. Duchet and D. Cornet, J. Catal., 87,
      283 (1984).
[64]  J. Bachelier, M. J. Tilette, J. C. Duchet, and D. Cornet,
      Ibid., 87, 292 (1984).
[65]  Yu. I. Yermakov, B. N. Kuznetsov, A. N. Startsev, P. A.
      Zhdan, A. P. Shepelin, V. I. Zaikovskii, L. M. Plyasova,
      and V. A. Burmistrov, J. Mol. Catal., 11, 205 (1981); Yu.
      I. Yermakov, Private Communications.

# Reactor Developments in Hydrotreating and Conversion of Residues

F. M. DAUTZENBERG AND J. C. DE DEKEN
Catalytica Associates, Inc.
Mountain View, California

## I. INTRODUCTION

Many industrialized western countries depend on imported foreign crude. In 1977, for example, the United States imported 46%
of its oil and petroleum products, chiefly from Saudi Arabia, Iran,
and other OPEC nations. By September 1, 1983, the United States
imported only 28% of its oil. The chief suppliers were Mexico, with
826,000 barrels per day, Canada, with 479,000, and Venezuela, with
419,000. Saudi Arabia is now seventh on the list of suppliers for

- ● **Hydrodesulfurization/denitrogenation**
- ● **Thermal and hydrocracking**
- ● **Hydrogenation**
- ● **Demetallization and coke formation**
- ● **Asphaltene conversion**

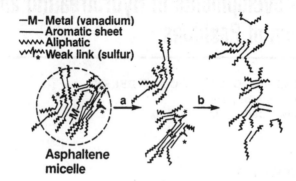

FIG. 1.  The most important residue upgrading reactions [2].

the United States.  This shift is responsible for the trend toward
a heavier crude supply mix on the international market.  Trans-
portation fuels and petrochemical feedstocks are in increasing demand; hence many refiners would like to get out of the fuel oil
business.  More "bottom-of-the-barrel" conversion capability will
be required in many refineries to efficiently process these heavier
feedstocks and maximize production of light products.

As illustrated in Fig. 1, the key challenge to manufacturing gasoline and middle distillate from atmospheric and vacuum residues is
how to handle the asphaltenes cost-effectively.  Asphaltenes are
complicated compounds in which heteroatoms such as nickel, vanadium, sulfur, nitrogen, and oxygen are built in a matrix of aromatic structures that have low hydrogen content [8, 10, 24].  Not
only must the sulfur, nitrogen, and oxygen components be removed, but the hydrogen/carbon ratio must be increased.  The
presence of trace metal components, of which nickel and vanadium
are the most abundant, does not facilitate upgrading.

Process selection generally consists of choosing a carbon-rejection technology or a hydrogen-addition process.  Typical processing alternatives are listed in Fig. 2.  In many cases a combination of these two approaches merits evaluation.  In such carbon-
rejection processes as visbreaking, solvent deasphalting, and de-
layed coking, a substantial part of the fuel is rejected as bottoms

| C-Rejection | C-Rejection | H-Addition |
|---|---|---|
| (Concentration into a residual by-product) | (Indirect conversion into low-Btu gas or steam) | (Direct conversion into oil products) |
| • Visbreaking<br>• Delayed Coking<br>• Deasphalting | • Fluid Coking<br>• Flexicoking<br>• Dynacracking<br>• Heavy oil catalytic cracking<br>• ART | • Shell-HDM/HDS<br>• Gulf-HDS<br>• H-Oil®<br>• LC-Fining<br>• UOP-Aurabon<br>• Etc. |

FIG. 2.   Disposition of asphaltenes—Processing alternatives.

or as coke. This results in the loss of potential liquid hydrocarbons. In the more modern carbon-rejection approaches, such as Flexicoking, Heavy Oil Cracking, the ART process, and Dynacracking, the asphaltenes are used to produce low-Btu gas, steam, and heat for internal or external use. This is reflected in the obtainable liquid yields. The hydrogen-addition technologies generally accomplish significant demetallization and Conradson carbon reduction, in addition to desulfurization and viscosity reduction. Relatively high yields are obtained compared to carbon-rejection processes.

A spectrum of hydrogen-addition processes is available. Many similar fixed-bed trickle reactor processes have been developed and can be licensed from such companies as Shell, Gulf, and Exxon. H-Oil and LC-Fining Processes, applying ebullating-bed technology, are also in an advanced stage of development. Some of the key technical challenges, and the merits of the trickle and ebullating reactor approaches in residue upgrading, will be discussed and evaluated. Such emerging slurry-phase technologies as M-Coke, CANMET, and Aurabon promise certain additional advantages. Areas requiring further development will be indicated.

## II. COMMERCIAL REACTOR SYSTEMS

### A. Fixed-Bed Trickle Reactors (FBR)

#### 1. Catalyst Contacting Efficiency

Even distribution of gas and liquid is of great importance in trickle-bed reactors. Figure 3 shows a poorly designed trickle

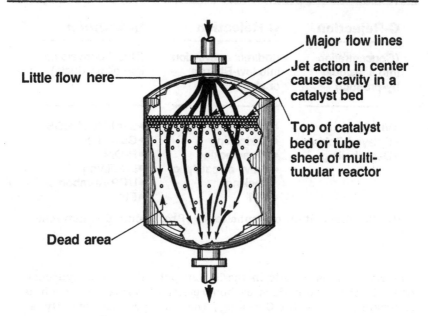

FIG. 3. Large diameter trickle reactor with poor flow distribution [18].

bed reactor, demonstrating the severe short circuits, dead areas, stagnant zones, etc. that may occur [18]. These deficiencies can be minimized by designing slim reactors with high length-to-diameter ratios (L/D). The pressure drop across the reactor will determine the effectiveness of the design. The scant literature on this issue has been reviewed systematically by Satterfield [20] in an effort to predict the effectiveness of contacting the catalyst. Satterfield developed the correlation shown in Fig. 4. The correlation predicts that in industrial trickle reactors, catalyst utilization is close to ideal if a sufficiently high superficial liquid velocity is applied. The liquid flow rates in experimental units, however, may be considerably lower. If not recognized, this factor can have serious consequences—such as the design of commercial reactors that are too large. An inexpensive way to assure efficient catalyst wetting in pilot plants is by diluting the catalyst bed with smaller inert particles [21].

Apart from assuring sufficient superficial liquid flow, special attention is paid to the design of the liquid distributors at the top of each catalyst level. Special baffles, radial diffusors, screens, perforated plates, and inert ceramic ball packings are used to

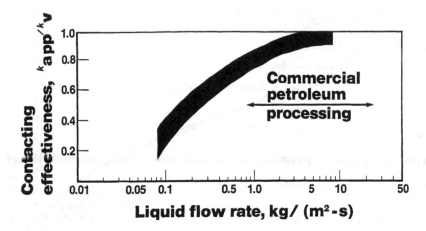

FIG. 4. Preliminary correlation of contacting effectiveness by Satterfield—Trickle bed data.

obtain optimum distribution of gas and liquid. With heavy residual feedstocks, bed plugging can occur and can lead to low contacting efficiencies. The first catalyst layer in most upstream reactors acts as a filter for such constituents as sand, salt, and corrosive products. Deep, two-stage crude oil desalting and special feedstocks filtration are often suggested as ways to avoid rapid bedplugging [22]. Replaceable top-bed filters and recycle gas scrubbing are also used as special means of protection.

## 2. Catalyst Deactivation

For both trickle and ebullating-bed reactors, catalyst deactivation is of great importance. The catalysts deactivate as metal is deposited in the pores. Thus catalyst deactivation is rapid early in the run, followed by a relatively long period of constant deactivation. The catalyst life is terminated after another rapid increase in deactivation. The generally observed S-shape deactivation curve can be described by the two-parameter model developed by Shell [5, 6]. According to this model, the hydrodemetallization activity and the overall catalyst metal load vary with the relative catalyst age according to

$$k_m/k_m^{\circ} = 1 - \theta \tag{1}$$

$$C/C_{max} = 2\theta - \theta^2 \tag{2}$$

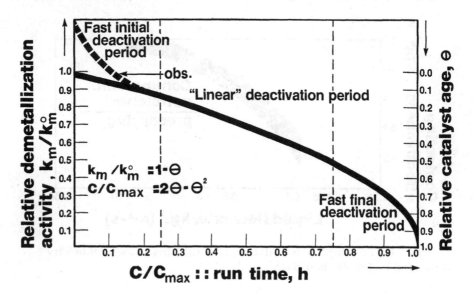

FIG. 5.  Catalyst aging—Pore mouth plugging model.

where $\theta = 0$ for a fresh catalyst and $\theta = 1$ for a catalyst in which the pore mouths are fully plugged.

Figure 5 represents Correlations 1 and 2 graphically. The catalyst performance must be evaluated experimentally at two catalyst ages to determine the initial HDM activity and the metal uptake capacity, $C_{max}$. These two important design parameters can be expressed in terms of the geometrical catalyst parameters and the physicochemical constants of the reaction system [22]. The relative catalyst age, $\theta$, can be found experimentally by applying Eq. (2). The metal load of the catalyst can be calculated at any time throughout the run from feed and product analysis. The experimental determination of $C_{max}$ requires running until the catalyst is completely deactivated. Alternatively, $C_{max}$ can be determined from the minimum life, $T_{min}$, and the initial HDM activity:

$$C_{max} = \frac{1}{2} k_m {}^{\circ}C_m T_{min}$$

(3)

The minimum life time, $T_{min}$, is reached when the catalyst is exposed to the full metal concentrations of the feed. Experimental values of $T_{min}$ can be found by plotting the actual run time, $t$ (in hours), against relative catalyst age, $\theta$, according to

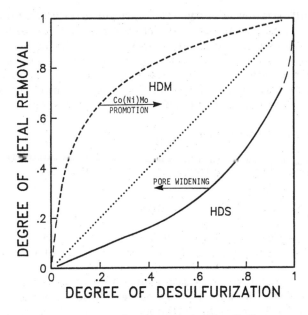

FIG. 6. Typical catalyst selectivity.

$$\frac{t}{T_{min}} = 1 + \frac{k_m^{\,\circ}}{SV}\,\theta - \frac{1}{2}\frac{k_m^{\,\circ}}{SV}\,\theta^2 \tag{4}$$

Once the pertinent activity decline parameters have been established, the model allows the correlation of the results of a large
number of experiments [6]. This confirms that the deactivation
of residue desulfurization catalysts is determined by the process
of metal deposition in the outer core of the particles, at least for
catalysts with not-too-large pores. If catalysts with large pores
(15-20 nm or more) are used, core poisoning becomes an important
issue.

Many oil companies and catalyst manufacturers have been active
in establishing catalyst design rules for heavy oil upgrading. The
ideal catalyst in terms of HDS activity and metal tolerance, however, appears to be unattainable. This is because desulfurization and hydrocracking occur mainly in the inner core of the catalyst particles, while metal deposition is concentrated at the pe-
riphery of the catalyst particles. Figure 6 shows that catalysts
with high desulfurization activity have limited demetallization activity and, conversely, effective demetallization catalysts have

lower desulfurization activity. Catalyst selectivity can be altered
by varying pore size to particle diameter ratio [9, 23] and by adding Co(Ni)/Mo sulfides instead of Ni/V subsulfides as active catalyst components. HDS and HDM are coupled, and empirical correlations over a wide range of conversions have been found to be
valid. For a typical HDS catalyst, the demetallization conversion
follows from the degree of desulfurization by applying

$$X_m = 0.055 \; X \; \exp \; (2.7 \; X \; X_{HDS})$$                    (5)

A similar equation is applicable for a typical HDM catalyst:

$$X_m = 0.215 \; X \; \log \; (105.26 \; X \; X_{HDS})$$                    (6)

In these empirical equations, 0.055, 2.7, 0.215, and 105.26 are the
fitting parameters applicable for the catalyst examples shown in
Fig. 6.

Combining catalysts with different HDM-HDS selectivity characteristics opens the path to reasonable solutions, especially for
the treatment of high-metal feedstocks with Ni + V contents of
more than 150-200 ppm. This catalyst tailoring principle is illustrated schematically in Fig. 7. The lower part of Fig. 7 gives the
V deposition profiles of the various catalysts, characterizing the
different accessibilities of the catalysts for metal penetration. The
wide-pore, front-end catalyst A has a high selectivity for HDM,
whereas the relatively small-pore catalyst C has good HDS characteristics.

## 3. Catalyst Requirements and Catalyst Handling

If catalyst life is not determined by metal deposition, the required catalyst volume for trickle-mode desulfurization in a fixed-
bed plug flow reactor can be found using Eq. (7), assuming
pseudo-first-order HDS kinetics:

$$V_{PFR} = \frac{F}{k_p} \; \log \; \frac{1}{1 - X_{HDS}}$$                    (7)

However, the catalyst life will be controlled in many residue upgrading operations by the amount of metal deposits that can be ac-
commodated by the catalyst. The metal deposition rate, $\dot{m}$, follows
from

$$\dot{m} = C_m X_m F$$                    (8)

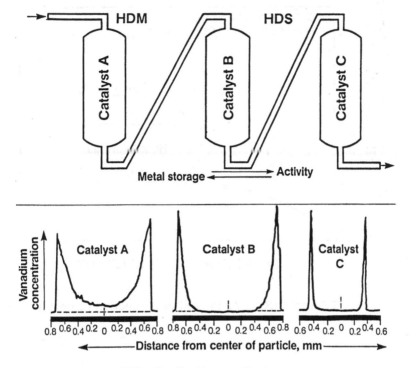

FIG. 7.  Catalyst tailoring.

For a required catalyst life, $L_0$, the minimum required catalyst volume can be calculated using

$$V_{min} = mL_0 \, \frac{1}{C^*} \, \frac{1}{\rho_{cat}}$$

(9)

In Eq. (9), $C^*$ represents the overall metal load on the catalyst, averaged over the length of reactor, which would render the catalyst inactive for the desired degree of desulfurization, $X_{HDS}$. As explained above, $X_m$ can be found using Eq. (5) or (6) at a given $X_{HDS}$.

Using the above equation, one can estimate the tolerable amount of Ni + V in the feedstock that would give, at a required degree of desulfurization, a minimum desired catalyst life. Figure 8 represents typical results for a mean catalyst life of 6 months, assuming catalysts with variable metal uptake capacities range from 10 to 25% w/w. Figure 8 illustrates that for a specific Ni + V content, a mean catalyst life of 6 months can be attained for a reasonable range of

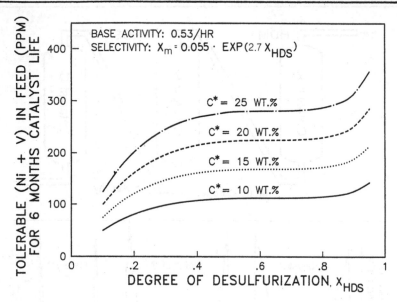

FIG. 8. Tolerable (Ni + V) in feed as a function of desulfurization conversion.

conversions—for 30-90% desulfurization—assuming a catalyst or catalyst combination with a sufficiently high metal uptake capacity is
used. For equivalent desulfurization and run-length requirements
with higher metal content, a catalyst with a higher metal uptake
capacity is required.

The above equations can also be used to estimate catalyst consumption. For a feedstock containing 100 ppm Ni + V, for example, 80% desulfurization can be maintained for about 1 year.
This corresponds to a catalyst consumption of 0.15 lb/bbl feedstock, assuming a catalyst with an overall metal uptake capacity
of 10%. At 200 ppm Ni + V, the catalyst life would drop to under
6 months and the catalyst consumption will be 0.31 lb/bbl of feedstock. The above examples are in general agreement with published catalyst consumption data [14]. These data illustrate that
for commercial fixed-bed residue upgrading units, catalyst cost
is a major operating cost item. Because of the high catalyst cost,
upgrading residues that contain more than 200 ppm Ni + V is
usually not feasible in trickle-bed operation using state-of-the-
art catalysts with metal uptake capacities ranging from 10 to 20
wt%.

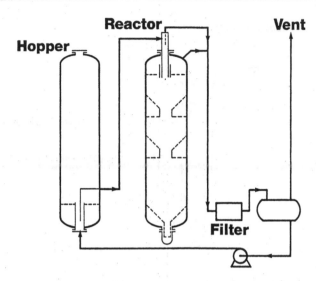

FIG. 9.  Catalyst loading system.

To achieve commercially acceptable run lengths, very large catalyst quantities must be installed.  A trickle-operation fixed-bed
reactor unit with a capacity of 50,000 bbl/day, designed for a feedstock of 200 ppm Ni + V and a run length of 6 months/cycle, will
require about 2.4 million pounds of catalyst (assuming C* = 10 wt%),
twice a year at 75% desulfurization.  With conventional catalyst handling facilities, it will take 10-20 days to unload spent catalyst and
load fresh catalyst, resulting in a loss of operating time of 10% or
more.  Downtime can be reduced with a loading/unloading system
that uses conical reactor catalyst support grids (Fig. 9) rather
than horizontal grids, and large interconnecting pipes between
the catalyst beds [16, 22].  These "quick catalyst replacement"
reactors allow loading rates of 40 m$^3$/h catalyst and unloading
rates of 20-30 m$^3$/h.  With conventional catalyst handling facilities, loading and unloading rates are generally of the order of 6
to 4 m$^3$/h.

### B.  Backmixed Ebullating-Bed Reactors (BMR)

### 1. General Characteristics

Ebullating-bed reactors—as in the H-Oil or the LC-Fining processes—are commercially applied in various residue upgrading proj-
ects.  The H-Oil process was introduced commercially 25 years ago

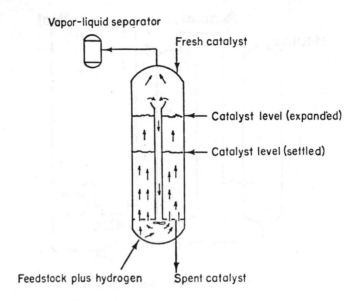

FIG. 10.  Ebullating-bed reactor.

and is presently licensed by Hydrocarbon Research, Inc. (HRI) and by Texaco Development Corporation [11, 12]. Like the H-Oil process, the LC-Fining process is based on technology developed by Cities Service Research and Development Company. A 18,500 bbl/day residue LC-Fining unit has been in operation at the Pemex Salamanca refinery since 1973 and several very high-metal content feedstocks have been processed successfully [17]. Since both reactor technologies are based on the same parent technology from Cities Service, the following discussions are thought to be equally valid for the LC-Fining and the H-Oil processes.

In ebullating-bed reactors, catalyst particles are held in suspension by the upward velocity of the liquid feedstock (Fig. 10). The indicated circulation pump assures that oil and catalyst particles are well mixed. Hydrogen gas bubbles through the catalyst suspension. Ebullating-bed reactors have a high dirt tolerance, feature ease of catalyst addition and withdrawal, and exhibit uniform reactor temperature. Relatively small catalyst particles can be applied without the pressure drop constraint of fixed-bed trickle reactors.

Due to the catalyst bed expansion, the volumetric catalyst holdup per reactor volume is smaller than in trickle-bed reactors. From a kinetics standpoint an ebullating-bed reactor behaves like a fully

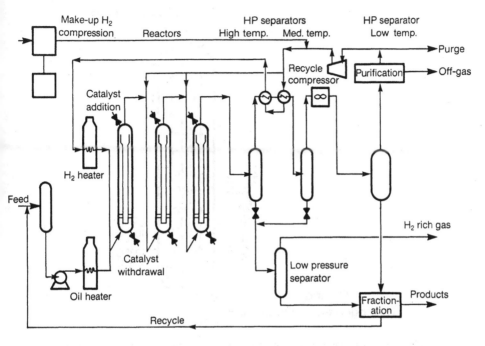

FIG. 11.  LC-Fining process flow sketch [19].

backmixed reactor (BMR). As a consequence, a series of reactors
is usually proposed where high conversions must be obtained. This
is reflected in Fig. 11, which shows a process flow sketch of a typical LC-Fining unit. Ebullating-bed reactors are usually operated
at higher temperatures than fixed-bed reactors. This limits the
total reactor volume, but enhances hydrogen consumption, as the
formation of light gases increases.

Reactor temperature control is important in ebullating-bed reactors, as is reflected in the great number of thermocouples installed in the reactor. Figure 12 shows the location of these thermocouples in one of the two reactors in operation at Salamanca.
The data illustrate the reactor's uniform temperature profile, which
is mainly controlled by the temperature of the feed. There are a
number of thermocouples installed in the three rings in the bottom
of the reactor wall (Fig. 12). These are mainly used as indicators
of maldistribution and formation of stagnant zones. Theoretically,
ebullating beds are considered to be more prone to thermal run-
away than are trickle-bed reactors [22], for which interbed quenching can be used to control exothermicity of the reactions.

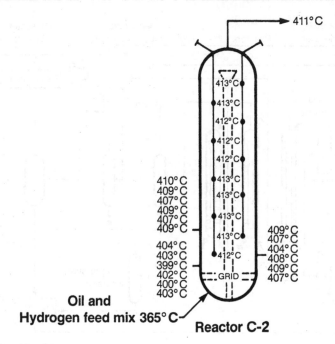

FIG. 12.  Petroleos Mexicanos reactors temperature profile,
1/12/76 [19].

Extensive analyses of the 1970 Baywater explosion have led to
improvements in the design for distributing the gas and feed in
the reactor.  These measures, combined with the use of cold-wall
reactor construction, can be regarded as efficient and reliable
protection against the temperature excursions that occurred in
the Baywater unit.  From an operational standpoint, the ebullating-
bed reactor system is no more complex than other residue hydroprocessing reactor systems.

## 2. Reactor Volume Requirements

Both trickle-bed and ebullating-bed reactors for residue upgrading applications are expensive capital cost items because of
the high temperatures and pressures applied and the huge catalyst volumes required.  To obtain high conversions, two or three
ebullating-bed reactors, in series, are usually used.  Because of
the required catalyst expansion, the catalyst hold-up per installed
reactor volume is lower than in fixed-bed reactors.  However,
smaller catalyst particles can be used in ebullating-bed reactors

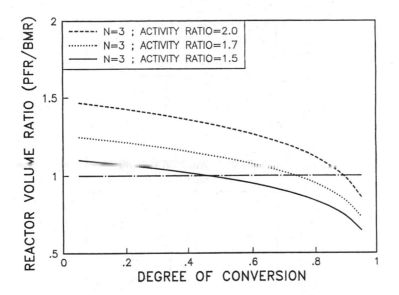

FIG. 13. Ratio of the reactor volumes as a function of conversion.

than in fixed-bed reactors. The average reaction temperature in ebullating-bed reactors is usually higher than in fixed-bed trickle reactors. Consequently, the apparent catalyst activity in ebullating-bed reactors may easily be a factor of 1.5-2.0 higher than in fixed-bed applications. Allowing for catalyst bed expansion, the required reactor volume for a backmixed reactor is given by

$$V_{BMR} = \frac{N}{(1 - \xi_\beta)} \frac{F}{k_B} \frac{1 - e^y}{e^y} \tag{10}$$

$$y = \frac{1}{N} \log (1 - X_{HDS}) \tag{11}$$

The ratio of the reactor volumes required in fixed-bed over backmixed operations has been plotted as a function of conversion, assuming three ebullating-bed reactors in series (Fig. 13). The activity of the catalyst in the ebullating-bed reactors was assumed to be 1.5, 1.7, or 2.0 times the activity in fixed-bed operation. The reactor volume for fixed-bed reactors was calculated using Eq. (7). The results show that only at very high conversion levels will the reactor volume in fixed-bed operation be smaller than in ebullating-bed reactors. No corrections were made for different

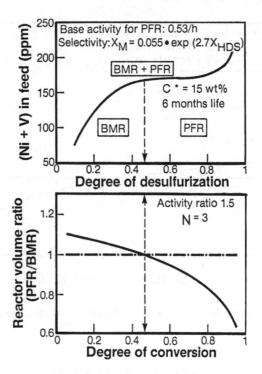

FIG. 14.   Selection of reactor technology.

kinetics in fixed-bed and ebullating-bed reactors.  It is well known
that desulfurization in fixed-bed operation can best be described
as a second-order reaction, whereas 1.5-order kinetics are applicable in backmixed reactor operation [6].  For simplicity the calculations were done assuming constant first-order kinetics.  This
is justified as a first-order approximation because desulfurization
of residues can be described as a combination of fast and slow first-
order desulfurization reactions.  Combining these results with the
data shown in Fig. 8 leads to the observation that for a feedstock
containing a given amount of Ni + V, a preferred reactor technology can be recommended depending on the required degree of desulfurization.  This is shown qualitatively in Fig. 14.

For this case a catalyst with a metal uptake capacity of 15% was
selected and the activity in ebullating beds was assumed to be 1.5
times that in fixed-bed reactors.  As can be seen in Fig. 14, a
backmixed reactor operation is preferred at low conversion, and
a fixed-bed technology is recommended for high conversion as
long as the Ni + V content of the feed is lower than about 150 ppm

in this specific case. For higher metal contents, a backmixed reactor followed by a plug-flow reactor (or reactors) may give the
best results.

Using an ebullating-bed reactor to remove the majority of the
metals also reduces catalyst consumption because the metal uptake
potential of the installed catalyst can be used to a greater extent
than in fixed-bed operation. Ideally a selective HDM catalyst that
performs according to Eq. (6) should be used. In such a case
high metal removal is achieved at low desulfurization conversion,
thus limiting the exothermicity of the reactions in the ebullating
bed and facilitating reactor temperature stability control. Once
the metal content of the feedstock has been lowered, high conversion operation can be executed in fixed-bed trickle operation
using a high-activity HDS catalyst. Clearly, this marriage of
technologies merits further investigation and would require the
cooperation of competing licensors. The potential rewards favor
the effort to overcome the inherent managerial problems.

## III. EMERGING SLURRY-PHASE TECHNOLOGY

### A. Technology Features

Slurry-phase technology, also called "liquid-phase" or "liquid-
fluidized" technology, consists of a gas/liquid/solid operation in
which the catalyst is fluidized (fully expanded) by the combined
motion of gas and liquid. Figure 15 shows how slurry-phase technology can be construed as a natural prolongation of ebullating-
bed technology. The average catalyst particle size is now much
smaller, resulting in a considerable increase in specific surface
area, even allowing for the much lower specific volume load of
the catalyst (1% versus 40%). The net result is a much higher
density of particles in the liquid medium ($2 \times 10^9$ versus 250 par-
ticles/$cm^3$), with drastic reduction in the interparticle distance.
This feature will prove advantageous in suppressing homogeneous (e.g., liquid polymerization) reactions.

Emerging technologies for hydroconverting heavy feedstocks
are based exclusively on slurry-phase operation. They include
Exxon's M-Coke process [3], UOP's Aurabon process [1], and the
Canadian Government's CANMET process [13, 15]. These processes differ mainly in the type of catalyst used. Their concep-
tual approach to hydroconversion is discussed below.

### B. Conceptual Approach

As opposed to the conventional approach to hydroconversion,
in which massive amounts of catalyst are used to directly attack

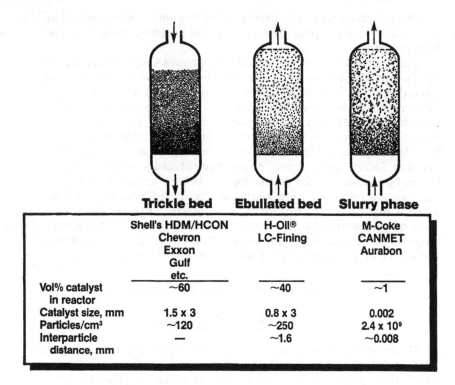

| | Trickle bed | Ebullated bed | Slurry phase |
|---|---|---|---|
| | Shell's HDM/HCON Chevron Exxon Gulf etc. | H-Oil® LC-Fining | M-Coke CANMET Aurabon |
| Vol% catalyst in reactor | ~60 | ~40 | ~1 |
| Catalyst size, mm | 1.5 x 3 | 0.8 x 3 | 0.002 |
| Particles/cm³ | ~120 | ~250 | 2.4 x 10⁹ |
| Interparticle distance, mm | — | ~1.6 | ~0.008 |

FIG. 15.   Comparison of reactor technologies.

the hydrocarbon feed molecules (Reaction 1 in Fig. 16), the hydroconversion mechanism is primarily conceived of as a thermally
induced free-radical cracking reaction of heavy materials such
as asphaltenes and resins (Reaction 2 in Fig. 16), with subsequent hydrogenation (inhibition) of the unstable radicals to oil
in the presence of hydrogen and a catalyst (Reaction 3 in Fig.
16).   Essential to this concept [3] is that the catalyst should be
highly dispersed in the reacting oil in order to engage the free-
radical intermediates as quickly as possible at any point in the
reactor to prevent further degradation to polymeric materials
(Reaction 4 in Fig. 16).   Equally important to obtaining high conversions to lighter oil is that the catalyst should also be active
toward Reaction 1, since it is possible that some heavy feed fractions (in particular some of the coke precursors or Conradson
carbon materials) may first necessitate some hydrogen input via
Reaction 1 before they can be cracked via Reaction 2' [3].   The

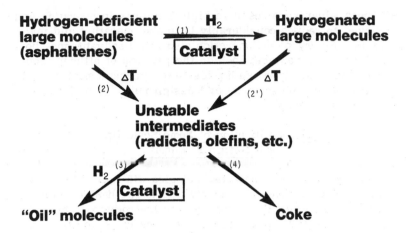

FIG. 16.  Slurry-phase heavy oil upgrading reaction mechanism.

need for a highly dispersed catalyst, in which catalyst particles
act as nuclei for metal deposition, led to the development of slurry-
phase operations.

## C.  Generic Advantages of Slurry-Phase Operation

Among the most attractive features of slurry-phase operation
are the limited amounts of catalyst required to achieve the desired
conversions, and the simplicity, high efficiency, and improved
temperature control possibilities of the reactor vessel.

Only a few hundred ppm of catalytically active species are
needed to achieve the desired conversion, since the small particle size provides a high specific external surface area [$m_{cat}^2$/
$m_{slurry}^3$] and reduced internal diffusion resistance.  Replenish-
ing the catalyst is easy, and continuous regeneration may be possible.  Compared to ebullating-bed technology, no special mechanical strength seems to be required since attrition of the catalyst is
not undesirable.  The minimum size of the catalyst particle is a
concern, however, in its efficient separation from the hydrocarbon liquids.  As with the ebullating-bed technology, slurry-phase
operation solves the problem of reactor plugging which is inherent to fixed-bed operation.  The large liquid-phase heat transfer coefficients and high liquid mass (which acts as a thermal
sink) diminish the risk for temperature excursions.  Heat recovery through external heat exchange is practical, and probably
more easily achieved than in ebullating beds, which use internal recycle.

All these advantages result in high flexibility in operation and
are believed to reduce considerably both the operating and investment costs. Slurry-phase operation also has disadvantages, however. These disadvantages include the problem of efficient separation of the solids from the liquids, and, perhaps most dramatically, a lack of reliable scale-up and design rules. The latter concern is discussed below.

### D.  Identification of Some Scale-Up and Design Problems

The emerging slurry-phase technologies will necessitate a few
more years of development, primarily due to the limited understanding of the factors that influence the hydrodynamic behavior
of the reactor system. Such knowledge is essential for the scale-
up and design of reactor systems, beginning with reliable model-
ing of their hydrodynamic behavior, including such phenomenon
as flow regime, backmixing, interfacial areas, sedimentation, and
convection heat transfer.

A major problem in the scale-up of slurry-phase bubble column
reactors is the assessment of the proper type of flow regime under which the reactor is operated. This decision is of critical
importance since it also determines the applicability of the correlation used for such design parameters as the gas-holdup, mass
transfer, and dispersion coefficients. Some insight into the flow
regimes that can be encountered in three-phase systems can be
gained from the better understood two-phase (gas/liquid) systems.
Three-phase systems behave in essentially the same way as pseudo-
two-phase systems, provided the catalyst concentration and size
are small [7]. For such systems it is known that at low superficial gas velocities, the gas flow is characterized by bubbles rising
in a rather undisturbed fashion in the column. This hydrodynamic
situation, in which the interaction between the bubbles is small and
their size distribution narrow, is usually referred to as homogeneous or bubbly flow (Fig. 17a). The bubbles start to coalesce as
the superficial gas velocity is increased at values depending on
several factors. An equilibrium between coalescence and break-
up of the bubbles is established, and the flow regime denoted as
heterogeneous or churn-turbulent (Fig. 17b). Among the pseudo-
homogeneous dispersion of small bubbles, a few large bubbles appear which rise through the column in a churnlike motion [7]. In
slim reactor columns the size of these bubbles can even reach the
diameter of the column, the gas flow in this situation being characterized by slugs (Fig. 17c). Deckwer and co-workers have
tried to characterize these various flow regimes as a function of

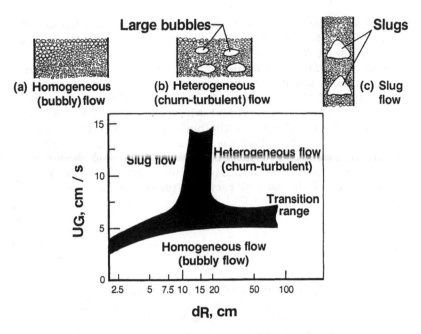

FIG. 17.   Assumed flow regions for slurry-phase reactors [7].

the superficial gas velocity and reactor diameter (Fig. 17). Although originally developed for two-phase systems, this flow diagram is claimed to be equally applicable to three-phase systems, provided the catalyst load and particle size are small. The transition range in this representation (shaded area) is believed to depend also on the dispersion height, the gas distribution system, the liquid velocity, and the physicochemical properties of the three phases involved (actually covering all unknown factors). Although the dependency on these factors has been recognized, little work has been done to quantify their influence on the flow pattern. A complicating factor is that the nature of the flow often changes with distance from the gas distributor.

If it is true that industrial applications imply operation at churn-turbulent conditions since high superficial gas velocities and reactor diameters are desired for high capacity, it is important to realize that most correlations for important design parameters have been derived only under bubbly flow conditions. The limited availability of reliable correlations appears especially crucial in the case of axial dispersion coefficients, and essential in determining the degree of phase backmixing and hence the product composition. In particular, the often postulated assumption of uniformity of the catalyst distribution is not always justified, but no correlations exist for predicting the axial dispersion coefficient of the solids. The degree of liquid-phase mixing, on the other hand, although usually assumed to be complete, essentially depends on the magnitude of the reactor $L/D_R$ ratio [4]. Finally, the limited availability of data is also evident when estimating the solubility and diffusivity of hydrogen in the hydrocarbon liquid mixture, a complicating factor being that the mixture composition may change with time on stream. Provided reliable scale-up and design of slurry reactors can be accomplished, this technology would appear to be one of the most attractive alternatives to conventional technology.

## SYMBOLS

| | |
|---|---|
| $C^*$ | averaged metal uptake capacity (kg/kg) |
| $C$ | overall metal loading of catalyst bed (kg/m$^3$) |
| $C_{max}$ | maximum value of $C$ (kg/m$^3$) |
| $C_m$ | metal content of feedstock (kg/m$^3$) |
| $F$ | feed rate (m$^3$/h) |
| $k_B$ | first-order desulfurization rate constant in backmixed reactor (h$^{-1}$) |
| $k_m$ | first-order demetallization rate constant (h$^{-1}$) |
| $k_m^\circ$ | initial value of $k_m$ (h$^{-1}$) |
| $k_p$ | first-order desulfurization rate constant in plug-flow reactor (h$^{-1}$) |
| $L_0$ | required run length (h) |
| $\dot{m}$ | metal deposition rate (kg/h) |
| $N$ | number of backmixed reactors in series |
| $SV$ | space velocity (kg/h·m$^3$) |
| $t$ | run time (h) |
| $V_{PFR}$ | reactor volume of plug-flow reactor (m$^3$) |
| $V_{BMR}$ | reactor volume of backmixed reactor (m$^3$) |
| $V_{min}$ | reactor volume of plug-flow reactor for metal storage (m$^3$) |
| $X_m$ | degree of metal removal |
| $X_{HDS}$ | degree of desulfurization |
| $\theta$ | dimensionless catalyst age |
| $\rho_{cat}$ | bulk density of catalyst (kg/m$^3$) |
| $\xi_\beta$ | bed expansion (m$^3$/m$^3$) |
| $k_{app}/k_v$ | contacting effectiveness |

REFERENCES

[1]   F. H. Adams, J. G. Gatsis, and J. G. Sikonia, in Pro-
      ceedings of First International Conference on the Future
      of Heavy Crude Oils and Tar Sands (R. F. Meyer and
      C. T. Steele, eds.), McGraw-Hill, New York, 1981, p. 632.
[2]   S. Asaoka, Ind. Eng. Chem., Process Des. Dev., 22, 242-
      248 (1983).
[3]   R. Bearden and C. L. Alderidge, 1981; Energy Prog.,
      1(1-4), 44 (1981).
[4]   D. B. Burkur, Chem. Eng. Sci., 38(3), 441 (1983).
[5]   F. M. Dautzenberg, S. E. George, C. Ouwerkerk, and
      S. T. Sie, Advances in the Catalytic Upgrading of Heavy
      Oils and Residues; Paper Presented at the Advances in
      Catalytic Chemistry Symposium, Salt Lake City, Utah,
      May 1982.
[6]   F. M. Dautzenberg, J. Van Klinken, K. M. A. Pronk,
      S. T. Sie, and J.-B. Wijffels, "Catalyst Reactivation
      through Pore Mouth Plugging during Residue Desulfuriza-
      tion, ACS Symp. Ser. 65, 254 (1978).
[7]   W. Deckwer, Y. Louisi, A. Zoudi, and M. Ralik, Ind.
      Eng. Chem., Process Des. Dev., 13, 633 (1980).
[8]   J. P. Dickie and T. F. Yen, Anal. Chem., 39, 1847 (1967).
[9]   C. T. Douwes, J. Van Klinken, J.-B. Wijffels, and W. C.
      Van Zijll Langhout, "Developments in Hydroconversion
      Processes for Residues," 10th World Petroleum Congress,
      Bucharest, Hungary, 1979, Panel Discussion 10.
[10]  H. V. Drushel, Analytical Characterization of Residuals
      and Hydrotreating Products, Paper Presented at Am.
      Chem. Soc., Div. Pet. Chem., New York, New York,
      August 27-September 11, 1972.
[11]  R. M. Eccles, A. M. Gray, and W. B. Livingston, H-Oil:
      Texaco's Design for Flexibility, Paper Presented at Natl.
      Pet. Refiners Assoc. Annu. Meet., San Antonio, Texas,
      March 21-23 (1982).
[12]  R. M. Eccles, A. M. Gray, and W. B. Livingston, Oil Gas
      J., 80(15), 121 (1982).
[13]  M. A. Menzies, T. F. Scott, and J. A. Denis, Proc.
      Intersoc. Energy Conv. Eng. Conf., 15(1), 26 (1980).
[14]  W. L. Nelson, Oil Gas J., 74, 72 (November 15, 1976).
[15]  D. J. Patmore, C. P. Khulbe, and K. Belinko, Am. Chem.
      Soc., Div. Pet. Chem. Prepr., 26(2), 431 (1981).
[16]  A. A. Pegels and J.-B. Wijffels, U.S. Patent 3,826,737
      (1976), assigned to Shell Oil Co.

[17]   J. D. Potts and H. Unger, Proc. Intersoc. Energy Conv. Eng. Conf., 15(3), 1832 (1980).

[18]   H. F. Rase, Chemical Reactor Design for Process Plants, Wiley, New York, 1977.

[19]   B. Robson, K. Chinnery, C. Herster, C. Viens, and J. Caspers, Proceedings of the C. E. Lummus Montreal Area Refiner's Seminar, May 21, 1980.

[20]   C. N. Satterfield, Am. Inst. Chem. Eng., 21, 209 (1975).

[21]   J. Van Klinken and R. H. Van Dongen, Chem. Eng. Sci., 35, 59-66 (1980).

[22]   W. C. Van Zijll Langhout, C. Ouwerkerk, and K. M. A. Pronk, Development of and Experiences with the Shell Residue Hydroprocesses, Paper Presented at 88th National AIChE Meeting, Philadelphia, Pennsylvania, June 1980.

[23]   R. H. Wolk and W. C. Robesti, Demetallization of Heavy Residual Oils, Phase I, U.S. Environ. Prot. Agency, U.S. Dept. Comm., Natl. Tech. Inf. Serv., December 1973, PB 277568.

[24]   T. F. Yen, The Role of Trace Metals in Petroleum, Ann Arbor Science Publishers, Ann Arbor, Michigan, 1975.

# Hydrotreatment of Cracked Light Gas Oil*

R. GALIASSO, W. GARCIA, M. M. RAMIREZ DE AGUDELO,
AND P. ANDREU
Intevep S.A.
Caracas, Venezuela

## I. INTRODUCTION

Since worldwide conversion processes are used to upgrade heavy oil to distillates, the hydrotreatment of light gas oil (LGO) as a downstream process has been used more extensively. This fraction (LGO) is produced from thermal or catalytic cracking or hydrocracking processes. It contains high amounts of unsaturates, nitrogen, and sulfur compounds which cause instability while in storage due to gum formation. The use of LGO as a fuel oil for diesel engines plugs the filter and produces sulfur and nitrogen

---

*This paper, scheduled for presentation, was not presented at the Conference due to the inability of the authors to attend.

emissions. These sulfur and nitrogen compounds arise from the cracking of heavy cuts and are aromatic-type molecules which are difficult to hydrogenate. This cut also possesses a low cetane index (CI) which must be increased (by aromatic hydrogenation) because of its poor motor performance. Color and color stability are associated with a high bromine number (BN, unsaturate content), nitrogen, and aromatic content. In order to improve these properties, a deep hydrogenation is sometimes required.

A modern refinery tries to optimize the diesel pool from different crudes. Thus, hydrotreatment, as a downstream process, must be adapted to use different types of feed, straight and cracked, by using appropriate catalysts and operating conditions. Thus different market requirements are met.

In this paper an attempt is made to study the parameters which influence hydrotreatment as variables used to yield diesel specifications. Special attention is paid to catalyst selection according to its surface composition and the objectives.

## II. EXPERIMENTAL

Materials. Six commercial hydrotreatment catalysts named A, B, C, D, E, and F were considered in this study. They may be recognized from the properties presented in Table 1. The four types of feed used in this study were two LGOs produced by catalytic cracking of two different crudes, a coker gas oil (CGO), and a hydrocracker gas oil (HCGO). The analyses of these four cuts are shown in Table 2.

Equipment. The flow diagram of the bench-scale unit used for the catalytic studies is shown in Fig. 1. The feedstock was filtered, pumped from the storage tank, and preheated in an electrically heated coil. Subsequently it was mixed with hydrogen and passed in a downflow manner through an isothermal fixed-bed reactor. The bed contained about 100 cc of catalyst diluted with inert material in order to assure proper fluid dynamics. The product leaving the reactor was separated in a high-pressure, low-temperature separator. The gas stream containing hydrogen sulfide, ammonia, and $C_1$-$C_4$ compounds was analyzed by gas chromatography. Liquid samples were stripped and analyzed as explained later. Mass balance was performed on stream every 24 h.

Catalysts were normally dried, weighed, and presulfided in situ using $H_2$ + $H_2S$ or $RS_2$ compounds at 250-350°C. Spent catalyst samples were washed with xylene and analyzed for metal content.

The unit was started in a hydrogen flow. The liquid was then fed to the unit and the reactor heated to the reaction temperature. The first sample was withdrawn after 24 h when it assumed that a steady state was achieved.

Previous fluid dynamic tests showed that regardless of the size and shape of the catalyst, the kinetic results were reproducible when the particles were diluted and filled the reactor properly.

Analysis of Liquid Samples. Most of the chemical analyses performed followed ASTM standard techniques, such as: Cetane Index (D-976), aniline point (D-611), distillation (D-86), sulfur (Leco Analysis), nitrogen (Kjeldahl for total content and perchloric for basic compounds), C-H-N analysis (Perkin-Elmer 240 Elemental Analyzer), aromatics (GC-MS), density (ASTM D-287), bromine number (ASTM D-159), and color (ASTM D-1500).

Analysis of Catalysts. Total metal content was determined by atomic absorption using a Varian Techtron Analyzer. Sulfur and carbon were determined using a Leco Analyzer. Physical properties were determined by conventional techniques. The results are reported in Table 1.

Photoelectron spectra (XPS) were obtained in a Leybold-Heraeus LHS-10 system using A1K$\alpha$ (1486.6 eV) with 50 eV pass energy and 300 W of power. The Al$_{2p}$ peak (74.6 eV) was used as reference for charge effect correction. Oxide and sulfide catalysts were considered, the latter transferred to the apparatus from sealed ampules to avoid contact with air. Atomic concentration was evaluated from the peak areas and sensitivity factors which account for the orbital ionization cross section, electron mean free path ($\lambda$), and spectroscopic effects. Curve fitting of the Mo$_{3d}$ doublet was performed according to Patterson et al. [1].

Raman spectra were obtained in a Cary 86 Spectrometer equipped with a triple monochromator. The 514.15 line of Ar$^+$ laser from Spectra Physics 165 was used for excitation. The typical slit which was 4 cm and the wavenumber accuracy was better than 2 cm$^{-1}$.

Diffuse reflectance spectra (DRS) were obtained in a Pye Unicam AEI-SP8-100 spectrophotometer using $\gamma$-Al$_2$O$_3$ as reference and a sample thickness of 1.2 mm which permits the R$_\infty$ determination in the UV-Vis region.

Electron spin resonance (ESR) was performed in a Varian E-12 dual cavity spectrometer using x-band microwave radiation and 100 kHz modulation in order to derive the absorption curve. Varian strong pitch was used as standard.

Magnetic susceptibility was measured using the Faraday method and a Sartorius or a Cahn microbalance. Diamagnetic and ferromagnetic contributions were eliminated from the measurements which were carried out from -198 to 130°C.

TABLE 1

Catalyst Properties

|  | A | B | C | D | E | F |
|---|---|---|---|---|---|---|
| Chemical composition: |  |  |  |  |  |  |
| $MoO_3$ | 16.0 | 15.5 | 12.4 | 19.8 | 18.4 | – |
| CoO | 2.0 | 5.5 | 3.8 | – | – | – |
| NiO | – | – | – | 4.1 | 6.8 | 3.4 |
| $WO_3$ | – | – | – | – | – | 20.9 |
| $P_2O_5$ | – | – | – | 6.8 | 7.3 | – |
| Calcination loss wt% (600°C) | 4.1 | 3.9 | 2.9 | 4.4 | 3.3 | 4.0 |
| Physical properties: |  |  |  |  |  |  |
| Average particle diameter (mm) | 1.28 | 1.63 | 1.61 | 1.59 | 1.63 | 1.20 |
| Shape | Extruded | Extruded | Extruded | Extruded | Extruded | Sphere |
| Apparent density (g/cm³) | 1.48 | 1.20 |  | 1.46 | 1.45 | 1.47 |

| | | | | | | |
|---|---|---|---|---|---|---|
| Real density (g/cm$^3$) | 3.50 | 5.51 | 3.00 | 3.63 | 3.84 | 3.78 |
| Pore volume (cm$^3$/g) | 0.55 | 0.79 | 0.67 | 0.58 | 0.67 | 0.34 |
| Surface area (m$^2$/g) | 200 | 285 | 283 | 133 | 135 | 197 |
| Average pore diameter (A) | 120 | 92 | 73 | 120 | 136 | 69 |
| Pore distribution (A), V%: | | | | | | |
| 20 - 30 | 4.3 | 2.8 | 5.7 | 0.0 | 3.3 | 12.0 |
| 30 - 60 | 8.6 | 40.3 | 75.0 | 12.5 | 26.1 | 1.5 |
| 60 - 90 | 19.9 | 51.6 | 11.6 | 28.0 | 48.6 | 20.3 |
| 90 - 150 | 23.7 | 6.3 | 3.8 | 54.4 | 18.1 | 10.3 |
| 150 - 300 | 19.4 | 1.6 | 2.0 | 3.3 | 2.0 | 13.2 |
| 300 - 500 | 4.6 | 0.0 | 0.0 | 0.5 | 1.5 | 2.4 |
| 500 - 10$^3$ | 1.7 | 0.0 | 0.0 | 0.3 | 0.4 | 0.5 |
| 10$^3$ | 2.8 | 0.0 | 0.0 | 1.0 | 0.0 | 0.0 |
| Mechanical strength (kg/mm) | 8.3 | 6.1 | 10.0 | 9.3 | 6.1 | 10.0 |

TABLE 2

Feed Properties

| Properties | Delayed coking feed I | | Catalytic cracking feed II | | Catalytic cracking feed III | | Hydro- cracking feed IV | |
|---|---|---|---|---|---|---|---|---|
| Sulfur (ppm) | 4500 | | 3500 | | 12500 | | 19400 | |
| Nitrogen (ppm) | 350 | | 515 | | 1238 | | 4810 | |
| Basic nitrogen (ppm) | 180 | | 217 | | - | | - | |
| Aromatic TLC (ndm) | 50/(40) | | 53(42) | | 77.5 | | (35.0) | |
| Bromine number | 35 | | 20 | | 6.4 | | 9 | |
| Density | 0.873 | | 0.8682 | | 0.9720 | | 0.8690 | |
| Cetane index | 39.0 | | 35 | | 27.0 | | 41 | |
| Molecular weight | 200.0 | | 212 | | 217 | | 203 | |
| Color | Black | | Black | | Black | | Brown | |
| Distillation ASTM (°F/%V) | 399 | 0 | 399 | 0 | 471 | 0 | 392 | 0 |
| | 410 | 10 | 446 | 10 | 529 | 5 | 662 | 95 |
| | 498 | 30 | 488 | 30 | 549 | 10 | | |
| | 523 | 50 | 530 | 50 | 582 | 30 | | |
| | 582 | 70 | 595 | 70 | 604 | 50 | | |
| | 603 | 90 | 618 | 90 | 634 | 70 | | |
| | 633 | 100 | 653 | 100 | 637 | 90 | | |
| C (%wt) | 86.5 | | 86.2 | | - | | 84.9 | |
| H (%wt) | 11.9 | | 12.3 | | - | | 12.7 | |
| Gums formation ASTM 2274 (mg/100 mL) | 7 | | 3 | | - | | - | |

Total acidity was measured using ammonia uptake by a gravimetric method utilizing a Sartorius balance. The weighed sample (about 50 mg) was saturated with 600 torr $NH_3$ at room temperature. The irreversibly adsorbed ammonia was measured at various temperatures.

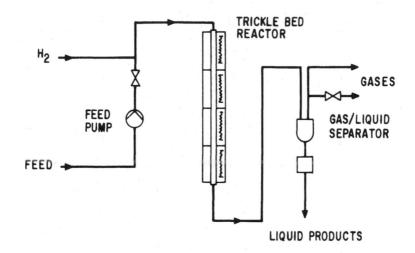

FIG. 1.  Bench-scale hydrotreating unit.

## III.  RESULTS

The properties of the four feedstocks shown in Table 2 are evidence that there exists a considerable difference among them. Feeds I and II, coming from the coking and cracking (respectively) of a paraffinic crude, contain low levels of sulfur and nitrogen but have high CI and BN.  Feed III was produced by the cat-cracking of a naphthenic crude.  It contains the highest amounts of nitrogen, sulfur, and aromatics, and the lowest CI and density.  Feed IV was the hydrocracked product of an extraheavy Orinoco Belt crude.  It contains the highest basic nitrogen content and its properties are intermediate between Feeds I and II.  The carbon and hydrogen contents and the gum formation for Feeds I and II are also indicated in Table 2.

In order to search for the operating parameter sensitivity, a previous kinetic study was carried out for Feed I using a CoMo-type catalyst (A) and a NiW-type catalyst (F).  Temperature was varied from 350 to 390°C, and pressure ranged between 30 to 50 bars while the space velocity ranged between 1 and 3 h$^{-1}$.  The H$_2$/feed ratio was held constant in all experiments.  Table 3 summarizes the results obtained with Catalyst A.

This study indicated that sulfur and nitrogen removal are sensitive to pressure and temperature.  The apparent orders of reaction for sulfur and hydrogen are about 2 and 0.75, respectively.

TABLE 3

Hydrotreatment of Feed I Using Catalyst A

| | | | | | | | | |
|---|---|---|---|---|---|---|---|---|
| **Operating conditions:** | | | | | | | | |
| Temperature (°C) | 370 | 370 | 370 | 370 | 370 | 350 | 390 | 350 | 390 |
| Pressure (psig) | 50 | 50 | 50 | 80 | 30 | 50 | 50 | 80 | 30 |
| LHSV ($h^{-1}$) | 3 | 1.5 | 1 | 1.5 | 1.5 | 3 | 3 | 1.5 | 1.5 |
| $H_2$/HC | 200 | 200 | 200 | 200 | 200 | 200 | 200 | 200 | 200 |
| **Analysis of the product:** | | | | | | | | |
| S (ppm) | 110 | 45 | 35 | 33 | 66 | 214 | 70 | 95 | 50 |
| $N_T$ (ppm) | 130 | 85 | 63 | 68 | 108 | 150 | 110 | 82 | 95 |
| $N_B$ (ppm) | 75 | 45 | 34 | 38 | 54 | 115 | 44 | 70 | 48 |
| C (%wt) | 85.41 | 86.15 | 84.95 | 85.95 | 85.30 | 85.60 | 85.30 | 85.17 | 85.31 |
| H (%wt) | 13.31 | 13.60 | 13.75 | 13.80 | 13.40 | 13.20 | 13.50 | 13.39 | 13.39 |
| Aromatic (wt%) | 36 | 30 | 38 | 25 | 34 | 37 | 32 | 30 | 34 |
| Naphthenic (wt%) | 19 | 19.5 | 20.50 | 18 | 19 | 18.50 | 19.50 | 19 | 19 |
| Bromine number | 30 | 0.8 | - | 0.1 | 2 | 5 | 0.3 | - | 1 |
| Density | 0.8682 | 0.8615 | 0.8605 | 0.8610 | 0.8645 | 0.8675 | 0.8633 | 0.8610 | 0.8640 |
| Molecular weight | 212 | 187 | 180 | 181 | 914 | 200 | 185 | 190 | 190 |
| Color | 5 | - | 3 | 2 | 3 | 2 | 2 | - | - |

TABLE 4

Apparent Kinetic Results

| | Catalysts A /F | | |
| --- | --- | --- | --- |
| | Hydrocarbon order | Hydrogen order | Activation energy (kcal/mol) |
| Sulfur removal | 2/1.8 | 0.75/1 | 25 |
| Nitrogen removal | 1.4/1.0 | 0.60/0.8 | 20 |
| Basic nitrogen removal | 1.2/1.0 | 0.5/0.5 | 20 |
| Aromatic saturation | 1.0/1.0 | 1.0/1.0 | 25 |
| Bromine number improvement | 1.5 | - | - |

The order in total content nitrogen was 1.4 and 1.2 for basic nitrogen. The sensitivity to hydrogen partial pressure for nitrogen removal was 0.8 and 0.5 for total and basic, respectively. The aromatic saturation, which is a complex reaction, followed apparent first order in both aromatic content and hydrogen partial pressure. The BN appeared to follow an order higher than 1, but the poor reproducibility of the data precluded further exploitation. The activation energy values are summarized in Table 4; they varied between 25 and 30 kcal/mol. The results obtained for Catalyst F are also included in Table 4.

Comparison of the results on reaction order obtained with these two catalysts permitted us to assume that the sensitivity to hydrocarbons and hydrogen are substantially similar in view of the reproducibility of the experimental results.

Sonnemans [2] reported a 1.5 order of reaction for the HDS and the HDN of a LGO, and stated also that this value was rather insensitive to the type of catalyst used. The activation energy values that he obtained were moderately lower than ours, 15 to 20 kcal/mol for HDS and 10 to 12 kcal/mol for HDN.

Table 5 shows the variation of product quality obtained for all the feedstocks, catalysts, and operating conditions. A comparative analysis of these results is presented in Figs. 2 and 3. Figure 2 is indicative of an interrelation between the type of catalyst and the type of feed, i.e., the effect on product quality of

TABLE 5

Product Quality vs Type of

| | I | | |
|---|---|---|---|
| | A | B | F |
| **Operating conditions:** | | | |
| Pressure (bar) | 30/50 | 50 | 50 |
| Temperature (°C) | 350/370 | 370 | 370 |
| Space velocity (LHSV) | 1.5 | 1.5 | 1.5 |
| $H_2$/HC (N1/1) | 200 | 200 | 200 |
| **Product quality:** | | | |
| Sulfur (ppm) | 66/45 | 50 | 55 |
| Nitrogen (ppm) | 108/85 | 90 | 70 |
| Basic nitrogen (ppm) | 54/45 | 50 | 38 |
| Aromatic wt% TLC (ndm) | (34/30) | (30) | (25) |
| Bromine number | 2/0.8 | - | - |
| Cetane Index | 40/43.5 | 42.5 | 44 |
| Molecular weight | 194/187 | 190 | 185 |
| Density (kg/1t) | 0.8645/0.8615 | 0.8620 | 0.8615 |

one depends on the other. The two CoMo catalysts did not exhibit a significant difference for Feed I at 50 atm and 370°C. Catalyst F (NiW type) showed a better activity for nitrogen and aromatic removal, but the CI improvement resembled that obtained for Catalyst A.

Color and color stability did not correlate well with severity. However, we have observed more stable products and a lower color number at high hydrogenation conditions.

Regarding Feed II, Catalyst D (NiMo type) was more active than Catalyst A (CoMo type) for aromatic removal and CI improvement. Likewise, for Feed III, NiMo catalyst (D) was more active than CoMo and NiW catalysts (A and F, respectively). These last two catalysts showed similar activities.

Catalyst and Feed

| II | | III | | | IV |
|---|---|---|---|---|---|
| A | D | C | E | F | D |
| 50 | 50 | 28/48 | 28/48 | 28/48 | 48.3 |
| 370 | 370 | 350/370 | 350/370 | 350/370 | 380 |
| 1.5 | 0.5 | 0.5 | 0.5 | 0.5 | 1.07 |
| 200 | 200 | 85/250 | 85/250 | 85/250 | 600 |
| 310 | 280 | 4260/520 | 2853/505 | 6300/830 | 700 |
| 60 | 45 | 1039/367 | - | 1083/270 | Nil |
| 45 | 30 | - | - | - | - |
| (33) | (29) | 76/70 | 676/577 | 74/71.5 | 1 |
| 1.5 | 0.5 | 2.4/0.9 | - | 3.5/1 | 1 |
| 39.2 | 40.5 | 29/32.7 | - | 287/32.5 | 50 |
| 188 | 184 | -/215 | 213/215 | - | 205 |
| 0.8620 | 0.8610 | 0.9300/0.95700 | - | 0.9580/0.9330 | 0.870 |

Sulfur removal was comparatively the same for all catalysts. Better activity of NiMo catalysts for nitrogen removal obtained in this work agrees with results reported previously by Sonnemans [2], Edgar [3], and others [4] (compare KF 165: HDS = 100, HDN = 100, HDBN = 100 with KF 153-S: HDS = 90, HDN = 137, HDBN = 110). The difference in activity between NiMo and CoMo catalysts may become smaller by controlling the operating conditions, as seen from Fig. 3. The higher the pressure and temperature, the higher the nitrogen removal and aromatic conversion. Furthermore the NiMo catalyst (E) was more sensitive to severity than CoMo catalyst; this sensitivity depends more on pressure than on temperature. Sonnemans [2] found similar activation energy values for KF 165 (CoMo) and KF 153-S (NiMo).

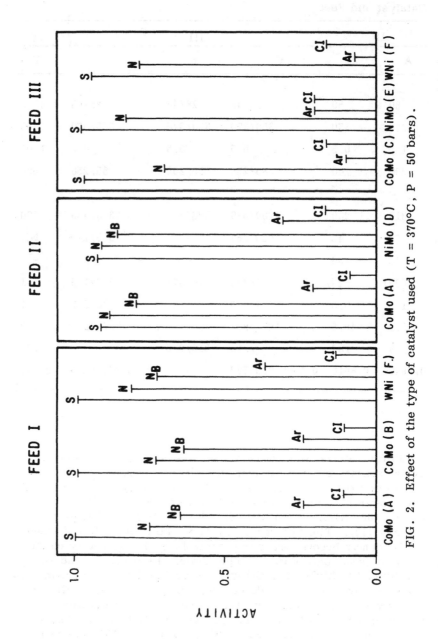

FIG. 2. Effect of the type of catalyst used (T = 370°C, P = 50 bars).

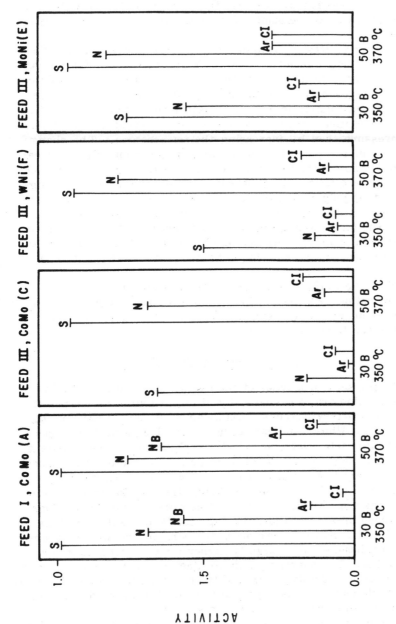

FIG. 3. Effect of pressure and temperature for different catalysts.

The effect of the type of sulfur compound present in the feed-stock was reflected by the fact that HDS activity for CoMo and NiMo catalysts became similar for a specific set of operating conditions.

In order to attempt to understand the hydrogenation reaction and the CI improvement, the aromatic and paraffinic distribution analysis were carried out by GC-MS. The results reported in Table 6 correspond to Catalysts A and D at two different severities. Hydrogenation increased the saturate content in Feed III. A stronger effect on CI improvement was observed by increasing the pressure than by increasing the temperature. The change detected in the paraffin content was in agreement with the measured CI improvement. Since the polyaromatic content decreased when monoaromatic and naphthenic contents increased, it is suggested that polyaromatic hydrogenation produces monoaromatic and naphthenic molecules. In addition, the polyaromatic rate of conversion is higher than the monoaromatic rate [5, 6]. Thus, the paraffin content increases and so does the CI. This phenomenon was more noticeable for NiMo (D) than for CoMo (C) catalysts.

Polycyclic sulfur compounds decreased simultaneously when hydrogenation took place. Increasing the pressure and the temperature decreased the sulfur content. NiMo catalysts were more active than CoMo catalysts in this case.

## A.  Catalyst Characterization

The physical properties of the catalysts are summarized in Table 1. They were prepared by different industrial companies using proprietary methods. Catalysts A, B, and C were of the CoMo type. While Catalysts A and C contained almost the same amount of metals, Catalyst B had a slightly higher metal content. Catalyst B showed the higher average pore diameter and surface area. Catalysts D and E were of the NiMo type, containing phosphorus in their formulation. Although these NiMo catalysts contained the same amount of metals, this metal content was higher than those present in the CoMo-type catalysts. Although Catalysts D and E had the same area, the pore volume and the average pore diameter were slightly higher for Catalyst E than for Catalyst D. The average particle diameter of these five catalysts was the same. Catalyst F was of the NiW type with a relative high amount of metals. This spherical Catalyst F presented low pore volume and intermediate surface area.

All the catalysts were submitted to a surface study. Our discussion will first deal with Ni and Co and then with Mo.

TABLE 6

Compound Distribution Using Feed III

| | Feed III | Catalyst A | | | Catalyst E | |
|---|---|---|---|---|---|---|
| | | (Expt. 1) | (Expt. 2) | (Expt. 3) | (Expt. 5) | (Expt. 6) |
| Operating conditions: | | | | | | |
| Temperature (°C) | – | 350 | 370 | 370 | 350 | 370 |
| Pressure (bars) | – | 30 | 30 | 50 | 30 | 50 |
| Space velocity | – | 0.53 | 0.53 | 0.46 | 0.5 | 0.5 |
| $H_2$/HC (N1t/1t) | – | 236 | 236 | 236 | 236 | 236 |
| Saturates (ASTM 2786) | 25.4 | 29.6 | 28.0 | 35.2 | | |
| Paraffins (iso type) | 11.9 | 13.8 | 13.8 | 16.1 | | |
| Naphthenes: | 13.5 | 15.8 | 14.2 | 19.1 | | |
| 1 ring | 3.9 | 4.6 | 4.1 | 5.3 | | |
| 2 ring | 3.8 | 4.7 | 4.1 | 5.5 | | |
| 3 ring | 3.1 | 3.6 | 3.1 | 4.8 | | |
| 4 ring | 2.7 | 2.9 | 2.9 | 3.5 | | |
| 5 ring + | 0.0 | 0.0 | 0.0 | 0.0 | | |
| Aromatics (ASTM 3234) | 74.6 | 70.4 | 72.0 | 64.8 | 67.6 | 52.7 |
| Mono-aromatics | 9.8 | 24.6 | 17.6 | 26.9 | 34.0 | 25.5 |
| Di-aromatics | 39.6 | 32.3 | 36.7 | 27.5 | 26.0 | 14.2 |
| Tri-aromatics | 15.4 | 10.9 | 14.8 | 8.6 | 6.8 | 5.3 |
| Tetra-aromatics | 2.0 | 1.1 | 1.7 | 0.8 | 0.3 | 1.2 |
| Penta-aromatics | 0.0 | 0.0 | 0.0 | 0.0 | 0.0 | 0.6 |
| Tiophene aromatics | 7.8 | 1.5 | 1.2 | 1.0 | 0.4 | 2.8 |
| Cetane index | 27.0 | 32.2 | 31.7 | 32.7 | | |

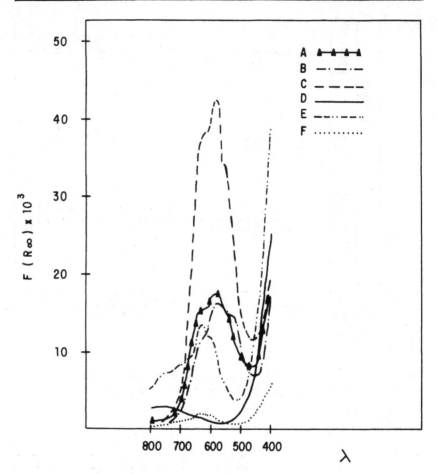

FIG. 4.  $F(R_\infty)$ for several catalysts (DRS).

## 1. Ni and Co Results

### Diffuse Reflectance Spectroscopy (DRS)

Figure 4 shows the visible region of the diffuse reflectance spectra plotted using the Kubelka-Munk function $F(R_\infty)$.

The spectra of CoMo-type Catalysts A, B, and C present the typical absorption between 600 to 800 nm. According to Ashley and Mitchell [7], the observed bands (558, 595, and 634 nm) can be assigned to $Co^{2+}$ in tetrahedral coordination. Thus, Catalyst C contains the highest amount of $CoAl_2O_4$. We could not find any evidence of the presence of $Co_3O_4$ in these catalyst (450-700 nm bands). The $Co^{2+}$ bands were not observed for the sulfided catalysts due to the overlapping of the $MoS_x$ specie bands.

The spectra of Ni-containing Catalysts E (NiMo) and F (NiW) present characteristic bands at 575 and 612 nm assigned to tetrahedral $Ni^2$ [8]. These bands were absent in Catalysts D.

Since the optical coefficients (scattering and absorption) are different for Co and Ni, it is not possible to compare the band intensities directly in order to withdraw quantitative conclusions. However, we point out that Catalysts D and F seemed to present less tetrahedral species than the CoMo catalysts.

## Magnetic Susceptibility

The temperature dependence of the magnetic susceptibility, which is shown in Fig. 5, was similar for all the catalysts. Although the ferromagnetic contribution was low, it was higher for Ni-containing catalysts than for CoMo catalysts. It is possible to determine the octahedral to tetrahedral ratios by using the known magnetic moment of those species and the measured Curie constant [8]. Thus, Catalyst C had the highest tetrahedral specie content while Catalyst D had the lowest, which is in good agreement with our DRS results.

If we assume that this tetrahedral configuration corresponds to sites of the spinel matrix, the major part of octahedral species must be at the surface.

## X-Ray Photoelectron Spectroscopy (XPS)

Co ($2p$) spectra are shown in Fig. 6. The oxide precursors showed the typical satellite structure of $Co^{2+}$ in an oxide matrix [9], i.e., the binding energy values were 780.2, 787, 798.3, and 804 eV. In general the main peaks, $2p_{1/2}$ and $2p_{3/2}$, were separated by 16 eV, while each satellite was separated by 5.7 eV from the main peak. Catalyst B presented the highest Co concentration at the surface (Table 7). The cobalt peaks were shifted 2.8 eV toward lower binding energies when the catalysts were sulfided. The satellite intensities decreased, and the separation of the main peaks increased to 19.4 eV. The changes observed on sulfiding for the shape and intensity ratios (satellite/main peaks) for different catalysts could be indicative of the number of atoms of sulfur, oxygen, and molybdenum neighboring at the surface.

Figure 6 also reports the Ni ($2p$) spectra. The $2p_{3/2}$ and its corresponding satellite peaks were located at 857 and 863 eV, respectively. The Ni content was higher on Catalyst E than on Catalyst D. The additional satellite structure observed might be

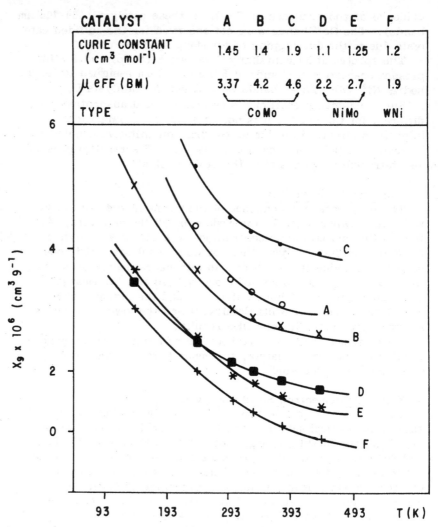

| CATALYST | A | B | C | D | E | F |
|---|---|---|---|---|---|---|
| CURIE CONSTANT ( $cm^3$ $mol^{-1}$ ) | 1.45 | 1.4 | 1.9 | 1.1 | 1.25 | 1.2 |
| $\mu$ eFF (BM) | 3.37 | 4.2 | 4.6 | 2.2 | 2.7 | 1.9 |
| TYPE | | CoMo | | | NiMo | WNi |

FIG. 5.   Gram susceptibility of the catalysts.

attributed to a modification of spin-orbit coupling of $Ni^{2+}$ located
in the Mo monolayer.  The peak position indicated the absence of
$NiAl_2O_4$ at the surface.  The shifting and changes in shape and
intensity ratios observed on sulfiding were similar to those already
mentioned for CoMo catalysts.  However, it was also observed that
an increase in Ni dispersion was obtained after sulfiding Catalysts
D and E.  Catalyst D showed the highest dispersion (Table 8).

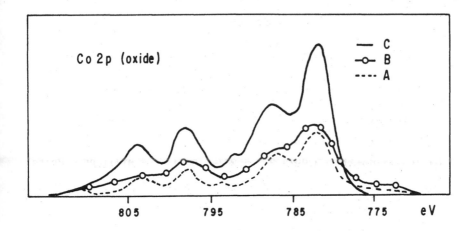

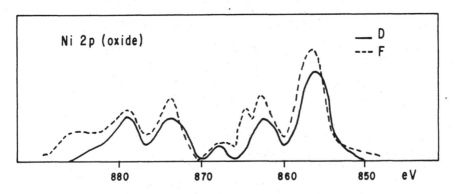

FIG. 6.  XPS spectra of Co (2p) and Ni (2p).

## 2. Mo Results

### DRS

All the Mo-containing catalysts showed the typical $Mo^{+6}$ band between 295 and 315 nm which corresponds to a ligand charge transfer.  Although the band position is almost the same in all catalysts, its intensity did not reflect the metal content.  This band is currently assigned to highly aggregated molybdenum located in an octahedral configuration [7].

TABLE 7

XPS Results

| | A | B | C | D | E |
|---|---|---|---|---|---|
| **Intensity ratios:** | | | | | |
| $I_{Ni}/I$ (Ni+Mo+Al) | - | - | - | 1.93/3.5 | -/5.36 |
| $I_{Co}/I$ (Co+Mo+Al) | 3.8/4.1 | 1.00/- | 6.29/- | - | - |
| $I_{Mo}/I$ (Mo+Mo+Al) | 3.00/3.3 | 5.23/- | 6.14/- | 6.81/11.07 | -/9.58 |
| $I_{Me}/I_{Mo}$ | 3.8/3.15 | 0.19/- | 1.02/- | 0.28/0.32 | -/0.55 |
| | | | | | |
| $Mo^{+5}$ (% of total Mo) | 39.75 | 19.97 | - | - | - |
| **Binding energy (eV):** | | | | | |
| Mo(IV) | 229.0 | 229.0 | - | 228.8 | 228.9 |
| Mo(V) | 230.6 | 230.9 | - | 230.2 | 230.4 |
| Mo(VI) | 233.0 | 233.0 | - | 232.1 | 232.0 |

TABLE 8

ESR Results

| | Catalyst | | | |
|---|---|---|---|---|
| Properties | A | B | C | D |
| g | 1.928/1.933 | 1.917 | 1.930 | 1.923 |
| Relative intensities | 1/300 | 105 | 80 | 75 |
| Hpp (Gauss) | 50/80 | 75 | 130 | 80 |

ESR

The ESR spectra of Catalysts A, B, D, and E are shown in Fig. 7. The Mo(V) ions present in an oxide matrix has been reported [11] to produce a signal centered at $g = 1.936$. The signal has

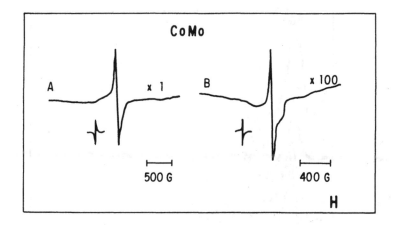

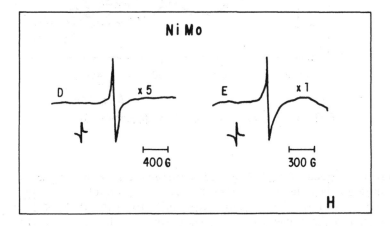

FIG. 7.  EPR analysis of Catalysts A, B, D, and E (oxide).

been attributed to molybdenum located in a tetragonally distorted
tetrahedra in a polymolybdate matrix.  From Table 8 it can be
seen that Catalyst B had the highest intensity and A the lowest.
The linewidth was larger for Catalyst C than for the others.  The
slight differences observed in the g-values could be attributed to
a different environment about the Mo(V) ions.  The signal intensity increased on sulfiding, as expected, since Mo(V) is an intermediate in the reduction of Mo(VI) to Mo(IV).  The relative
changes in intensity which occurred on sulfiding indicated a

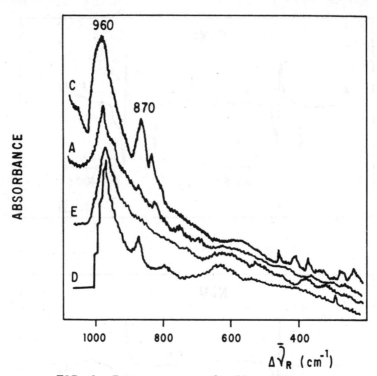

FIG. 8. Raman spectra of oxide catalysts.

different degree of sulfiding and a different reorganization of the catalyst monolayer. Furthermore, it is possible to suggest that the Mo(V) species present after sulfiding are different from those originally present in the precursors: oxide (g = 1.928) sulfide (g = 1.933). This fact has been previously reported [12].

### Raman Spectroscopy

The Raman spectra of CoMo and NiMo catalysts are shown in Fig. 8. Resolution was not very good. The 950-960 cm$^{-1}$ absorption, which has been assigned to polymolybdate species [10], was always detected. The other band of these species at 870 cm$^{-1}$ only appeared in some catalysts. The shoulder present at 600 cm$^{-1}$ is also due to these species. While Catalyst D presented the most intense signals, Catalyst A exhibited the smallest.

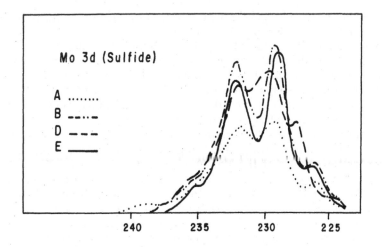

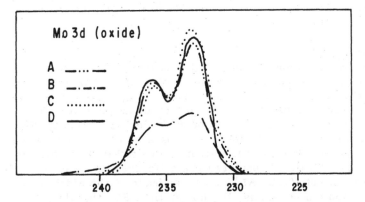

FIG. 9.  XPS spectra of $Mo_{3d}$ oxide and sulfide.

Medema et al. [10] concluded that Ni and Co improved the molybdenum dispersion at the surface of promoted catalysts, based on the lack of bulk bands of $MoO_3$ (998 and 821 $cm^{-1}$) in their Raman spectra. Our results indicated that the catalysts having the highest Co and Ni dispersions also presented the largest 960 $cm^{-1}$ band.

### XPS

The XPS spectra of the $Mo_{3d}$ doublet of the oxide and sulfide catalysts are shown in Fig. 9.  The evaluated binding energy of Mo(VI) gave the same values in all catalysts (233 and 236 eV).

The molybdenum dispersion in the oxide precursors increases
as follows: D > C > E > B > A (Table 7).

Even though the shape of the 3d envelope changed after sulfiding and was different for each catalyst, the binding energies
of Mo(VI), Mo(V), and Mo(IV) can be thought of as constant.
This difference in shape exhibited by the catalysts is due to different contributions of the oxidation state species mentioned on
each catalyst. According to Jepsen and Rase [13], the composition ratio of the Mo species depends on the degree and procedure
of sulfiding. Curve fitting and integration yielded the relative
concentration results of the Mo(V) reported in Table 7. The sulfiding of Catalysts D and E increased the molybdenum dispersion
originally present in the oxide catalysts: Catalyst D showed the
highest dispersion among all catalysts. We are inclined to suggest that this improvement in metal dispersion on Catalysts D
and E could be due to a certain role of phosphorus, present in
the samples. Some suggestions have been pointed out in the
literature [14] regarding the possible role of phosphorus in
NiMo catalysts.

### Acidity

Acidity measurements indicated primarily that all the catalysts
presented higher acidity than that typical of $\gamma$-Al$_2$O$_3$ but much
lower than that obtained for SiO$_2$-Al$_2$O$_3$. Figure 10 shows the
irreversibly adsorbed NH$_3$ as a function of temperature. This
type of desorption isobar could be regarded as an acidity distribution curve since the weakest sites will desorb the ammonia
at low temperatures while the strong sites will do so at high temperatures. We have considered the difference between the acidity
at 100°C and at 400°C (ASD) as an estimate of the width of the
distribution curve. Thus, the larger the ASD, the wider the
distribution of acidity strength. Catalyst A exhibited the highest acidity strength (i.e., acidity at 400°C).

We have measured the acidity of the precursor oxides. Thus,
we have to point out that the acidities of precursor catalysts
have been reported to depend on the pretreatment conditions
[15]. The results of Laine et al. [16] indicated that NiMo-type
catalysts possess higher acidity than CoMo-type catalysts. Pre-
sulfiding the catalysts changed the acidity distribution as shown
by Gajardo et al. [17] as well as by Ayerbe [18] while studying
two sulfided CoMo catalysts when they found that the catalyst
with highest ASD also had the highest HDN activity. Hence,
we have to admit that the acidity of the working catalysts must
be different from what we have shown here.

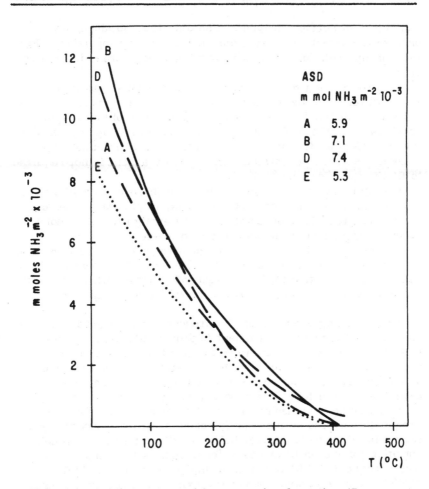

FIG. 10. Acidity measured by ammonia adsorption ($P_{NH_3}$ = 600 mmHg).

## IV. DISCUSSION

Generally, the foremost consideration when a hydrotreatment improvement must be considered is the selection of the best catalyst for particular applications. Since the working elements of a catalyst are located at its surface, an approach to knowledge of these active sites must be provided from the use of fundamental

techniques of surface analysis. The data thus obtained may be applied to the search of correlations with catalytic properties. However, many limitations arise when trying to simulate industrial conditions.

Most refineries use hydrotreatment as a downstream process.
Thus, different feeds for different purposes have to meet the requirements dictated by the market. In order to obtain the cut
specifications, the pool of diesel is usually formed by the combination of different streams, i.e., straight or cracked runs. The
cracked product incorporated after hydrotreatment is not required
to fit the specifications by itself.

The hydrogenation of a cycle diesel oil in order to achieve the
Cetane Index specifications (30/35-45) is, in general, expensive.
Meanwhile, hydrogen consumption is low enough to obtain color,
color stability, and sulfur and nitrogen levels.

The available new generation of denitrogenation catalysts seems
to be very active for aromatic saturation and nitrogen removal but
selectively less active for sulfur removal. Hence, when both nitrogen and sulfur are to be removed, a combination of catalyst is suggested.

The operating variables of the process can be regarded as the
second problem in hydrotreatment improvement. Most of the installed hydrofinishing units were designed for low pressure operation (30-50 bars) since they were planned for the treatment
of straight run cuts or more saturated feeds than they are actually being used for. The maximum reactor temperature is another limitation. Typically, it must be below 380°C. Since hydrogenation causes an increase in temperature, the maximum entry temperature must be kept below 360°C. The hydrogen consumption for treatment of heavy feedstocks is high enough that
using frequent $H_2/HC$ ratios of around 80 reduces the hydrogen
partial pressure in about 50% at the exit. At this partial pressure all catalysts perform quite similarly.

The hydrogenation of aromatic molecules is a reversible reaction under the conditions commonly used in industrial reactors.
High temperature and low hydrogen partial pressure will reduce
the conversion. Thus, the operating conditions have to be selected in order to decrease the contaminants (sulfur and nitrogen)
without completing hydrogenation of polyaromatics.

The reactivity is associated with the molecular weight. The
different feedstocks contain not only different amounts of sulfur, hydrogen, and aromatics, but also different types of molecules. The molecular distribution can be regarded as a function
of distillation temperature. Feed II, for instance, has the nitro-
ten, sulfur, and aromatic distribution presented in Table 9. This

TABLE 9

| Fraction | S (ppm) | N (ppm) | Aromatics (%) |
|----------|---------|---------|---------------|
| 399-488°F | 450 | 150 | 30 |
| 488-580°F | 3300 | 680 | 42 |
| 580-653°F | 6500 | 662 | 52.3 |

distribution indicates an enrichment of unwanted compounds at the tail of the cut. Therefore, not only the activity but also the stability of the catalyst depends on the distillation cut used as feed.

Hydrogen solubility is highly affected by the operating conditions, and consequently its partial pressure is also affected. The temperature, the total pressure, and the $H_2/HC$ ratio control the rate of vaporization of the feed. Sonnemans [2] demonstrated that the reaction rate increased with increasing $H_2/HC$ ratio until vaporization was completed. In our case, for instance, the percentage of vaporization for Feed I is 50% at 30 bars and 350°C but only 30% at 50 bars and 370°C.

In accordance with the wetting properties of the bed, the diffusion control, and the reaction types occurring, the reactivity in the gas phase could be different than that in the liquid phase. Hence, the coverage of the catalytic active surface by hydrogen and hydrocarbon species and the reactivity of the molecules involved might also be different whether or not they come from the liquid or the gas phase. Aromatic, sulfur, and nitrogen molecules might compete for the active centers, but since, on the other hand, their relative amounts in the liquid and gas phase are different, it is difficult to decide about reaction mechanisms using apparent kinetic data. This type of data is even less useful if one considers several catalysts with different coke and sulfur contents and at different steady states. In this sense the analysis of spent Catalysts C and E after 180 h of operation with Feed III proved that coke and sulfur accumulation depends on the type of catalyst used:

|  | C | E |
|----|-----|-----|
| %C | 6.7 | 8.4 |
| %S | 3.9 | 4.7 |

The reaction order for sulfur removal was higher than those for the other reactions. However, this reaction order decreased when the feed became more and more aromatic. This seems to indicate that sulfur removal in light molecules occurs by a simple C–S cleavage but in heavy molecules a previous aromatic ring saturation might be needed. The kinetic results obtained for Feeds I and II must be the addition of the individual kinetic parameters of each of the component cuts. The observed results for the different catalysts are also in agreement with those found by Satterfield et al. [19] and López et al. [20] who explained them in terms of the differences in hydrogenating power of NiMo, CoMo, and NiW catalysts.

Nitrogen removal and saturation of aromatics followed reaction orders close to 1. The nitrogen removal reaction has been reported to occur in a consecutive path which involves ring saturation prior to C–N bond breaking [21]. It was also observed for those catalysts having high hydrogenation activity (i.e., NiMo type) and a better performance than that obtained with catalysts having less hydrogenation activity (i.e., CoMo type).

The saturation of polynuclear aromatics normally requires moderate operating conditions to saturate the first two rings, but high severity is required if C–C bond scissions take place to form naphthene- or benzene-type molecules. This phenomenon is in part due to equilibrium control but also to the reactivities of the molecules that depend on the aromatic ring. A first-order reaction was estimated from our results which was independent of the cut used, i.e., the molecular weight of the reacting molecules. Probably the presence of sulfur and nitrogen polyaromatic compounds (changing the type of adsorption and introducing a new reaction path) would also change the global kinetics.

The selection of the tail of the cut would define the type and the amount of aromatics present in the feed. Their analysis could help in understanding how to improve CI, color, and color stability. At high conversion, the higher the polyaromatic content is, the higher the initial hydrogenation rate but the lower the total rate of hydrogenation.

Unfortunately, as may be seen from Table 10, high quality products are always obtained at the expense of utilizing high severity which involves a high hydrogen consumption. The hydrogenation severity must then be adjusted to an economic scenario taking into account operating costs vs the advantages of incorporating more cracked products in the diesel pool for selling it as a diluent for heating oil.

The catalysts studied lost their initial activity fairly quickly. This, in turn, changes selectivity. For CoMo and NiMo catalysts,

TABLE 10

Hydrogen Consumption vs Product Quality (Catalysts A, Feed I)

| Conditions | Hydrogen consumption $(Nm^3/m^7)$ | Color stability[a] | Cetane Index[b] | N Content | Density |
|---|---|---|---|---|---|
| Low Severity $(350°C/30 \text{ bar}/3 \text{ h}^{-1})$ | 15 | Poor | 1.25 | High | No change |
| Medium severity $(370°C/50 \text{ bar}/1.5 \text{ h}^{-1})$ | 41 | Good | 14.5 | Low | Small change |
| High severity $(390°C/80 \text{ bar}/1 \text{ h}^{-1})$ | 63 | Good | 20.0 | Very low | Important change |

[a]Storaged in $N_2$ at 50°C during 1 month.
[b]$CI = (CI)_0 - (CI)_f/(CI)_0 \times 100$.

sulfur removal was less sensitive than nitrogen removal. The loss of activity depends on the type of feed and the operating conditions. A coked product containing more basic nitrogen and olefins than a catalytic cracked product would also affect the catalyst stability more. Up to now there is no direct way of predicting catalyst life when processing different cracked feeds.

We will attempt to analyze the catalytic activity through the interpretation of some of the surface properties. We will use relative parameters utilizing Catalyst A as the basis of comparison. Figure 11 shows the relative activity for sulfur, nitrogen, basic nitrogen, aromatic, and CI as a function of the relative molybdenum dispersion. Two other abscissas are included in the same figure for comparison purposes: relative intensity of the 960 cm$^{-1}$ Raman band and relative Mo(V) (XPS) concentration. The relative activity varied in the range 0.8 to 1.3. The relative molybdenum dispersion ranged between 1 and 2.5 for the oxide precursors and between 1 and 3.25 for the sulfided catalysts. The NiMo catalysts not only exhibited a higher metal dispersion but also the highest activity for HDN, aromatic saturation, and CI modification. However, the change of dispersion observed in the two catalysts was much larger than the corresponding change in the measured catalytic activity. If one considers the 960 cm$^{-1}$ Raman band, in spite of observing that the general trend and the ranking of the catalysts agreed well with the XPS findings, the variation of this parameter itself differs among the catalysts. For instance, while the change in molybdenum dispersion between Catalysts B and D was 1.28, the corresponding change in the Raman band was only 1.03. The situation did differ when considering the XPS Mo(V) concentration. In this case the relative trend presented by the catalysts did not agree with the other two parameters already considered. A final remark with respect to this figure is the fact that no straight correlation between the parameters measuring the surface molybdenum concentration and the catalytic activities evaluated in this work have been found.

Figure 12 shows the same type of plot as Fig. 11 but considers Co and Ni dispersion on the abscissa. Our analysis that the activity must pass through a maximum at a given value of metal dispersion satisfactorily explains catalyst behavior. It can be seen, for instance, for the CoMo catalysts, that Catalyst B, having less dispersion than Catalyst A, also has less activity, but Catalyst C has less activity even though it had a better Co dispersion. This finding agrees perfectly with what has been stressed in the literature: There is an optimum promoter concentration for which maximum activity can be achieved.

Figure 13 summarizes the changes of some Co and Ni surface parameters. It is observed that good agreement exists between

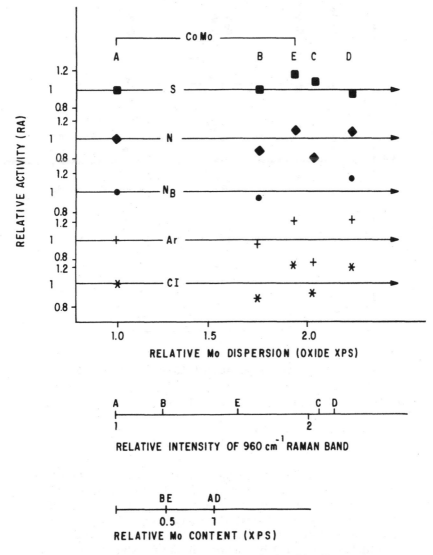

FIG. 11. Relative activity vs relative Mo dispersion.

DRS and magnetism. NiMo catalysts had the lowest content of tetrahedral species and also the highest activity for hydrogenation reactions (HDN, aromatic saturation, and CI improvement). The ranking obtained when comparing the XPS metal concentration was also the same as that obtained from DRS and magnetism.

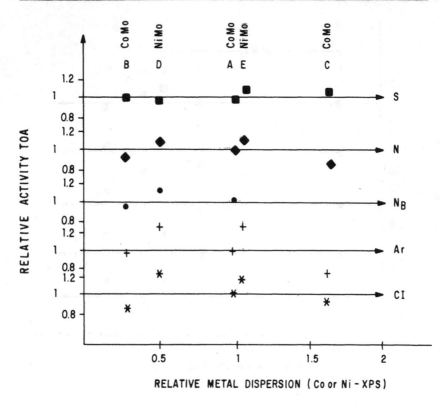

FIG. 12.   Relative activity vs relative metal dispersion (Co or Ni XPS).

The relative acidity of the precursors showed a quite different ranking, but we are inclined to suggest that this is due to the fact that acidity must change on sulfiding [17].  The degree of sulfiding might be estimated from the sulfur content of the catalyst related to the metal content.  This value was almost the same for Catalysts A and B but much higher for Catalysts D and E.

Up to now, several theoretical models have been proposed in the literature [22, 23] to explain the active surface of NiMo and CoMo catalysts.  No attempt is made here to support any of them, mainly because we have considered catalysts prepared by different companies using different methods, different supports, different types of metals, and different metal loads.

A, B, C : CoMo
D, E, F : NiMo

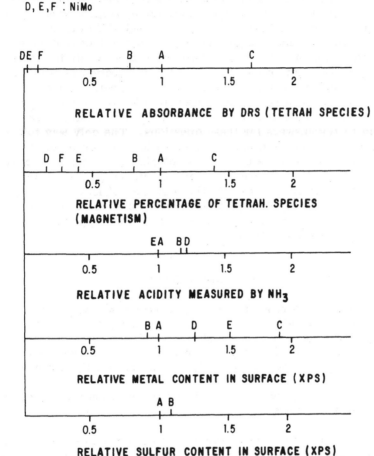

FIG. 13. Relative properties for Catalysts A, B, C, D, and E.

In this work we have tried to emphasize the difficulties encountered when attempting to select a catalyst and to correlate activity data obtained under industrial conditions with measurable surface parameters. The main difficulty arises from the complexity of the reaction networks taking place on the catalysts under these conditions. Trying to visualize or interpret the surface panorama may be too difficult a task. The wide controversy found in the

literature when considering ideal (or simple) situations is of no
help for this type of work. No accepted proposal for the nature
of the active sites (for hydrogenation, for hydrogenolysis, etc.)
nor for the role of acidity, as well as a model of deactivation, exists. However, general agreement points to the fact that some
parameters play a very important role in catalytic activity and
should be taken into account when catalyst selection is needed.
They are type of feedstock, metal dispersion, acidity distribution, and amount of octahedrally coordinated promoter ions. Finally, we emphasize the need for further and extensive work,
leading to the answers for these questions. The only way to
carry out this task is through a systematic research project
where model solids were deeply characterized not only physically
and chemically but also catalytically. Particularly, very little information is available to explain the high rate of loss of activity
which occurs during the hydrotreatment of cracked feeds. Future work must be centered on this feature; one approach might
be the study of spent catalysts.

## V. CONCLUSIONS

In order to optimize the operating conditions of the hydrotreatment of cracked light gas oil to satisfy a particular market scenario,
several parameters have to be adjusted and many features have to
be considered. For instance, if the feed and the target have been
selected, then the operating conditions depend on the catalysts
used.
The following generalities can be extracted from the trends observed in our results.

The higher the molecular weight and the aromatics in the cut,
the more difficult the HDS, HDN, and hydrogenation are. Consequently, it is also more difficult to meet the CI and color stability specifications.

The complexity of the events taking place in the system feed-
catalyst leads to the fact that the apparent kinetics determined
(reaction order, activation energy, etc.) not only depend on the
type of those two components but are also affected by feed vaporization and mass and heat transfer.

The apparent kinetics derived from our results for Feeds I and
II were

$$HDS = \frac{1}{(S)} - \frac{1}{(S)_0} = K_0 S e^{(-25,000/RT)} p_{H_2}^{0.75}$$

$$HDN_T = \frac{1}{(N_T)^{0.4}} - \frac{1}{(N_T)_0^{0.4}} = K_{0N_T} e^{(-28,000/RT)} p_{H_2}^{0.6}$$

$$HDN_B = \frac{1}{(N_B)^{0.2}} - \frac{1}{(N_B)_0^{0.2}} = K_{0N_B} e^{(-30,000/RT)} p_{H_2}^{0.5}$$

Aromatic saturation: $\ln \frac{Ar}{(Ar)_0} = K_{0Ar} e^{(-20,000/RT)} p_{H_2}$

CI improvement for a highly aromatic cracked feed (higher than 5) might be achieved by using a NiMo-type catalyst at 50 bars and 360-380°C.

The activity for nitrogen removal and aromatic saturation is higher for NiMo-type catalysts than for CoMo-type catalysts.

Hydrogen consumption is higher at higher severity, but color and color stability is proportionally improved.

Although there is no particular surface property of the catalysts which could be used directly for catalyst selection, it seems that metal dispersion, octahedral coordinated promoter, and acidity distribution must help in this decision. However, any systematic study should consider the influence of the method of preparation and a full characterization of the oxide precursor, the sulfided catalysts, and the spent catalysts. In this way an integral scheme of correlation can be drawn.

Catalyst deactivation (and/or stability) has to be considered a very important parameter for catalyst selection.

## REFERENCES

[1]  T. A. Patterson, J. C. Carver, D. E. Leyden, and D. M. Hercules, J. Phys. Chem., 80(15), 1700 (1976).

[2]  J. Sonnemans, From Ketjen Catalysts 1982, Akzo Chemie.

[3]  M. D. Edgar, Oil Gas J., p. 63 (November 16, 1981).

[4]  M. O. Rosenheimer and J. R. Kiosky, Preprints, American Chemical Society Meeting, Chicago, September 1967, B-147.

[5]  R. Voorhoeve and J. C. M. Sturven, J. Catal., 23, 228 (1971).

[6]  A. V. Sapre and B. C. Gates, Ind. Eng. Chem., Process Des. Dev., 20, 68 (1981).

[7]  J. H. Ashley and P. C. H. Mitchell, J. Chem. Soc., A, p. 2781 (1968); p. 2730 (1969).

[8]   J. Abart, E. Delgado, G. Ertl, H. Jeziorowski, H.
      Knozinger, N. Thiele, X. Zh. Wang, and E. Taglauer,
      Appl. Catal., 2, 155 (1982).
[9]   Y. Okamoto, H. Nakamura, T. Imanaka, and S. Teranishi,
      Bull. Chem. Soc. Jpn., 48, 1163 (1975).
[10]  J. Medema, C. Van Stam, V. H. J. de Beer, A. J. A.
      Konings, and D. C. Koningsberger, J. Catal., 53, 386
      (1978).
[11]  M. Lo Jacono, M. Schiavello, and A. Cinimo, J. Phys.
      Chem., 75, 1044 (1971).
[12]  A. J. A. Konings, W. L. Bentjens, D. C. Koningsberger,
      and V. H. J. de Beer, J. Catal., 67, 145 (1981).
[13]  J. S. Jepsen and H. F. Rase, Ind. Eng. Chem., Prod.
      Res. Dev., 20, 487 (1981).
[14]  J. J. Stanulonis and L. Pedersen, Preprints, American
      Chemical Society Meeting, Houston, March 1980, p. 255.
[15]  P. Ratnasami, A. V. Ramaswamy, and S. Sivasanker, J.
      Catal., 61, 519 (1980).
[16]  J. Laine, J. Brito, and S. Yunes, Preprints, Pet. Chem.
      Div., American Chemical Society Meeting, August 1980,
      p. 438.
[17]  P. Gajardo, A. Ayerbe, J. M. Cruz, and L. Gonzalez,
      Symp. Ibero. Catal. 8th, Huelva-España, 1982, p. 624.
[18]  A. Ayerbe, MSc Thesis, IVIC, Venezuela, 1982.
[19]  C. I. Chu and I. Wang, Ind. Eng. Chem., Process Des.
      Dev., 21, 338 (1982).
[20]  F. J. Gil Llambias, S. Mendioroz, F. Anía, and A. López
      Agudo, Symp. Iberoamericano Catal., 7th, La Plata,
      Argentina, 1983, p. 513.
[21]  C. N. Satterfield and J. F. Cocchetto, Ind. Eng. Chem.,
      Process Des. Dev., 20, 53 (1981).
[22]  P. Grange, Catal. Rev.-Sci. Eng., 21(1), 135 (1980).
[23]  B. S. Clausen, H. Topsoe, R. Candia, and B. Lengeler,
      Am. Chem. Soc. Symp. Ser., Catal. Mater., San Francisco,
      California, June 13-15, 1983.

# Catalyst Preparation

# Preparation Techniques for Hydrotreating Catalysts and Their Influence on the Location of the Metal Oxides and Performance

W. M. KEELY, P. JERUS, E. K. DIENES, AND
A. L. HAUSBERGER
United Catalysts Inc.
Louisville, Kentucky

## I. INTRODUCTION

There have been many studies in the last 30 years on hydro-treating catalysts. The early studies involve various metals and supports and their influence on activity. Later studies have applied bulk and surface analytical techniques to defining the structure of catalysts and to activity [1-4]. Few of the papers try to correlate the bulk and surface properties to the preparation method for a fixed composition, and the resulting performance observed [5].

A program was undertaken at United Catalyst Inc. (UCI) to study the effect that various preparation techniques have on hydrotreating catalyst activity and performance. The initial phase of this study involved the preparation of cobalt and molybdenum oxides on an extruded alumina support that is currently used for one of UCI's commercial hydrotreating catalyst.

These were made with various salts of cobalt and molybdenum and applied to a carrier together or in separate applications and alternating the order of addition.

In order to explain how these preparative techniques influence activity, various bulk chemical and physical analyses were run and correlated with the preparation method and activity. These bulk analyses are conventional to and well known by the catalytic industry. To supplement these analyses, surface analytical methods, ESCA, SEM, and TEM were used to further the understanding of the effect of the preparation method on the catalyst performance and the location of the metal oxides.

The ESCA, TEM, and SEM analyses were made by and through cooperation with the University of Kentucky Institute for Mining and Minerals Research. A selected group of these catalysts was tested in the laboratory for hydrodesulfurization activity using a light cycle oil feedstock. Finally, the bulk and surface properties were correlated with the preparative techniques and resulting catalytic performance. This report covers the data and results obtained in this study and the correlation made from those results.

## II. EXPERIMENTAL APPROACH

The approach to this study was to prepare catalysts from the same carrier material and vary the type of metal salts and the order of addition of these salts; one compounded catalyst was also made for comparison. In this study, cobalt oxide/molybdenum oxide on alumina catalysts were prepared and activity tested for the hydrodesulfurization of a light cycle oil feedstock. Each catalyst preparation was analyzed by various bulk chemical and

physical analytical techniques and by various surface techniques
(ESCA, TEM, and SEM). These bulk and surface properties were
then correlated to activity or performance of the catalyst for hydrodesulfurization of a light cycle oil.


## III.  CATALYST PREPARATION

The cobalt oxide/molybdenum oxide catalysts were prepared by
impregnation of a carrier material with various salts of cobalt and
molybdenum. The carrier selected for this study was United Catalysts Inc. (UCI) CS331-3 high surface area gamma alumina carrier. The physical properties of the carrier are given in Table 1.

For the application of the metal oxides to the carrier, various
salts of cobalt and molybdenum were selected (Table 2).


### TABLE 1

#### Carrier Properties

Carrier type:   UCI's CS331-3 gamma alumina
        Form:   Extrudates
        Size:   1/16 in. diameter

Physical properties:
   Density, lb/ft$^3$:  37
   Crush, lb, DWL:  16
   Attrition, wt% loss:  1.5
   Surface area, m$^2$/g:  223
   Pore volume, cm$^3$/g:  0.65


### TABLE 2

#### Metal Salts

| Cobalt salt | Molybdenum salt |
| --- | --- |
| Cobalt nitrate hexahydrate<br>$Co(NO_3)_2 \cdot 6H_2O$ | Ammonium molybdate<br>$(NH_4)_2MoO_4(2.5NH_3/MoO_3)$ |
| Cobalt ammine carbonate<br>$[Co(NH_3)_6]_2(CO_3)_3$ | Ammonium heptamolybdate<br>$(NH_4)_6Mo_7O_{24} \cdot 4H_2O$ |

TABLE 3

Catalyst Preparations

| Prepara-tion | Order of addition of salt | Cobalt salt | Molybdenum salt |
|---|---|---|---|
| 5194 | Mo/Co | Cobalt nitrate | Ammonium molyb-date |
| 5195 | Mo/Co | Cobalt ammine car-bonate | Ammonium molyb-date |
| 5196 | Mo/Co | Cobalt nitrate | Ammonium hepta-molybdate |
| 5197 | Mo/Co | Cobalt ammine car-bonate | Ammonium hepta-molybdate |
| 5198 | Co/Mo | Cobalt nitrate | Ammonium molyb-date |
| 5199 | Co/Mo | Cobalt nitrate | Ammonium hepta-molybdate |
| 5200 | Co/Mo | Cobalt ammine car-bonate | Ammonium molyb-date |
| 5201 | Co/Mo | Cobalt ammine car-bonate | Ammonium hepta-molybdate |
| 5202 | Simultaneous | Cobalt ammine car-bonate | Ammonium molyb-date |
| 5203 | Compounded | Cobalt ammine car-bonate | $MoO_3$ |

The cobalt and molybdenum oxides were applied to the carrier by impregnation with aqueous solutions of soluble salts of cobalt and molybdenum by complete submersion of the carrier in the metal salt solution.

The salt solutions were prepared in concentrations to obtain 5.0% CoO and 20.0% $MoO_3$ by co-dipping and/or sequence dipping and alternating the order of the Co or Mo application. After each impregnation the catalyst was calcined for 2 h at 850°F to decompose the metal salts to the oxides of cobalt and molybdenum. After the calcination the other salt and/or dip was applied and then calcined again to the same temperature, 850°F for 2 h.

The resulting catalyst preparations from the various salts above with alternating addition of the salts are outlined in Table 3.

## TABLE 4

### Chemical and Physical Properties

| Chemical properties | Method | Equipment |
|---|---|---|
| % CoO<br>% MoO$_3$<br>% Al$_2$O$_3$ | Atomic absorption<br>Atomic absorption<br>by difference | Instrumentation Laboratory Model No. 551 |

**Physical Properties**

| | | |
|---|---|---|
| Color | Visual inspection | - |
| Surface area, m$^2$/gm | N$_2$ adsorption | Micromeritics Model No. 2100D |
| Pore volume, cc/gm | Hg porosimetry | Aminco 60,000 psig Model No. 5 - 7125B |
| Compound identification | X-ray diffraction | Norelco Type 1221570 |

## IV. BULK ANALYSES OF CATALYSTS

The catalysts outlined above were analyzed for thsir bulk chemical and physical properties by analytical techniques well knowń to the industry. These analyses and methods are outlined in Table 4.

## V. SURFACE ANALYSES OF CATALYSTS

In addition to these bulk properties, which frequently do not correlate with the activity of the catalysts, surface properties were determined by ESCA, TEM, and SEM.

The principle of the techniques, the depth of penetration, the property measured, and the detection limits for these surface techniques are outlined in Table 5. X-ray fluorescence, when run on a pressed, unground sample, is also a surface technique and is shown with the other techniques for comparison [6].

There are several equipment manufacturers that now provide instruments to measure these properties. Equipment for these analyses was available to us through cooperation with the University of Kentucky Institute for Mining and Minerals Research. The equipment used is outlined in Table 6.

TABLE 5

Surface Analyses Techniques

| Surface analyses | Principle | Beam width | Depth of penetration | Property measured | Detection limits |
|---|---|---|---|---|---|
| ESCA | X-rays in<br>Electrons out | Few mm | 10-30 Å | Elemental composition | 0.1 to several percent |
| SEM | Electrons in<br>Back-scattered electron out<br>X-rays out | 70 Å | - | Elemental composition | 0.1 to 1% |
| TEM | Electrons through | 7 Å | <1000 Å | Particle shape and size | - |
| EDX | X-rays in<br>X-rays out | 5-250 µm | | Elemental composition | 50-100 ppm |

## TABLE 6

### Surface Analyses Equipment

| Surface analyses | Equipment name, model, or supplier |
|---|---|
| ESCA | AEI ES 200 Electron Spectrometer with Digital PDP 8/e Computer System |
| SEM | ETEC Omniscan Scanning Electron Microscope |
| TEM | Hitachi Hu-11-A Transmission Electron Microscope |
| XRF | EG&G ORTEC Energy Dispersive X-ray Model HTECH 6110 |

## VI. CATALYST TESTING

The activities of the various catalysts were determined in an integral laboratory test reactor. Each catalyst was loaded separately to the reactor tube, presulfided, and tested for hydrodesulfurization of a light cycle oil. Tests were run at a fixed pressure, liquid hourly space velocity, and $H_2$/barrel of feed. Only the temperature was varied to determine the relative activities of the catalyst. The loading, sulfiding, testing, and analytical procedures to determine the hydrodesulfurization activity are outlined below.

### A. Activity Test Procedure

The experimental catalysts were tested side-by-side in a dual tube reactor. The catalysts were tested with a light cycle oil feedstock.

The feedstock properties (Table 7), catalyst loading, sulfiding, and testing conditions (Table 8) are summarized.

#### Catalyst Loading

25 $cm^3$ of 20 X 40 mesh catalysts were used and mixed (diluted) with 75 cc of Alundum grid 30 (20 X 40 mesh). Glass wool was used as a plug below the catalyst bed to avoid catalyst slip through the support plate. 10 $cm^3$ of >20 mesh inert alumina was loaded and packed before the mixed catalyst was loaded and packed in the reactor. Another 10 $cm^3$ of >20 mesh inert alumina was then loaded on the top of the mixed catalyst.

## TABLE 7

### Feedstock Properties

| Feedstock | Light cycle oil |
|---|---|
| API gravity (60°F) | 22.3 |
| Boiling range | IBP, 281°F |
|  | 50%, 518°F |
|  | EBP, 761°F |
| % S | 1.41-1.65 |
| Total $N_2$ (ppm) | 350 |

### Sulfiding

The catalysts were purged with $H_2$ (about 15 min) at ambient temperature.

The catalysts were then heated to 400°F and sulfided with 10% $H_2S$ in $H_2$ for 2 h at 100 psig and 27 L/h exit gas rate. The temperature was then increased to 450°F and sulfiding continued for a total of 7 h. The temperature and pressure were adjusted to operating conditions. The $H_2S/H_2$ mixture was replaced with $H_2$ and the feed started.

The feed and product were analyzed for sulfur content by x-ray fluorescense. The activity ($K_S$) for hydrodesulfurization is calculated based on a 1.5 order reaction and is defined for this study as

$$K_S = LHSV \ [(1/S_O)^{1/2} - (1/S_i)^{1/2}]$$

where $S_O$ = S product and $S_i$ = sulfur in feed [7].

Since the S content of the feed, $H_2$ partial pressure, and catalyst particle size are constant, no corrections are made to the equation for these factors [7].

## VII. RESULTS AND DISCUSSION

All of the catalysts were visually inspected for color and analyzed for % CoO, % $MoO_3$, surface area, pore volume, and x-ray diffraction. Each catalyst was also tested for its HDS activity expressed as $K_S$. The bulk chemical/physical properties and activities are tabulated in Table 9.

TABLE 8

Testing Conditions

| | |
|---|---|
| Hydrogen partial pressure (psig) | 600 |
| Temperature (°F) | 550, 600 |
| $H_2$/oil (outlet) (SCF/bbl) | 2000 |
| LHSV[a] | 3.0 |

[a]LHSV = volume of oil/hour/volume of catalyst.

It is readily seen that the CoO and $MoO_3$ concentrations on the impregnated catalysts are essentially equal, as intended. The pore volume and surface areas are also essentially equivalent. The compounded preparation is an exception, having a higher surface area. The activities of the catalysts vary substantially when considering the metals level and pore volumes are essentially equal.

By reorganizing the data (see Table 10) and ranking the catalysts by $K_S$ in descending order, we can correlate the salts used, order of addition of the salts, and resulting color of the catalyst to their relative performance.

The activities of these preparations using cobalt ammine carbonate are superior to those using cobalt nitrate. No differentiation could be made in the activity with the two molybdate salts used. The compounded catalyst with $MoO_3$ and $Co(NH_4)_4CO_3$ gave less activity. The other observed influence on the activity is the order of addition of the Co or Mo salts with addition of cobalt first being preferred over addition of molybdenum. The major factor seemed to be that the cobalt ammine carbonate is much superior to $Co(NO_3)_2$ with the order of addition playing a secondary role. The resulting color does not correlate with activity, but there is some trend with blue to gray-green color being the most active, blue to blue-gray giving intermediate activity, and gray or gray-black giving poor activity.

Finally, the x-ray results show that all impregnated samples have $Al_2O_3$ crystallite size of 40-50 Å, which is expected since they were prepared on the same carrier. Most samples do not show $CoMoO_4$, which means it probably is at too low a concentration or very small crystallite size (amorphous). The few samples

## TABLE 9

Catalyst Preparation, Chemical and Physical Properties, and Activity

| Prepara-tion no. | Preparation | % CoO | % MoO$_3$ | Color | Surface area (m$^2$/g) | Pore volume (cc/g) 29.2 Å | XRD | Activity Ks 550°F | Activity Ks 600°F |
|---|---|---|---|---|---|---|---|---|---|
| 5194 | a. Ammonium molybdate<br>b. Cobalt nitrate | 5.14 | 20.7 | Blue-gray | 200 | 0.45 | 1. γ-Al$_2$O$_3$ 44 Å | 2.46 | 5.65 |
| 5195 | a. Ammonium molybdate<br>b. Cobalt ammine carbonate | 5.25 | 20.8 | Blue | 198 | 0.46 | 1. γ-Al$_2$O$_3$ 44 Å | 2.78 | 5.72 |
| 5196 | a. Ammonium hepta-molybdate<br>b. Cobalt nitrate | 5.19 | 20.3 | Blue-gray | 214 | 0.48 | 1. γ-Al$_2$O$_3$ 53 Å<br>2. CoMoO$_4$ 187 Å | 2.61 | 5.21 |
| 5197 | a. Ammonium hepta-molybdate<br>b. Cobalt ammine carbonate | 5.22 | 20.7 | Blue | 182 | 0.48 | 1. γ-Al$_2$O$_3$ 45 Å<br>2. CoMoO$_4$ 94 Å | 2.81 | 5.89 |
| 5198 | a. Cobalt nitrate<br>b. Ammonium molybdate | 5.13 | 20.4 | Gray-black | 197 | 0.43 | 1. γ-Al$_2$O$_3$ 40 Å<br>2. Co$_3$O$_4$ 515 Å | 1.89 | 3.48 |

| No. | Preparation | | | Color | | | Phases identified | | |
|---|---|---|---|---|---|---|---|---|---|
| 5199 | a. Cobalt nitrate | 5.50 | 19.6 | Gray-black | 193 | 0.47 | 1. γ-Al₂O₃ 40 Å<br>2. Co₃O₄ 321 Å<br>3. MoO₃ ?000 Å<br>4. Al₂(MoO₄)₂ | 2.35 | 3.89 |
| 5200 | a. Cobalt ammine carbonate<br>b. Ammonium molybdate | 5.10 | 20.4 | Gray-green | 191 | 0.46 | 1. γ-Al₂O₃ 53 Å | 3.70 | 6.73 |
| 5201 | a. Cobalt ammine carbonate<br>b. Ammonium heptamolybdate | 5.12 | 20.4 | Gray-green | 198 | 0.45 | 1. γ-Al₂O₃ 50 Å | 3.75 | 6.68 |
| 5202 | a. Cobalt ammine carbonate and ammonium molybdate<br>b. Cobalt ammine carbonate and ammonium molybdate | 4.73 | 21.0 | Blue | 191 | 0.45 | 1. γ-Al₂O₃ 47 Å<br>2. CoMoO₄ (?) | 3.75 | 7.01 |
| 5203 | Compounded molybdenum oxide, alumina, and cobalt ammine carbonate | 4.68 | 20.3 | Blue | 280 | 0.46 | 1. γ-Al₂O₃ 41 Å | 2.33 | 4.49 |

TABLE 10

Catalyst

| Activity ranking | $K_S$ (average) | Preparation no. | Color | Co salt |
|---|---|---|---|---|
| 1 | 5.38 | 5202 | Blue | Ammine carbonate |
| 2 | 5.22 | 5201 | Gray-green | Ammine carbonate |
| 3 | 5.22 | 5200 | Gray-green | Ammine carbonate |
| 4 | 4.35 | 5197 | Blue | Ammine carbonate |
| 5 | 4.25 | 5195 | Blue | Ammine carbonate |
| 6 | 4.06 | 5194 | Blue-gray | Nitrate |
| 7 | 3.91 | 5196 | Blue-gray | Nitrate |
| 8 | 3.41 | 5203 cpd | Blue | Ammine carbonate |
| 9 | 3.12 | 5199 | Gray-black | Nitrate |
| 10 | 2.69 | 5198 | Gray-black | Nitrate |

that do have detectable $CoMoO_4$ show increasing activity with decreasing $CoMoO_4$ crystallite size. Two preparations show the presence of $Co_3O_4$ and these are the least active, gray to black in color. Apparently this is due to decomposition of $Co(NO_3)_2$ to $Co_3O_4$ and not forming cobalt aluminate or cobalt molybdate.

The samples were also evaluated by ESCA, SEM, and TEM techniques. These results are tabulated in Table 11 with the activity ($K_S$) for each sample. The data show that the uniformity or surface enrichment of cobalt and molybdenum varies on the samples. The ESCA results show Mo enrichment on two samples and Co depletion on two samples, one sample being in common to both effects.

Properties

| Mo salt | Order of addition | S.A. | P.V. | ($Al_2O_3$) | ($CoMoO_4$) | ($Co_3O_4$) |
|---|---|---|---|---|---|---|
| Ammonium molybdate | Together | 191 | 0.45 | 47 | - | |
| Ammonium hepta-molybdate | Co/Mo | 198 | 0.46 | 50 | - | |
| Ammonium molybdate | Co/Mo | 191 | 0.46 | 53 | - | |
| Ammonium hepta-molybdate | Mo/Co | 182 | 0.48 | 45 | 94 Å | |
| Ammonium hepta-molybdate | Mo/Co | 198 | 0.45 | 44 | - | |
| Ammonium molybdate | Mo/Co | 200 | 0.45 | 43 | 172 | |
| Ammonium hepta-molybdate | Mo/Co | 214 | 0.48 | 53 | 187 Å | |
| Molybdenum oxide | Together | - | - | - | - | |
| Ammonium hepta-molybdate | Co/Mo | 193 | 0.47 | 40 | - | 321 |
| Ammonium molybdate | Co/Mo | 197 | 0.43 | 40 | - | 515 |

The remaining samples are essentially uniform. The two Mo-enriched samples are two of the four which have the Mo salt added last. The Co depletion is on the two samples which were made with cobalt nitrate followed by the molybdenum salt. This cobalt nitrate is apparently converted to $Co_3O_4$ rather than reacting with the $Al_2O_3$ and with the subsequent impregnation with the molybdenum salt leaves the surface cobalt depleted. There is no correlation of activity with the $MoO_3$-enriched samples versus the uniform samples but the samples impregnated with cobalt nitrate first are the least active and show Co depletion on the surface.

TABLE 11

Catalyst Surface

| Activity ranking | $K_s$ (average) | Prepara- tion no. | ESCA | SEM whole particle | SEM interior Al | (20 X 40 mesh) Mo | Co |
|---|---|---|---|---|---|---|---|
| 1 | 5.38 | 5202 | Uniform | Uniform | 0.74 | 0.17 | 0.08 |
| 2 | 5.22 | 5201 | Mo enriched. Al depletion | Strong Mo Co depletion | 0.72 | 0.21 | 0.07 |
| 3 | 5.22 | 5200 | Uniform | Strong Co Weak Mo | 0.76 | 0.19 | 0.05 |
| 4 | 4.35 | 5197 | Uniform | Weak Co Weak Mo | 0.76 | 0.18 | 0.06 |
| 5 | 4.25 | 5195 | Uniform | Uniform | 0.72 | 0.20 | 0.08 |
| 6 | 4.06 | 5194 | Uniform | Weak Co Weak Mo | 0.74 | 0.19 | 0.07 |
| 7 | 3.91 | 5196 | Uniform | Strong Co Strong Mo | 0.72 | 0.19 | 0.09 |
| 8 | 3.41 | 5203 cpd | Uniform | Uniform | 0.74 | 0.19 | 0.06 |
| 9 | 3.12 | 5199 | Mo enriched. Al and Co depletion | Strong Co Strong Mo | 0.75 | 0.18 | 0.07 |
| 10 | 2.69 | 5198 | Co depletion | Strong Co Weak Mo No conclusion | 0.72 | 0.22 | 0.07 |

'roperties

| Exterior | | | TEM | | | |
|---|---|---|---|---|---|---|
| Al | Mo | Co | (1) Round | (2) Rodlike | (4) | (3) Dense Coating |
| .61 | 0.30 | 0.09 | 30-75 | 250-400 | 150-200 | Most |
| .40 | 0.53 | 0.07 | 75-125 | Many 250-400 | | |
| .59 | 0.29 | 0.13 | 100-200 | 1000 (?) | | Most |
| .67 | 0.25 | 0.07 | 100-200 | Many 280-450 | | |
| .72 | 0.23 | 0.05 | 75-150 | 200-300 | | Most |
| .62 | 0.27 | 0.11 | Many 100-250 | Few | | |
| .66 | 0.28 | 0.06 | Many 100-250 | Few | | |
| .60 | 0.31 | 0.09 | 30-75 | 250-400 | | Half |
| .45 | 0.48 | 0.07 | 100-250 | Few | Needle 80-100 X 750-950 $Co_3O_4$ 100-300 Å | |
| .76 | 0.18 | 0.06 | 34-75 | 145-735 | | <Half |
| ome concentration of $MoO_3$ at the surface | | | More active catalysts have more rodlike particles on the surface. Less active catalysts have more round particles | | | Dense coating gives more activity than less coating |

In the SEM study the original scan starting with whole particle gave essentially random results. There are uniform and Co- or Mo-enriched samples, regardless of the salts used the order of impregnation, and therefore no correlation with activity. On running the 20 X 40 mesh samples, the size loaded to the activity test reactor, the interior, and exterior of Co, Mo, and Al relative concentration were determined. The interior of all of the samples is relatively uniform, with the exterior SEM results showing more variation. The exterior is Mo enriched and sometimes Co enriched compared to the interior, with the exception of one sample which showed lower surface Co and Mo, and this is the least active sample.

We are unable to find a clear correlation by ESCA and SEM. There does seem to be some surface enrichment of Mo and Co, with Mo enrichment preferred to cobalt enrichment [1, 2]. The lack of a correlation of Mo and Co distribution to activity suggests the metals level is very near the maximum, and uniformity or surface Mo enrichment is preferred at this metals level. At some other metals level a different distribution may be desirable [1, 2].

In the TEM results three particles types or size are observed. These are (a) round-looking particles ranging from 30 to 300 Å in size, (b) rod-shaped particles 20 to 35 Å in width and 50-735 Å in length, and (c) dense-looking irregular shaped particles which are 100-500 Å in diameter. Although there is no correlation with size of the particles either with the method of preparation or activity, there does appear to be some correlation with the amounts or surface density of each type of particle and the activity. The more active catalysts are mostly covered by the dense-looking material and the less active catalysts are less than half covered. The more active catalysts also have more rodlike particles and fewer round particles while the less-active catalysts conversely have more round and fewer rodlike particles.

Combining the observations of x-ray, TEM, and SEM, it is speculated that the rodlike and dense phases are some Co-Mo complex which is fairly uniform on all the samples, apparently due to selecting near a near optimum metals level [3, 4, 8, 9]. There seems to be a concentration of this Co-Mo phase on the surface of the most active catalyst with less Co-Mo phase in the middle activity samples and actually separate phases of Co and Mo on the least active catalyst. The colors tend to support the TEM conclusion; CoMo complex and $CoAl_2O_4$ phases giving a blue to gray-green color and separate $Co_3O_4$, $MoO_3$, and $Al_2O_3$ phases giving a gray-to-black color.

## VIII. CONCLUSIONS

From this study several conclusions were drawn on the effect of the catalyst preparation method on catalyst activity. These are

(1) Cobalt ammine carbonate is preferred over cobalt nitrate as the cobalt salt.
(2) No appreciable difference in activity or properties was observed between ammonium molybdate ($2.5NH_3/MoO_3$) and ammonium heptamolybdate.
(3) Simultaneous addition of Co and Mo or Co first is preferred over addition of molybdenum first.

There appears to be a relationship between these preparative methods and the resulting color and surface properties. First, the metals level appears near optimum because there is no gross or maldistribution of active metals. There is a dense phase material which is more active and gives a blue to gray-green color. This material appears to be some Co-Mo phase and is of smaller crystallite size and more dense on the more active catalysts.

This phase becomes less apparent on the middle activity samples and disappears on the least-active samples. Although Co and Mo are concentrated on the surface of both the most-active and least-active sample, this is apparently due to some Co-Mo phase on the surface of the very active samples while the least-active materials have $Co_3O_4$. Laser Raman studies are recommended on these samples to try and verify these observations of the surface properties [4, 5].

## REFERENCES

[1] F. E. Massoth, "Characterization of Molybdena Catalysts," Adv. Catal., 27, 265-310 (1978).

[2] J. S. Brinen and W. D. Armstrong, "Surface Chemistry of Activated Hydrodesulfurization Catalysts by X-Ray Photoelectron Spectroscopy," J. Catal., 54(1), 57-65 (1978).

[3] P. Ratnasamy and S. Sivasanker, "Structural Chemistry of Co-Mo-Alumina Catalysts," Catal. Rev.-Sci. Eng., 22(3), 401-429 (1980).

[4] A. Morales et al., "Correlation between HDS or HDN Activity with Spectroscopic Characterization of Molybdenum-Alumina and Cobalt-Molybdenum-Alumina Catalysts Modified by Extraction," Appl. Catal., 6(3), 329-340 (1983).

[5]   S. Kasztelan et al.   "Preparation of Co-Mo-Al$_2$O$_3$ and Ni-Mo-
       Al$_2$O$_3$ Catalysts by pH Regulation of Molybdenum Solution.
       Characterization of Supported Species and Hydrogenation
       Activities," Ibid., 7(1), 91-112 (1983).
[6]   C. Cornelius, "Determination of Catalyst Surface Contami-
       nants by X-Ray Fluorescence Spectrometry," Anal. Chem.,
       53(14), 2361-2363 (1981).
[7]   M. L. Vrinat, "The Kinetics of the Hydrodesulfurization
       Process—A Review," Appl. Catal., 6(2), 137-158 (1983).
[8]   H. Topsøe et al., "In situ Mössbauer Emission Spectroscopy
       Studies of Unsupported and Supported Sulfided Co-Mo Hy-
       drodesulfurization Catalysts: Evidence for and Nature of
       a Co-Mo-S Phase," J. Catal., 68(2), 433-452 (1981).
[9]   C. Wivel et al., "On the Catalytic Significance of a Co-Mo-S
       Phase in Co-Mo/Al$_2$O$_3$ Hydrodesulfurization Catalysts: Com-
       bined in situ Mössbauer Emission Spectroscopy and Activity
       Studies," Ibid., 68(2), 453-463 (1981).

# A Novel Catalyst Geometry for Automobile Emission Control

C. J. PEREIRA, G. KIM, AND L. L. HEGEDUS
W. R. Grace & Co.
Columbia, Maryland

## I. INTRODUCTION

The catalytic control of automobile exhaust pollutants (hydro-carbons, carbon monoxide, nitrogen oxides) is now a common prac-tice in the United States and Japan, and there are strong indica-tions that the technology may be introduced in other parts of the world in the coming years. The simultaneous conversion of all three of the above pollutants is achieved by the near-stoichiomet-ric operation of noble metal catalysts which are supported either

on alumina particles or on alumina-washcoated cordierite monoliths. A summary of the developments which lead to this technology can be found, e.g., in Hegedus and Gumbleton [1] and
references therein.

Automobile exhaust catalysts have to satisfy a wide range of
chemical and physical requirements. Besides their conversion
performance, of importance are catalyst durability (chemical,
thermal, mechanical), converter weight and volume, converter
backpressure, converter shape and positioning flexibility, thermal inertia, ease of catalyst replacement, manufacturability, and
cost.

The gradual drift toward smaller and lighter cars over the
past several years resulted in a gradual drift toward monolith-
type converters at the expense of pellet-type converters. If
pellet-type converters were to be improved beyond their current
state of development, a new catalyst geometry could be of potential interest. This, indeed, will be the subject of this paper.

In what follows we will propose a new catalyst particle geometry, discuss its theoretical advantages, optimize its shape and
pore structure with the help of appropriate mathematical models,
and show first evidence of their improved performance in laboratory-scale reactors.

Ongoing work is aimed at developing the new catalyst to the
stage where full-scale testing in automobile exhaust can be conducted.

## II. OPTIMUM DESIGN OF CATALYST PELLET GEOMETRY

Besides the chemical composition of catalyst particles, several physical aspects have important effects on their performance. These include their pore structure (which determines
their internal surface area, compressive strength, and density), and their size and shape (which determine their geometric surface area per unit particle volume, their bulk or packed
density, and the backpressure characteristics of their assembly).

Monolithic catalysts (e.g., Johnson et al., Ref. 2) are characterized by a high geometric surface area per unit volume, resulting from their small flow channels and thin wall structures.
However, monoliths operate in the laminar flow regime (e.g.,
Hegedus, Ref. 3) where their transport coefficients have relatively low values when compared with the turbulent transport
properties of particulate-type systems (e.g., Oh et al., Ref.
4). The prime advantage of monoliths lies in their low thermal

mass and in their flexible shape; both of these are favorable for efficient packaging in small vehicles.

Pellet-type catalysts are easier to replace in their converter and have a favorable resistance to thermal excursions due in part to the higher melting point of alumina compared to that of cordierite (Mg-Al-silicate).

Although it has been experimentally shown (e.g., Oh et al., Ref. 5; or Adomaitis et al., Ref. 6) that smaller catalyst pellets provide superior performance to larger ones, too small particles result in a dramatic increase in converter backpressure; this would dictate even flatter converter shapes which are not practical for a variety of reasons. Thus, an interest developed in some new catalyst pellet geometry which provides a high geometric surface area per reactor volume, turbulent transport characteristics, but with a relatively low backpressure.

Catalyst pellet geometries can be conveniently discussed in the context of a tradeoff between their heat and mass transfer performance and their backpressure in a catalytic converter.

For a first-order reaction controlled by extraparticle mass transfer (corresponding to the highest achievable reaction rate in a given catalytic converter),

$$\phi = 1 - \exp\left[\frac{-k_m s_g V_r}{Q}\right] \tag{1}$$

which shows that the conversion performance increases with increasing mass transfer coefficient and with increasing geometric surface area. Both of these quantities depend on the shape of the catalyst particles.

The pressure drop in a packed bed is well described by the Ergun equation (e.g., Bird et al., Ref. 7):

$$\Delta P = \left(\frac{L_r}{D_h}\right)\left(\frac{1-\epsilon}{\epsilon^3}\right)G^2\left(150\frac{(1-\epsilon)}{\frac{(D_h G)}{\mu}} + 1.75\right) \tag{2}$$

Accordingly, the pressure drop increases as the hydraulic diameter of the particle decreases.

Figure 1 shows this tradeoff relationship between geometric surface area and backpressure for a 2600-cm$^3$ catalytic converter filled with a variety of catalyst particle sizes and shapes under equivalent conditions. (The contributions of the converter to the system's backpressure have been left out of consideration; only the $\Delta P$ associated with the packing material is shown.)

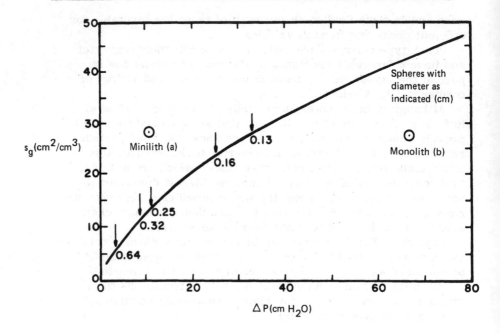

FIG. 1.  Geometric surface area vs pressure drop.  Pelleted converter: $L_r = 5.1$ cm, $D_r = 12.8$ cm, $Q = 10^5$ cm$^3$/s, T = 500°C, P = 1 atm; only the $\Delta P$ contribution of the packing is computed. (a) $L_p/D_r = 1.33$, $D_p = 0.25$ cm, w = 0.038 cm, $N_s = 4$.  (b) For comparison, $V_r = 2786$ cm$^3$, $L_r = 28$ cm, $D_r = 11.2$ cm, w = 0.015 cm, $N_c = 62$/cm$^2$.

In computing the points in Fig. 1, the hydraulic particle diameter for solid particles (spheres) was defined as six times the ratio of the particle volume to its external surface area.  For complex particles with internal flow channels, to be discussed later, the hydraulic diameter was defined as four times the cross-sectional area of the particle (along the flow direction) divided by its total wetted perimeter.

In Fig. 1, smaller spheres show a larger geometric surface area per unit converter volume (which, according to Eq. 1, should result in improved catalyst performance), but at the expense of an increased backpressure.

The point in Fig. 1 labeled as Minilith refers to the new geometry we are proposing for spoked extrudates.  (The term Minilith is a W. R. Grace trademark.)

This new geometry is shown in Fig. 2; it consists of cylindrical extrudates with a hollow interior and with spokes to reinforce the

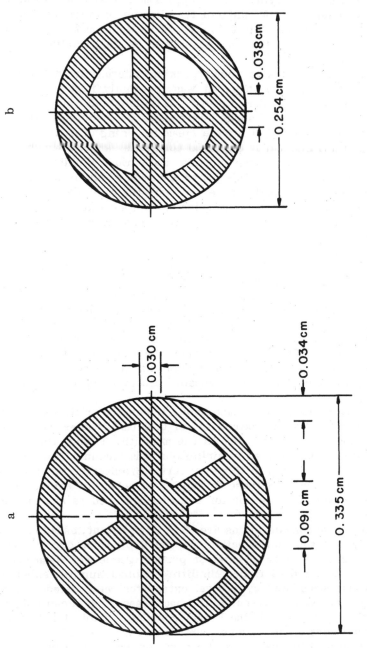

FIG. 2. Six-spoke (a) and four-spoke (b) Minilith cross sections. The extrudates are cut into approximately regular cylinders.

external rim. The cylinders are approximately as long as wide;
their diameter is in the range of the diameter of currently used
catalyst particles (0.25 to 0.30 cm). This would make the retention of the spoked extrudates possible in existing catalytic converter designs.

Spoked extrudates as column packing materials are not new.
As a catalyst particle shape, it was apparently first proposed
by Lundsager [8]. (Very large (1.55 cm long and wide) catalysts with the shape of spoked extrudates were subsequently
patented by Villemin [9] for the steam reforming of hydrocarbons.) While Lundsager employed cordierite particles which
were washcoated by porous alumina, the various Minilith spoked
extrudates discussed in this paper are made of highly porous
alumina which is directly impregnated by the catalytic materials.
Alumina-spoked extrudates promise several advantages with respect to cordierite-spoked extrudates, including a higher melting point.

As Fig. 1 shows, the spoked extrudate geometry, with the
same diameter as a spherical catalyst pellet, has approximately
twice the external geometric surface area without any increase
in backpressure. (This property of the Minilith shape could, of
course, be utilized in another way: larger spoked extrudates
could be made to match the geometric surface area of spherical
pellets of 0.25 cm diameter, resulting in a significantly reduced
backpressure. This could then allow the reshaping of the converter to have a smaller frontal area and increased depth; this
would, among others, have favorable effects on lightoff performance.)

For comparison, Fig. 1 also shows the characteristics of a typical monolith. Spoked extrudates of 0.25 cm diameter have the
same geometric surface area as the monolith. Even if the backpressure of the louvered retaining plates is accounted for (about
20 cm $H_2O$), the Minilith-filled converter promises backpressures
which are lower than those for spherical pellets of equivalent
geometric surface area, or for the monolith used as an example
in Fig. 1.

The above considerations are, to a large extent, based on the
validity of the Ergun equation to predict pressure drops. To
check this, a variety of catalyst particle geometries were prepared or acquired (4-ribbed Minilith, 6-ribbed Minilith, spheres
of various sizes, and cylindrical extrudates) and tested in the
laboratory. As Fig. 3 shows, the backpressure data can indeed
be well correlated with the Ergun equation. Table 1 lists the
characteristics of the catalyst particles we have tested.

Encouraged by the above, we have conducted some paramet-
ric sensitivity calculations with respect to the optimum number

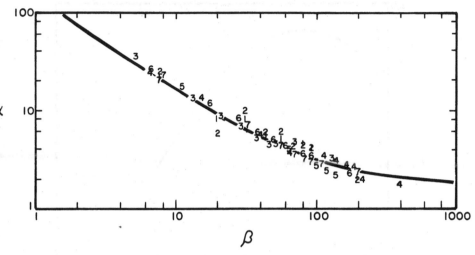

FIG. 3. Backpressure measurements on a variety of catalyst particles compared with the prediction of the Ergun equation. See Table 1 for symbols.

$$\alpha = \frac{\Delta P \rho}{G^2} \frac{D_h}{L} \frac{\epsilon}{1-\epsilon}; \quad \beta = \frac{D_h G}{\mu} \frac{1}{1-\epsilon}$$

TABLE 1

Explanation of Symbols for Fig. 3

| Symbol | Description (dimensions in cm) |
|--------|-------------------------------|
| 1 | Minilith, $N_s = 4$, $w = 0.038$, $D_p = 0.25$, $L_p/D_p = 1.33$, $\epsilon = 0.54$ |
| 2 | Minilith, $N_s = 6$, $w = 0.038$, $D_p = 0.34$, $L_p/D_p = 1.33$, $\epsilon = 0.63$ |
| 3 | Sphere, $D_p = 0.10$, $\epsilon = 0.37$ |
| 4 | Sphere, $D_p = 0.30$, $\epsilon = 0.37$ |
| 5 | Cylinder, $D_p = 0.160$, $L_p/D_p = 3$, $\epsilon = 0.40$ |
| 6 | Cylinder, $D_p = 0.085$, $L_p/D_p = 5$, $\epsilon = 0.43$ |
| 7 | Sphere, $D_p = 0.15$, $\epsilon = 0.37$ |

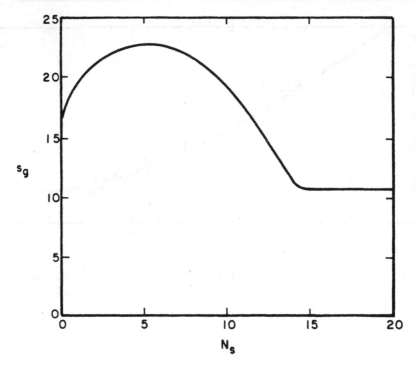

FIG. 4. Geometric surface area as a function of the number of spokes. $D_p = 0.25$ cm, $L_p/D_p = 1.33$, w = 0.038 cm, $\varepsilon = 0.40$.

of spokes giving a maximum geometric surface area for the Minilith. Figure 4 shows the results for the Minilith with fixed external dimensions and wall thickness. The optimum is five spokes, with four spokes representing a reasonable compromise for manufacturing ease.

Having selected the four-spoked Minilith for further consideration, we undertook a sensitivity analysis with respect to a trade-off between external geometric surface area (determining mass transfer-limited conversion) and solid pellet volume (determining the available support internal surface area). Figure 5 shows the results, expressed in the form of mass transfer-limited conversion (a) or kinetically limited conversion (which was taken to be proportional to the internal surface area), (b), as a function of $w/D_p$. A value of $w/D_p = 0.15$ appears to be a reasonable compromise, yielding, for $D_p = 0.25$ cm, w = 0.038 cm (for $L_p/D_p = 1.0$).

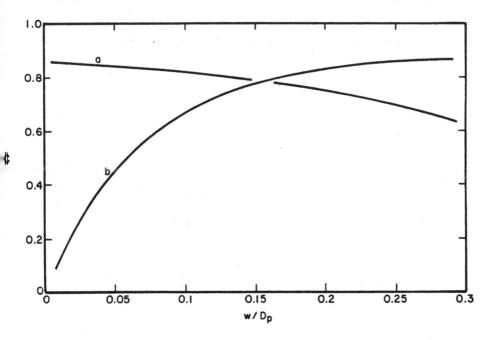

FIG. 5. Effect of wall thickness of the Minilith on its computed conversion performance under the assumptions that conversion is proportional to the external (geometric) surface area (a) or that it is proportional to the internal (support) surface area (b). $w/D_p = 0.15$ represents a reasonable compromise.

Thus, to summarize our efforts in designing a spoked extrudate, we have selected $D_p = 0.25$ cm, $L_p/D_p = 1.33$, $w = 0.038$, and $N_s = 4$ for further consideration.

### III. OPTIMUM DESIGN OF THE INTERNAL PORE STRUCTURE

The pore structure of automobile exhaust catalysts has a large effect on their performance. Especially important are large surface area (to disperse the noble metals, and to scavange the poisons which enter the catalyst pores) and carefully tailored macro- and microporosities for optimal diffusive transport characteristics.

Hegedus [10] and references therein discuss how the poisoning process can be approximated and simulated by phosphorus

## TABLE 2

Catalysts Employed in the Propylene Oxidation Experiments

| | Shape | | |
|---|---|---|---|
| | Spheroid | Cylinder | Minilith[a] |
| $L_p$ (cm) | 0.32 | 0.36 | 0.35 |
| $D_p$ (cm) | 0.26 | 0.25 | 0.26 |
| $L_p/D_p$ | 1.21 | 1.42 | 1.33 |
| $\rho_p$ (g/cm$^3$) | 0.765 | 0.932 | 0.964 |
| $V_{macro}$ (cm$^3$/g) | 0.404 | 0.125 | 0.133 |
| $V_{micro}$ (cm$^3$/g) | 0.645 | 0.625 | 0.618 |
| $\bar{r}_{macro}$ (Å) | 4500 | 500 | 500 |
| $\bar{r}_{micro}$ (Å) | 110 | 113 | 105 |
| $s_i$ (m$^2$/g) $N_2$ BET | 117 | 118 | 110 |
| Pt (wt%) (uniformly impregnated) | 0.055 | 0.044 | 0.054 |
| $\epsilon$ | 0.39 | 0.39 | 0.52 |

[a] $w = 0.038$ cm, $N_s = 4$.

poisoning, and how a simple model reaction, such as propylene oxidation, can be used to simulate catalyst performance. Other useful simplifications for catalyst pore structure optimization are net oxidizing, fully warmed-up conditions, and the use of one noble metal (Pt) to simulate the transport-limited behavior of partially poisoned catalysts.

Thus, we have prepared Pt-impregnated catalysts for propylene oxidation experiments in a laboratory-scale reactor. The catalysts were investigated in their fresh stage and after accelerated poisoning in a pulse-flame apparatus [11, 12].

Three catalyst geometries were considered (Table 2): sphere, cylinder, and spoked extrudate.

The propylene oxidizing reactor had a volume of 10.0 cm$^3$ and was operated at near atmospheric pressure. Its inner diameter was 2.68 cm, wide enough to accommodate about 10 catalyst particles along its diameter and thus minimizing flow channeling. The feedstream (at 57,000 h$^{-1}$, STP) consisted of 50 ppm (by volume) of propylene, 2% (by volume) of O$_2$, with N$_2$ as the balance. The low propylene concentration ensured isothermal operating conditions.

The poisoning experiments with the pulse-flame apparatus employed temperature cycling (566°C for 75 min, 732°C for 15 min) for a total duration of 42.5 h. The fuel to the combustor consisted of n-hexane in which tricresyl phosphate (for P) and thiophene (for S) were dissolved. The feedstream entering the catalyst had a phosphoric acid-equivalent phosphorus concentration of 1.5 X 10$^{-10}$ mol/cm$^3$, expressed at STP. The space velocity (STP) was 16,400 h$^{-1}$, and the air-fuel ratio of the feed mixture oscillated about the stoichiometric midpoint at the natural amplitude and frequency (1 Hz) of the pulse-flame combustor.

The purpose of these experiments was to calibrate our mathematical models and to check their ability in predicting catalyst performance (for propylene oxidation) and catalyst poisoning (by phosphorus).

The propylene conversion-inlet temperature data, before and after phosphorus poisoning, are displayed in Figs. 6 (spheres), 7 (cylinders), and 8 (spoked extrudates). These figures show both the measured data points and the predictions of a mathematical model which had noble metal dispersions, first-order kinetic preexponentials, and propylene oxidation activation energies as adjustable parameters.

The resulting propylene oxidation kinetic rate expression is as follows:

$$R = 1.0 \times 10^{25} \exp\left[\frac{-18.79 \times 10^3}{T}\right] c_{O_2} c_{C_3H_6} \left(\frac{mol}{cm^2\ Pt\ s}\right) \qquad (3)$$

This was applicable to all three catalysts tested. The best-fit noble metal dispersions were used in computing reaction rates per unit catalyst volume.

The conversion-inlet temperature curves in Figs. 6, 7, and 8 show not only that the model can be used to describe reasonably well the propylene conversion characteristics of fresh and aged catalysts of varying shape and pore structure, but also that the poisoning model (Cho et al., Ref. 13) correctly predicts the time-dependent poison deposition process without any adjustable

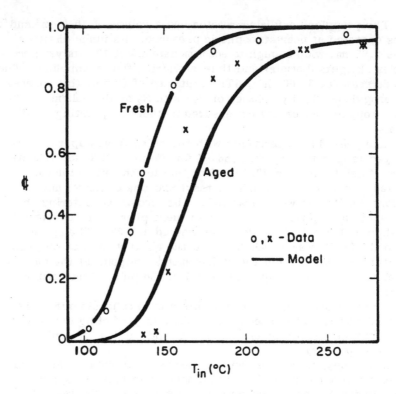

FIG. 6. Propylene conversion over spherical catalyst particles.

parameters, using poison inlet concentration, feed gas flow rate, reactor and catalyst dimensions, inlet temperature and pressure, and catalyst pore structure as pre-specified parameters.

Since part of the phosphorus which was combusted in the pulse-flame apparatus deposited within the system and thus never reached the catalyst, the actual catalyst-inlet phosphorus concentration was backtracked from chemically measured phosphorus deposition levels on the catalysts.

The poisoning model [13] employs a pore-mouth poisoning scheme; this approximation has been proven to be reasonable for spherical pellets (e.g., Hegedus and Baron, Ref. 14), should hold equally well for cylinders, and is used here without further proof for the spoked extrudates, accepting the fact that the model's predictions (Fig. 8) are reasonable.

A flat-slab geometry was used for spoked extrudates, neglecting curvatures due to the relative thinness of their walls

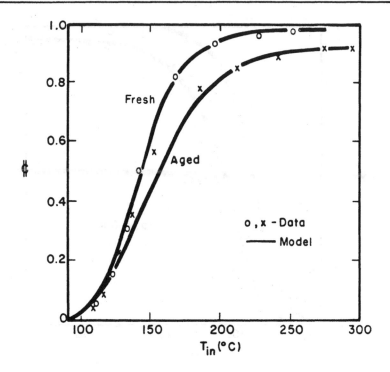

FIG. 7. Propylene conversion over cylindrical catalyst particles.

with respect to the curvature radius. The geometric approximation of Aris [15] was then used both for cylinders and spoked extrudates for the calculation of effectiveness factors.

Transport properties were estimated from catalyst geometries, hydrodynamic conditions, gas properties, and pore structures, as described by Hegedus [10].

Electron microprobe analysis of the poisoned spoked extrudates showed that their internal channels were somewhat less exposed to phosphorus than their outer surfaces. This implies that the gas flow through their interior channels may be slightly different from the gas flow field surrounding them. This aspect is currently being investigated in more detail; for the purposes of this analysis, average and symmetric phosphorus profiles were considered in the calculations.

Figures 6, 7, and 8 show only minor differences among the three catalyst geometries. This is due to the fact that all three

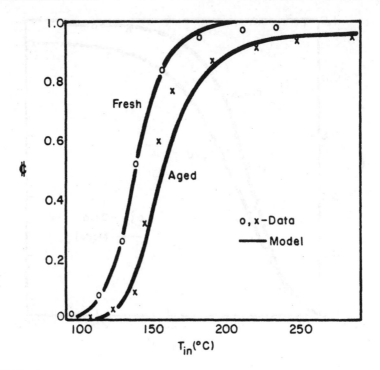

FIG. 8.  Propylene conversion over spoked extrudate catalyst particles.

catalysts had different pore structures and it so happened that they tended to cancel out the differences among their geometries.

Thus, it became evident that, to make best use of the spoked extrudate geometry, its pore structure needs to be optimized.

For the optimization procedure, the analysis of Hegedus [10] was employed for our four-spoke extrudate.  The conversion of propylene after 1000 h of simulated vehicle service was taken as the objective function, to be maximized as a function of catalyst pore structure parameters.

Table 3 shows the values of the floated (and constrained) pore structure parameters and computed diffusion-limited performance at 1000 h for two cases; a base case (spoked extrudate pore structure shown in Table 2) and the spoked extrudate with a computed optimal pore structure.

## TABLE 3

### Minilith Pore Structure Optimization Calculations

| Parameter | Constraints | Base case[a] | Optimum |
|---|---|---|---|
| $V_{macro}$ (cm$^3$/g) | <2 | 0.133 | 0.730 |
| $V_{micro}$ (cm$^3$/g) | <2 | 0.618 | 0.390 |
| $\bar{r}_{macro}$ (Å) | 300...10000 | 500 | 7900 |
| $\bar{r}_{micro}$ (Å) | 50...300 | 112 | 50 |
| $s_i$ (m$^2$/g) | <200 | 110 | 165 |
| $\rho_p$ (g/cm$^3$) | >0.7 | 0.964 | 0.700 |
| $\phi$ (at 1000 h) X 100 | | 75.1 | 93.9 |

[a]Identical to the Minilith displayed in Table 2.

With respect to the base case, the optimal Minilith shows a larger macropore volume, a smaller micropore volume, a larger macropore radius, and a smaller micropore radius; this results in a larger surface area, a lower density, and a substantially improved diffusion-limited conversion performance after the simulated 1000-h aging process.

The optimized pore structure converged to the lowest permissible density; it remains to be seen whether spoked extrudate structures with such a low density can be made to have an acceptable compressive strength; if not, a new pore structure may have to be computed at a larger value for the minimum permissible density.

## IV. EXPERIMENTS IN SIMULATED AUTOMOBILE EXHAUST

Even though we have not yet completed the experimental implementation of our pore structure optimization project, we have

TABLE 4

Comparison of Fully Formulated Three-Way Catalysts[a]

|                                      | Spheroid | Minilith |
|--------------------------------------|----------|----------|
| $V_{macro}$ $(cm^3/g)$               | 0.380    | 0.037    |
| $V_{micro}$ $(cm^3/g)$               | 0.614    | 0.551    |
| $\bar{r}_{macro}$ $(\overset{o}{A})$ | 4200     | 500      |
| $\bar{r}_{micro}$ $(\overset{o}{A})$ | 108      | 80       |
| $\rho_p$ $(g/cm^3)$                  | 0.766    | 1.150    |
| $s_i$ $(m^2/g, N_2$ BET)             | 114      | 137      |
| $L_p$ (cm)                           | 0.302    | 0.335    |
| $D_p$ (cm)                           | 0.262    | 0.254    |
| $L_p/D_p$                            | 1.15     | 1.32     |
| w (cm)                               | -        | 0.038    |
| $s_g$ $(cm^2/cm^3)$                  | 15       | 28       |
| Shrinkage $(\%)$[b]                  | 0-3      | 3-7      |
| Crush strength (kg):                 | 2.3      |          |
|   Between spokes           | -        | 1.6      |
|   Along spokes             | -        | 3.3      |
|   Longitudinal             | -        | 12.5     |
| Pt $(wt\%)$                          | 0.165    | 0.159    |
| Rh $(wt\%)$                          | 0.0099   | 0.0095   |
| Ce $(wt\%)$                          | 3.0      | 3.0      |

[a]Both catalysts were impregnated nonuniformly along the depth of the particle. Impregnation patterns are not disclosed but comparable for both geometries.

[b]985°C, 2 h, in air.

nevertheless decided to prepare a fully formulated three-way spoked extrudate and test it against a fully formulated spherical-type, commercial three-way catalyst. The characteristics of these catalysts are displayed in Table 4.

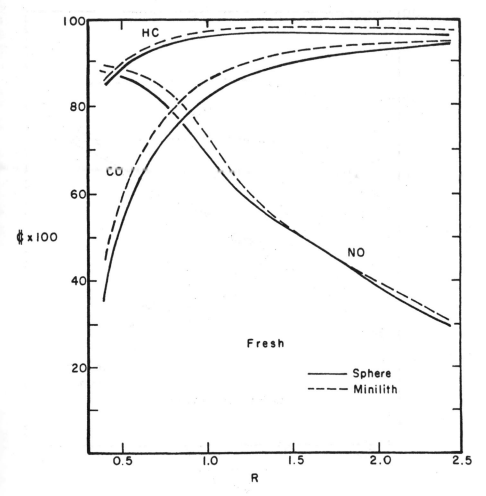

FIG. 9.  Three-way performance comparison of fresh sphere and Minilith.

The three-way reactor's dimensions were identical to the one described for propylene oxidation, and the phosphorus poisoning experiments were essentially identical to those described in conjunction with the propylene oxidation runs, except that, in addition to P and S, the feed also contained Pb in a P/Pb molar ratio of 2.5, and that the poisoning experiment was longer (90 h instead of 42.5 h). The poison exposure (inlet concentration times exposure time) remained the same, however.

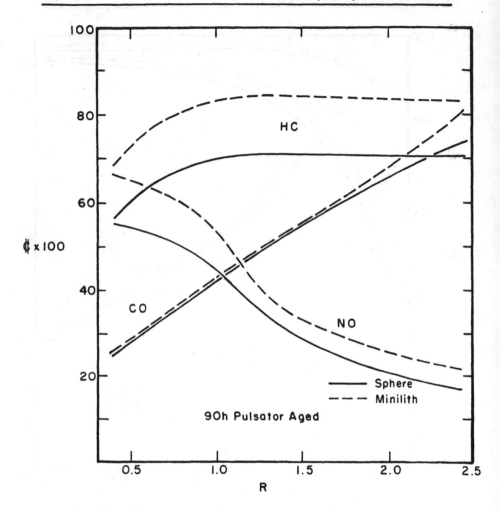

FIG. 10.   Three-way performance comparison of aged sphere and Minilith.

Before and after poisoning, the catalysts were tested for their three-way conversion performance at 60,000 h$^{-1}$ space velocity (STP), scanning the air-fuel ratio range by modulating CO and $O_2$ at a frequency of 1 Hz, with an air-fuel ratio amplitude of ±0.5 about the midpoint.  The feedstream also contained propylene, propane, NO, $SO_2$, $CO_2$, and $H_2O$.  The test was described by Ernest and Kim [12].

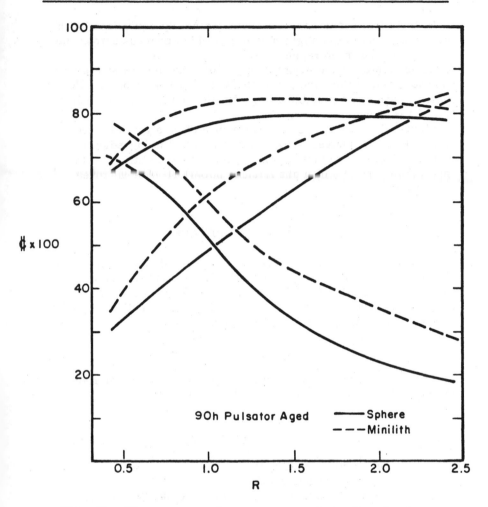

FIG. 11. Three-way performance comparison of aged sphere and Minilith. Formula includes a proprietary additive for both geometries.

Figure 9 shows the fresh catalysts against a scale (R) which indicates the oxidizing or reducing character of the feedstream:

$$R = \frac{[O_2] + 0.5[NO]}{\frac{2}{3}[CO] + 5[C_3H_8] + 4.5[C_3H_6]} \qquad (4)$$

In their fresh state, the catalysts are nearly equivalent, with the Minilith showing a slight advantage which prevails across the scanned air-fuel ratio range.

After 90 h of pulsator aging (Fig. 10), the Minilith shows clear superiority over the spherical catalyst for hydrocarbon and NO control, with a marginal advantage for CO at the lean (oxidizing) end of the scale.

Figure 11 shows the aged performance of catalysts (spheres and spoked extrudates) which, in addition to the ingredients listed in Table 4, contain another, proprietary additive. With this formula, the Minilith shows clear superiority for all three pollutants.

## V.  CONCLUDING REMARKS

While our initial work, as discussed above, shows promise for the Minilith, much work remains until its commercial viability can be established.  Ongoing investigations are aimed at crush strength and vibration resistance, and at transient lightoff performance using fully formulated samples.

### Acknowledgment

The authors thank J. E. Kubsh for the electron microprobe analysis of the phosphorus-poisoned catalyst samples.

### SYMBOLS

| | |
|---|---|
| $\mathcal{C}$ | conversion (fractional) |
| $d$ | catalyst dispersion (%) |
| $D_h$ | hydraulic diameter of catalyst particle, see text (cm) |
| $D_p$ | particle diameter (cm) |
| $D_r$ | reactor diameter (cm) |
| $G$ | mass feed flux, superficial ($g/cm^2/s$) |
| $k_m$ | mass transfer coefficient (cm/s) |
| $L_p$ | particle length (cm) |
| $L_r$ | reactor length (cm) |
| $N_c$ | number of monolith channels per unit cross section ($1/cm^2$) |
| $N_s$ | number of spokes in the Minilith |
| $\Delta P$ | pressure drop ($g/cm/s^2$) |
| $Q$ | feed flow rate at reactor inlet conditions ($cm^3/s$) |
| $\bar{r}_{macro}$ | integral-averaged macropore radius (see Ref. 10) (Å) |

$\bar{r}_{micro}$   integral-averaged micropore radius (see Ref. 10) ($\overset{o}{A}$)

R   parameter characterizing the oxidizing or reducing nature of the feedstream, see Eq. (4)

$s_g$   geometric surface area per unit reactor volume ($cm^2/cm^3$)

$s_i$   internal surface area per unit weight ($m^2/g$)

$V_{macro}$   macropore volume (see Ref. 10) ($cm^3g$)

$V_{micro}$   micropore volume (see Ref. 10) ($cm^3/g$)

$V_r$   reactor volume ($cm^3$)

w   wall thickness (for spoked extrudates and monoliths) (cm)

$\varepsilon$   void fraction of catalyst particle assembly (excluding pore volumes, including flow volumes)

$\rho_p$   particle density (excluding flow channels ($g/cm^3$)

$\mu$   gas viscosity ($g/cm/s$)

## REFERENCES

[1]   L. L. Hegedus and J. J. Gumbleton, Chemtech., 10, 630 (1980).

[2]   L. L. Johnson, W. C. Johnson, and D. L. O'Brien, Chem. Eng. Prog. Symp. Ser., 35, 55 (1961).

[3]   L. L. Hegedus, Prepr., Div. Pet. Chem., Am. Chem. Soc., 18, 487 (1973).

[4]   S. H. Oh, J. C. Cavendish, and L. L. Hegedus, AIChE J., 26, 935 (1980).

[5]   S. H. Oh, K. Baron, E. M. Sloan, and L. L. Hegedus, J. Catal., 59, 272 (1979).

[6]   J. R. Adomaitis, J. E. Smith, and D. E. Achey, SAE, Paper No. 800084, Detroit, Michigan, February 25, 1980.

[7]   R. B. Bird, W. E. Stewart, and E. N. Lightfoot, Transport Phenomena, Wiley, New York, 1960.

[8]   C. B. Lundsager, U.S. Patent 3,907,710 (September 23, 1975).

[9]   B. Villemin, U.S. Patent 4,089,941 (May 16, 1978).

[10]  L. L. Hegedus, Ind. Eng. Chem., Prod. Res. Dev., 19, 533 (1980).

[11]  K. Otto, R. A. Dalla Betta, and H. C. Yao, J. Air Pollut. Control Assoc., 24, 6 (1974).

[12]  M. V. Ernest and G. Kim, SAE, Paper No. 800083, Detroit, Michigan, February 25, 1980.

[13]  B. K. Cho, L. L. Hegedus, and R. Aris AIChE J., 29, 289 (1983).

[14]  L. L. Hegedus and K. Baron, J. Catal., 54, 115 (1978).

[15]  R. Aris, Chem. Eng. Sci., 6, 262 (1957).

# Preparation and Properties of Fluid Cracking Catalysts for Residual Oil Conversion

JAMES M. MASELLI AND ALAN W. PETERS
Davison Chemical Division
W. R. Grace & Co.
Columbia, Maryland

## I.  INTRODUCTION

Catalytic cracking of petroleum to produce gasoline began in
about 1912.  The early pioneering work was carried out by Eugene
Houdry [1].  Modern fluid catalytic cracking (FCC) was conceived
at Exxon and commercially developed in about 1940 [2] using amorphous catalysts.  Fluid catalysts are small spherical particles
ranging from 40 to 150 μm in diameter with acid sites capable of
cracking large petroleum molecules to products boiling in the
gasoline range.  One advantage of the FCC process is the absence of the diffusion limitations present in conventional gas oil
cracking due to the small size of the catalyst particle.  Since 1964
virtually all catalysts contain faujasite, a stable, large pore, Y-
type zeolite dispersed in a silica/alumina matrix [3].  The catalytic aspects of contemporary FCC processes have been reviewed
by Venuto and Habib [4], Gates, Katzer, and Schuit [5], Magee
and Blazek [6], and Magee [7].  A more recent update of refinery trends has been made available by Blazek [8].

In a conventional FCC process, Fig. 1, preheated oil at $\sim 300°C$
is sprayed into the riser where it mixes with a hot (700°C) catalyst to produce a reaction temperature of about 550°C.  The vaporization and cracking of the oil provide a threefold volume expansion.  This expansion plus the introduction of steam provides
a gas flow that transports the catalyst and oil/gasoline/gas mixture up the riser into a reactor zone.  A series of cyclones with
steam stripping disengages the catalyst from the petroleum products.  The catalyst is then transported to the regenerator where
$\sim 1\%$ carbon and a small amount of hydrogen as hydrocarbon on the
catalyst is burned off.  The reactions shown below, which occur
in the regenerator, are exothermic and produce temperatures of
about 700°C and a $\sim 20\%$ steam environment.  Most of the steam
comes from burning hydrogen derived from coke or entrained
hydrocarbons.

$$CH_n + \frac{(n + 1)}{2} O_2 \rightarrow CO + \frac{n}{2} H_2O$$

$$CO + \frac{1}{2}O_2 \rightarrow CO_2$$

The unit is maintained in heat balance and is nearly adiabatic.
There is some control over the degree to which the feed is preheated, but in general the heat for the overall unit operation is
supplied by burning coke in the regenerator.  If too much coke

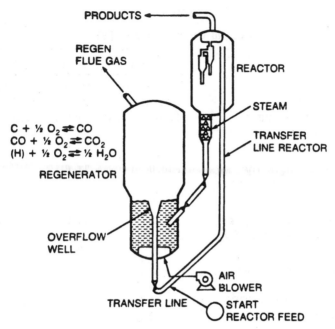

$$C + \tfrac{1}{2} O_2 \rightleftharpoons CO$$
$$CO + \tfrac{1}{2} O_2 \rightleftharpoons CO_2$$
$$(H) + \tfrac{1}{2} O_2 \rightleftharpoons \tfrac{1}{2} H_2O$$

Source: Davison *Catalagram* #65.

FIG. 1.  Regenerator and reactor: FCCU Exxon Flexicracker.

is made, the regenerator temperature increases and the amount
of feed must be reduced.  Since this results in diminished unit capacity, a requirement for a good catalyst is that it optimizes coke
production without sacrificing activity.

Although in the reactor there is usually five times as much catalyst by weight as there is oil, typical catalyst consumption in conventional fluid cracking is about 1 kg per 3000 kg of oil.  Fresh
catalyst is continuously being added while older catalyst is withdrawn or is lost through attrition.  On the average, each catalyst particle goes through about 15,000 cycles before it is dis-
carded.  A given particle may spend 1 to 3 months in the unit
where metallic poisons accumulate.  Much of that time is spent
in a severe hydrothermal environment.  There are therefore additional requirements that the catalyst must be hydrothermally
stable and not lose activity and selectivity after accumulation of
contaminants such as Ni, V, Fe, Cu, and Na.

TABLE 1

Typical Crude Assays [9]

| Boiling range | Volume % of total crude | | |
|---|---|---|---|
| | Arab Light (Berri) | Arab Medium (Khursaniyah) | Arab Heavy (Safaniya) |
| Naphtha, 20 to 150°C | 18.8 | 16.6 | 14.7 |
| Kerosine and light fuel oil (some FCC feed), 150 to 340°C | 39.9 | 32.6 | 28.9 |
| Heavy gas oil (conventional FCC feed), 340 to 565°C: | 30.3 | 30.9 | 30.6 |
| % Sulfur | 1.8 | 2.9 | 2.9 |
| VAC residual oil (565°C+): | 9.0 | 18.7 | 23.2 |
| % Sulfur | 2.9 | 5.4 | 6.0 |
| % Coking carbon (CCR) | 16 | 23 | 28 |
| Vanadium, ppm | 9 | 96 | 205 |
| Nickel, ppm | 6 | 32 | 64 |
| Iron, ppm | - | 30 | 30 |

## II.  RESID PROCESSING

Until recently, residual oil represented a small ($\sim$10%) portion of the crude oil.  Relatively inexpensive, light, low sulfur feedstocks were available for cracking.  Naphtha, kerosine, and light fuel oil obtained by simple distillation are usually processed to improve product quality without molecular weight reduction.  In a typical refinery the heavy fuel oil (340-565°C) is obtained as a product from vacuum distillation.  The FCC unit reduces the molecular weight of this fraction primarily to gasoline.  The vacuum residual oil is processed to make coke plus coker gas oil or a residual fuel oil.  Recently this alternative has become less attractive because the high sulfur content of the residual oil remains in the products and can create $SO_x$ emissions during burning of the coke or oil.  Furthermore, economic and political pressures have forced refiners to process some of the heavier feedstocks such as Arab Medium and Arab Heavy, Table 1.  These crude oils contain more vacuum residual fraction and less light products [9]. The residual oil contains increased amounts of sulfur and contaminant metals. As the amount of residual oil in feedstocks has increased, quality has decreased and the market for conventional products such as coke or heavy fuel oil has declined. As a result, some refiners have developed processes to convert the vacuum resid to gasoline and other valuable products.

Figure 2 depicts a refinery design developed 20 years ago by Phillips Petroleum and now commercial at two Phillips locations and at Saber Refining [10]. Atmospheric resid is hydrotreated to remove sulfur, some contaminant metals, and some of the coke precursors. After vacuum distillation the vacuum residual oil is sent to the HOC (heavy oil cracker) unit, a modified FCC unit of proprietary design. In another example, Fig. 3, atmospheric resid may be sent directly to an FCC unit specifically developed to process the very heavy feed. If the contaminant level is high, the resid may be solvent or otherwise deasphalted or hydrotreated to remove metal contaminants and coke precursors. Resid cracking processes of this type are presently being operated by Ashland Oil [11] and Total Petroleum [12, 13]. It is interesting to note that these three resid cracking processes have completely different operating philosophies and, not surprisingly, use three distinctly different types of cracking catalysts.

A recent survey of resid cracking in the United States [14] has shown that between 2/81 and 10/82 the number of resid cracking operations doubled. By 10/82 37% of United States refiners were processing at least some resid in their FCC units. Most of these refiners are adding atmospheric bottoms along with their conventional gas oil feedstocks. The resid part of the feed (Table 2) contains much higher levels of poisons such as nickel and vanadium, potential pollutants such as sulfur, and coke precursors identified as Conradson or Ramsbottom carbon [15]. For the refiners these properties lead to increased catalyst usage rates, higher metals (Ni, V) content on catalyst, and an observable decline in equilibrium activity. The increase of metals in the feed is partly responsible for the higher catalyst addition rates, higher metal on circulating catalyst, and a lower equilibrium activity compared to operation on heavy gas oil. Increased amounts of coke have caused a lower feed preheat by $\sim$50°C, and the use of antimony to reduce coke and hydrogen due to nickel is now common.

Although not apparent from the resid survey, the addition of vacuum bottoms (bp > 500°C) to the feed suggests that some very large asphaltene molecules may be present as a liquid aerosol, at least during the initial stages of the reaction. While diffusion

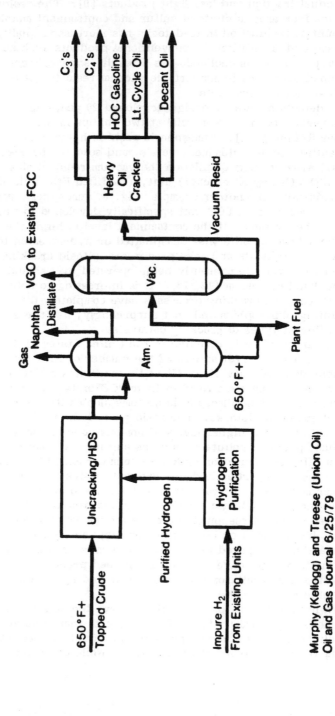

FIG. 2. HDS and HOC combined at Phillips refinery.

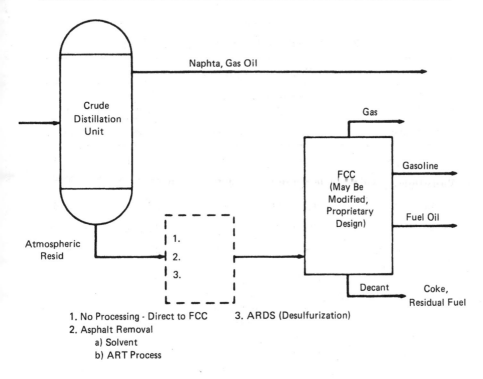

1. No Processing - Direct to FCC     3. ARDS (Desulfurization)
2. Asphalt Removal
   a) Solvent
   b) ART Process

FIG. 3.  Resid FCC refinery (Total, Ashland).

limitations do not occur in conventional gas oil FCC units, in resid
operations it is necessary to crack these large nonvolatile components under conditions where diffusion limitations may play a role
in the overall reaction rate.

### III.  CATALYST COMPOSITION, STRUCTURE, AND PREPARATION

FCC catalyst compositions have been described in detail in various patents and have been reviewed by Magee and Blazek [6].  A
typical commercial fluid cracking catalyst is composed of about 5
to 40% of a Y-type, faujasitic zeolite, a silica/alumina binder or
matrix, and a clay filler.  These materials are slurried together
at a controlled concentration and pH, and spray dried.  The matrix provides porosity and may contain a catalytically active silica/
alumina component.  In some families of catalysts the matrix also

TABLE 2

Resid Feedstock and Catalyst Quality [14]

| Feedstock Quality | | | | |
|---|---|---|---|---|
| Feedstock: | Resid portion of feed | | Total resid feed | |
| Properties: | Range | Median | Range | Median |
| Nickel, ppm | 0.6-50 | 5.0 | 0.1-7.0 | 1.4 |
| Vanadium, ppm | 0.5-130 | 4.0 | 0.15-25.0 | 1.5 |
| Sulfur, wt% | 0.04-3.0 | 0.7 | 0.04-3.0 | 0.6 |
| CCR, wt% | 0.25-15 | 3.6 | 0.25-6.0 | 1.25 |
| API, gravity | 8-37 | 21 | 20-37 | 25 |

Catalyst Usage Rates and Quality

| | Atmospheric bottoms | Vacuum bottoms | Heavy gas oils |
|---|---|---|---|
| Total catalyst additions: | | | |
| lb/bbl | 0.412 | 0.423 | 0.231 |
| % of inventory | 1.93 | 3.01 | 1.80 |
| Equilibrium catalyst quality: | | | |
| Microactivity, vol% | 64.5 | 68.7 | 72.0 |
| Ni, ppm | 1550 | 1400 | 650 |
| V, ppm | 2200 | 2100 | 900 |
| Cu, ppm | 60 | 40 | 50 |
| Fe, wt% | 0.64 | 0.62 | 0.57 |
| Na, wt% | 0.59 | 0.64 | 0.58 |

serves to bind the ingredients together. In others a separate
binder is used. The zeolite is normally either a rare earth exchanged Y or an ultra-stable dealuminated Y. The zeolite typically forms agglomerates of 1 to 2 μm in size and may be loosely
or tightly embedded in the matrix as shown in Fig. 4. This figure shows SEM micrographs of the individual components and of
the interior of a typical commercial catalyst particle. An essential operation and in some cases the final step in catalyst manufacture is the removal of soda by extensive exchanging and

washing with ammonium salts and water. Efficient resid processing may require specific adjustments in preparation, especially
in the nature of the components and in the extent of soda removal.


## IV. CATALYST REQUIREMENTS

A summary of new challenges faced by the catalyst manufacturer introduced by resid cracking includes increased levels of
poison (Ni, V), higher coke levels resulting in higher regenerator temperatures, and the introduction of very high boiling components to be converted, if possible, to gasoline. Another issue
not to be discussed here is the increased amounts of $NO_x$ and
$SO_x$ pollutants which result from the higher nitrogen and sulfur
levels in the resid fraction of the feed. These processing changes
translate into the following catalyst requirements that may be met
in part by adjusting the catalyst properties, especially the properties of the zeolite and the matrix.

Catalyst Requirements
    Resistance to poisoning by metal (vanadium and nickel)
    Hydrothermal stability at high regenerator temperatures
    Low coke and gas make, especially in a high metals
        environment
    Good activity for very large molecules
    Low cost due to increased catalyst usage rates


## V. TOLERANCE TO NICKEL AND VANADIUM

As shown in Table 3 and pointed out in a previous paper [16],
the major effect of nickel is to produce additional gas and coke,
while vanadium deactivates the zeolite-containing catalyst. The
relative effects of nickel and vanacium on activity, gas, and
coke show that both nickel and vanadium increase gas and coke,
and that vanadium (but not nickel) decreases activity. Gas make
is additive but coke is less so.
The vanadium directly attacks the zeolite during steam deactivation [17]. After impregnation by a vanadium naphthenate and
subsequent steam deactivation of the catalyst, the electron microprobe trace for vanadium coincides with the trace of the lanthanum
exchanged into the zeolite, Fig. 5. However, the presence of a
high surface area, catalytically active, silica/alumina additive as
a matrix or binder serves to disperse the vanadium more uniformly
throughout the particle, Fig. 6. Besides silica/alumina, materials

Catalyst Components

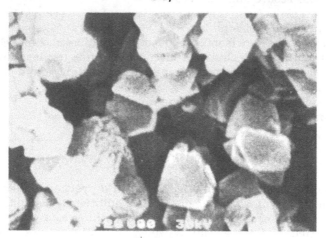

Clay

Faujasite

FIG. 4. Micrographs of interior of a commercial catalyst (DA-300).

suggested to scavenge vanadium in cracking catalysts include alumina [18], phosphorus [19], and titanium [20].

Nickel passivation techniques were developed by Phillips Petroleum in the mid 1970s using proprietary antimony compounds [21]. Results on hydrogen yield are illustrated in Fig. 7. An

Finished Catalyst

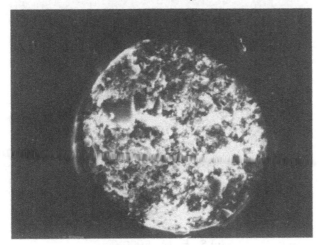

(a) S. E. M. 500X

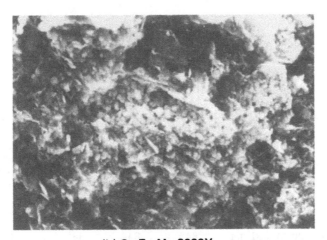

(b) S. E. M. 2000X

FIG. 4. (continued)

improvement in coke yield is also observed. These benefits can mean that a refiner processing resid can increase his yield of gasoline and increase his product value as well as increase capacity due to lower coke make. Other proposed nickel or metal passivators include tin [22], barium [23], lithium [24], and tungsten [25].

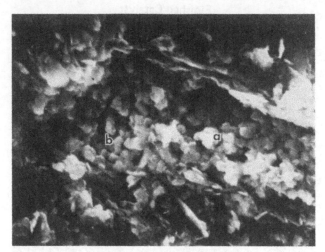

(c) S. E. M. 5000X

FIG. 4. (continued)


TABLE 3

Nickel and Vanadium Contributions to Catalyst
Deactivation and Cracking Selectivity

| Catalyst | Davison residcat | | | |
|---|---|---|---|---|
| Contaminant metal level | % Metals | 0.33% Ni | 0.67% V | 1% Metals (0.33% Ni + 0.67% V) |
| X-ray crystallinity: | | | | |
| Metal impregnated material, deactivation at 733°C, 8 h, 2 atm steam (% of base, 3 h at 680°C) | 84% | 84% | 38% | 38% |
| MAT after deactivation: | | | | |
| Vol% conversion | 80 | 82 | 61 | 61 |
| $H_2$ (wt% FF) | 0.14 | 0.274 | 0.109 | 0.244 |
| Coke (wt% FF) | 3.3 | 5.8 | 1.9 | 2.5 |

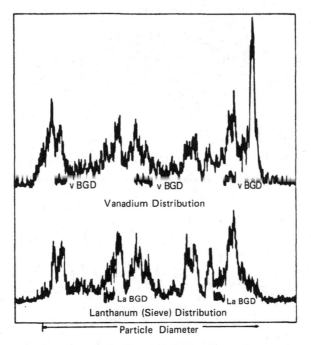

FIG. 5.  Zeolite (lanthanum) and vanadium distributions by
EPA on a zeolite bound with an inert matrix (clay plus silica
binder).  After steaming for 8 h at 2 atm and 733°C.


## VI.  HYDROTHERMAL STABILITY

Laboratory procedures for measuring hydrothermal stability involve submitting the catalyst to a steam environment at ~700 to
800°C for several hours.  By using these tests it has been shown,
Fig. 8, that both catalyst and zeolite hydrothermal stability decreases in the presence of Group Ia and IIa ions [26, 27].  Conversely, zeolite stability is increased by rare earth exchange [28].
Since the activity of zeolites [29] and their resistance to deactivation by vanadium [30] is increased with lower sodium concentrations, there is an incentive to produce low soda, stabilized, or
rare earth exchanged zeolites.  Stable rare earth or ammonium
(hydrogen) zeolites can be prepared by exchange to ~4% $Na_2O$,
further calcination, and additional exchange to even lower soda.
The rare earth exchanged version [31] is typically referred to as
CREY (Calcined Rare Earth Y), and USY (Ultrastable Y) is used
as the designation for the ammonium or hydrogen form [32].

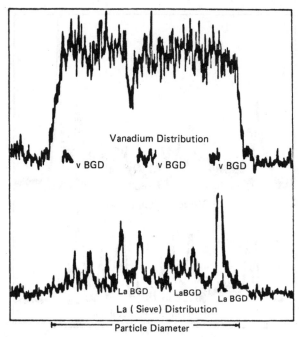

FIG. 6.  EPA zolite and vanadium (0.67%) distributions after steaming for 8 h at 2 atm and 733°C.  Zeolite is bound with a silica/alumina binder.

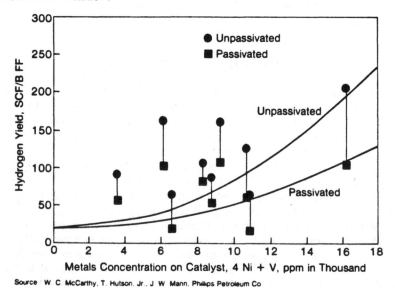

Source  W. C. McCarthy, T. Hutson, Jr., J. W. Mann, Phillips Petroleum Co

FIG. 7.  Licensee hydrogen yield decrease.

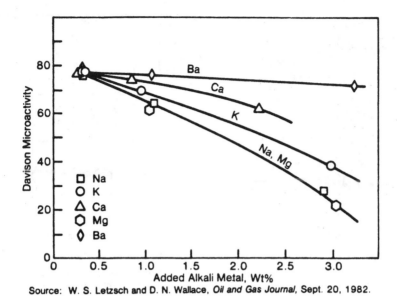

Source: W. S. Letzsch and D. N. Wallace, *Oil and Gas Journal*, Sept. 20, 1982.

FIG. 8.  Fresh catalyst activity after steaming (8 h, 733°C, 2 atm steam).

## VII.  DIFFUSION CONSIDERATIONS

In order to be effectively converted to more valuable products, oil molecules must reach the external surface of the catalyst particle and diffuse through the pore structure to an active site. The reactants must then adsorb and react on the site.  The products must then desorb, diffuse to the outer surface, and desorb from the particle into the bulk gas stream.

Residual oil is typically composed of very large molecules, Fig. 9, a fraction of which boil above about 500°C [33, 34]. Some may even be present in the form of an aerosol liquid during introduction and reaction in an FCC unit.  This presents two problems.  Both liquid and gas molecules must diffuse through the matrix to the zeolite, and then must react.  Molecules of 20 Å or greater in diameter cannot be cracked easily inside the zeolite pore structure which is only accessible through 8-9 Å openings.  Instead they are restricted to reaction on the external surface of the zeolite particles or to cages very close to the external surface.  As a result, there are observable diffusion restrictions within the zeolite and a loss of activity for even moderate size molecules [35], Table 4.

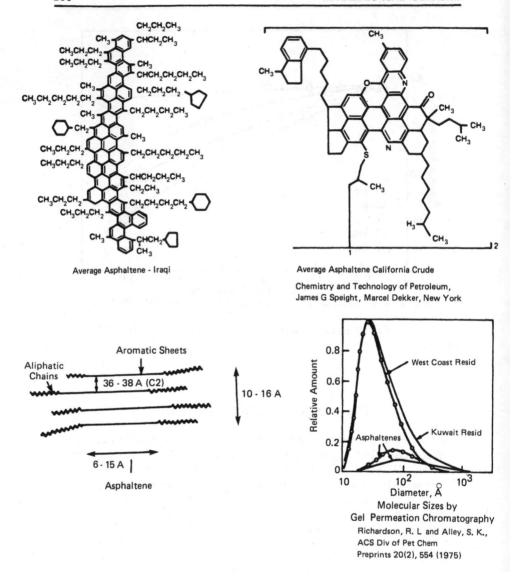

FIG. 9.  Residual oil molecular sizes.

There are two possible changes that can be made.  One is to reduce the size of the zeolite particle so as to increase external surface area.  The small particle zeolite shown in Fig. 10 contains crystallites with an external surface area (defined from BET data

TABLE 4

Cracking Rate Constants at 900°F[a]
(2 min on-stream instantaneous value)

| Reactant hydrocarbon | $SiO_2/Al_2O_3$ | REHX | Ratio $\dfrac{k_{REHX}}{k_{SiO_2/Al_2O_3}}$ |
|---|---|---|---|
| $nC_{18}H_{3-4}$ | 60 (0.1%C) | 1000 (1.4%C) | 17 |
| (benzene with two $C_2H_3$ and one $C_2H_3$ substituents) | 140 (0.4%C) | 2370 (2.0%C) | 17 |
| (decalin/naphthene with $CH_3$, $CH_3$, $CH_3$ substituents) | 190 (0.2%C) | 2420 (0.7%C) | 13 |
| (three-ring fused naphthene) | 205 (0.2%C) | 953 (1.0%C) | 4.7 |
| (four-ring fused naphthene) | 210 (0.4%C) | 513 (1.6%C) | 2.4 |

[a]Comparison of cracking rates of n-hexadecane and a series of naphthenes of progressively larger ring structure over amorphous silica-alumina and zeolitic catalyst [35].

as surface area in pores > 100 Å) of about 45 m$^2$/g, equivalent to a cube 0.07 μm on a side. The larger size sieve has an external surface area of only about 4 m$^2$/g which is equivalent to a cube 0.7 μm on a side. Microactivity and selectivity studies using a heavy resid feed show the small particle sieve catalyst is both more active and makes more liquid product. It also leaves less heavy bottoms and makes less gas, Fig. 11. However, after hydrothermal or metals deactivation the smaller size sieves lose their selectivity and activity advantages, possibly due to an inherent instability relative to larger size zeolite particles, Table 5.

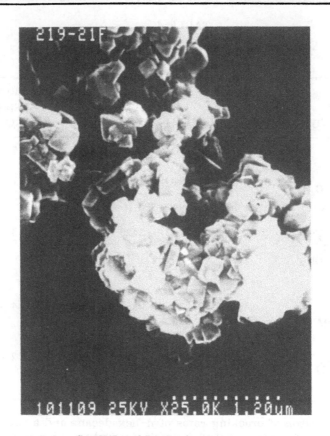

Small Particle Faujasite

Size                    0.07 μm

External SA             45 M$^2$/G
(0.57-0.87 P/P$_0$)

Total SA                960 M$^2$/G

FIG. 10. Properties of large and small sieves.

## VIII. MATRIX ACTIVITY

Large molecules can also be converted into liquid product by
the incorporation of an active, large pore amorphous matrix along
with the sieve in the catalyst particle. Amorphous matrix activity
is usually derived from a mixed oxide, porous material containing

Large Particle Faujasite

| | |
|---|---|
| Size | 0.7 μm |
| External SA<br>(0.57-0.87 P/P$_O$) | 4.1 M$^2$/G |
| Total SA | 930 M$^2$/G |

FIG. 10. (continued)

relatively large amounts of both silica and alumina. Systematic procedures for preparing silica, alumina, or binary oxides of defined pore structure have been described in the literature [36]. The matrix should be active enough to provide synergism between matrix and sieve, but not so active as to make significant amounts of gas and coke. The properties of some experimental matrices discussed below are listed in Table 6.

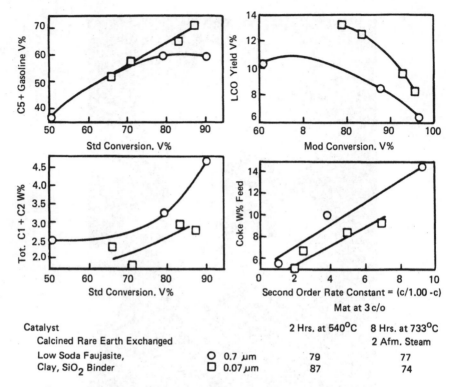

| Catalyst | | 2 Hrs. at 540°C | 8 Hrs. at 733°C |
|---|---|---|---|
| Calcined Rare Earth Exchanged | | | 2 Afm. Steam |
| Low Soda Faujasite, | ○ 0.7 μm | 79 | 77 |
| Clay, SiO₂ Binder | ▢ 0.07 μm | 87 | 74 |

FIG. 11. Selectivity and activity comparisons of a small (0.07 μm) and large (0.7 μm) zeolite containing cracking catalyst on West Texas GO: 500°C, 16 WHSV, C/O = 1.5-4.5.

TABLE 5

Effect of Deactivation on Small and Large Zeolite Catalysts

| Formulation: | Matrix + 20% Sieve | |
|---|---|---|
| Sieve: | (SP) Small (0.07 μM) | (LP) Large (0.7 μM) |
| Microactivity (8 h, 733°C, 2 atm steam) vol% conversion: | | |
| 0 | 81 | 76 |
| 0.2 | 52 | 67 |
| 0.4 | 33 | 49 |

TABLE 6

Experimental Matrix Properties

| Additive | A | B | C |
|---|---|---|---|
| % $Al_2O_3$ | 70-75 | 65-72 | 30-40 |
| Surface area, $m^2/g$ | 350 | 250 | 100 |
| Pore volume, $N_2$ $cm^3/g$ | 0.5 | 0.4 | 0.3 |
| Microactivity (8 h, 733°C, 2 atm steam) vol% conversion | 70 | 60 | 40 |

Synergism between matrix and sieve exploits the activity and selectivity of the sieve for cracking 300 to 500°C boiling range gas oil to gasoline, and the selectivity and activity of the matrix to crack larger molecules boiling at 500°C+ into the gas oil region. Fortunately, these differences in selectivity are known to exist and are illustrated in Tables 4 and 7. These differences have been shown to lead to synergism in the case of a reaction network

$$\text{Heavy oil} \xrightarrow{k_1} \text{light oil} \xrightarrow{k_2} \text{gasoline}$$
$$\downarrow k_3$$
$$\text{gasoline}$$

for a catalyst consisting of two types of activity: sieve activity for which $k_3$ and $k_2$ are large and large pore matrix activity for which $k_1$ is relatively large [37]. The existence of this phenomenon can be used to explain the results obtained in Fig. 12; when one adds an active matrix to a catalyst particle containing zeolite, the ability of the catalyst to convert the heavy portion of the feed improves.

With the still more active matrix A, it is possible to observe the effect of combining different amounts of the matrix with a sieve catalyst. While the matrix by itself has poorer selectivity

TABLE 7

Relative Rates of Cracking N-Hexadecane, a Vacuum Gas Oil,
and a 500°C+ Residual Oil by an Active Amorphous
Matrix and a Sieve/Inert Mixture

|  | Amorphous matrix catalyst (100% basis) | | | 15% Sieve/inert mixture | | |
|---|---|---|---|---|---|---|
|  | N-C$^{16}$ | VGO | 500°C+ Resid | N-C$^{16}$ | VGO | 500°C+ Resid |
| % Conversion (to coke and lighter products) | 14 | 58 | 70 | 70 | 75 | 62 |
| Relative rate constant (1st order) | 1 | 5.8 | 8.0 | 8 | 9.3 | 6.5 |

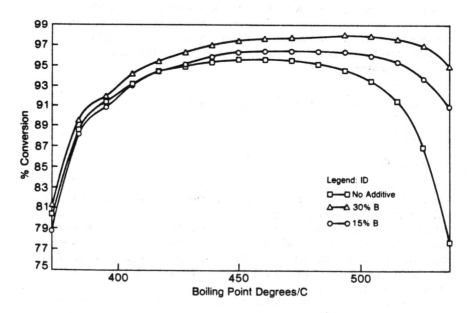

FIG. 12.   Conversion of heavy feed by matrix-active components.

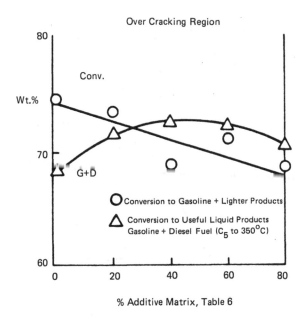

FIG. 13. Very heavy gas oil synergistic selectivities of additive A + Super D after steam deactivation (MAT test).

for G + D, the combination has improved selectivity over either the sieve or the amorphous matrix, Fig. 13.

## IX. MATRIX DIFFUSION

Typically, FCC catalysts are not diffusion limited for vapor-phase gas oil cracking because of the small size of the catalyst particle. However, in residual oil cracking the situation is not so clear. Residual oil can contain a substantial amount of 500°C+ material that according to a flash equilibrium calculation [38] can exist as a liquid in the reactor, Fig. 14. Liquid-phase diffusion is slower by four orders of magnitude [39] and involves larger size molecules than vapor-phase diffusion. The Wiesz-Prater criterion [40] illustrates the differences between gas- and liquid-phase diffusion, Table 8.

If a diffusion limitation does exist, depending on the size of the molecule one is interested in reacting, there is an incentive to increase pore size, for example, to $\sim 300$ Å diameter to convert a 30-Å size liquid molecule. A catalyst with a small pore

ASTM D - 1160 Distillation

Vapor Temperature, °C

| Vol. % | (10 MM Hg) | (Corrected to Atm.) |
|--------|-----------|----------------------|
| 10 | 267 | 427 |
| 20 | 297 | 461 |
| 30 | 323 | 490 |
| 40 | 344 | 514 |

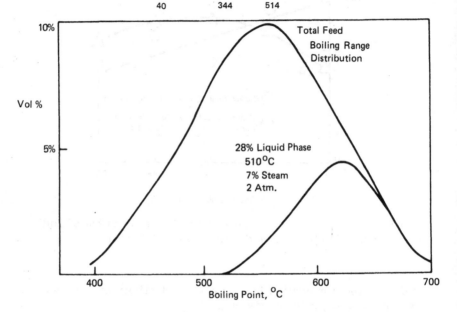

FIG. 14. Relative amounts of liquid and vapor phases in an
FCC unit: flash (equilibrium) distillation of a resid feed. Flash
program from Simulation Sciences, Inc., Fullerton, California.

size of 45 Å would not be expected to do as well, Table 9. The
pore structures of commercial equilibrium fluid cracking catalysts listed in Table 10 show that if there is a liquid-phase diffusion limitation, there may be observable performance differences among these catalysts. A less porous, smaller pore catalyst will be less effective, other things being equal. This result has been experimentally observed at temperatures above
450°C [41]. However, attempts in our laboratory to experimentally demonstrate diffusion limitations in a fixed-bed microactivity test have so far failed, Table 11. Particle size experiments have not shown significant improvements in microactivity
for smaller particle catalysts in resid cracking. These experiments are continuing utilizing circulating, riser type pilot units.

## TABLE 8

### Matrix Diffusivity

Weisz-Prater criterion for diffusion control:

$$\theta = \frac{R^2}{D_e C_o} \frac{DN_v}{DT} \sim \frac{10^{-5}}{D_e}$$

$D_e$ = diffusivity of reactant through catalyst (liquid $10^{-6}$ to $10^{-7}$; vapor $10^{-3}$ to $10^{-4}$)

$C_o$ = molar concentration of reactants

$DN/DT$ = reaction rate

$$\frac{1}{C_o} \frac{DN}{DT} = \sim 0.5$$

R = catalyst radius $\sim 5 \times 10^{-3}$

$\theta$ = liquid, 10-100, may be diffusion limited
vapor, $10^{-2}$-$10^{-1}$, not diffusion limited

## TABLE 9

### Parallel Pore Structure Diffusivity Sensitivity

| Pore diameter Å | 45 | 90 | 90 | 180 | 180 | 300 |
|---|---|---|---|---|---|---|
| Pore volume $cm^3/g$ | 0.2 | 0.2 | 0.4 | 0.2 | 0.4 | 0.2 |
| Relative diffusivity 30 Å liquid molecule | 0.26 | 1.0 | 1.5 | 2.2 | 3.3 | 2.9 |
| Relative rate constant | 0.5 | 1.0 | 1.2 | 1.5 | 1.8 | 1.7 |

## X. CONCLUSIONS

The result of this work suggests that catalysts can be prepared with improved selectivity and metals tolerance using selected matrix components. Table 12 shows some selectivity

TABLE 10

Pore Structures of Some Equilibrium Catalysts

| | SA, $m^2/g$ | Pore volume $cm^3/g$ Hg/N$_2$ | Porosity $cm^3/cm^3$ | Matrix (Hg) av pore diameter, Å | Relative reaction rate parallel pore model, molecular size 30 Å | 100 Å |
|---|---|---|---|---|---|---|
| A | 104 | .15/.39 | .50 | 1200 | 1.0 | .46 |
| B | 106 | .20/.25 | .44 | 1100 | .82 | .35 |
| C | 73 | .19/.42 | .50 | 700 | .97 | .41 |
| D | 116 | .25/.42 | .51 | 600 | .90 | .37 |
| E | 196 | .34/.39 | .55 | 600 | .80 | .32 |
| F | 152 | .29/.25 | .44 | 400 | .66 | .22 |

TABLE 11

Effect of Equilibrium Super DX Catalyst Particle Size on
Activity for Cracking a Residual Feed

| APS, μm | 126 | 90 | 68 | 53 |
|---|---|---|---|---|
| % NA$_2$O | .61 | .61 | .62 | .62 |
| Ni, ppm | 32 | 29 | 43 | 33 |
| V, ppm | 449 | 481 | 475 | 486 |
| SA, $m^2/g$ | 90 | 83 | 81 | 76 |

Residual oil:
  Reactor test conditions:      500°C, WHSV = 16, C/O = 2 (domi-
                                nant vapor phase cracking)
  Standard conversion, vol%:   67.1      65.0      64.2      63.0

490°C + Fraction of a VGO
  Reactor test conditions:      430°C, WHSV = 16, C/O = 2 (domi-
                                nant liquid phase cracking)
  Standard conversion, vol%:   53.5      54.0      50.6       -

## TABLE 12
### Selectivity Performance of B Matrix

| | Resid feed (MAT) 500°C/16 WHSV/ 3 C/O | | VGO feed (riser, 56 WHSV, 7 C/O, 520°C) | |
|---|---|---|---|---|
| | DA | DA + B | DA | DA + B |
| Conversion, vol% | 71.6 | 74.0 | 71.1 | 71.1 |
| C$^{5+}$ Gasoline, vol% | 60.0 | 64.0 | 58.5 | 62.5 |
| Diesel fuel, vol% | 15.6 | 16.1 | 16.2 | 17.8 |
| G + D, vol% | 75.6 | 80.1 | 74.7 | 80.3 |
| Coke, wt% | 4.8 | 4.3 | 4.5 | 4.8 |
| Slurry, vol% | 12.8 | 9.9 | 12.7 | 10.1 |

## TABLE 13
### Metal Tolerance Performance of Catalysts with Special Matrices (listed in Table 6)

| | | | |
|---|---|---|---|
| % V | 0 | 0.33 | .67 |
| % Ni | 0 | 0.17 | .33 |
| MAT (16 WHSV, 3 C/O, 500°C, VGO); 6 h, 733°C, 2 atm steam) | | | |
| Matrix C (Table 6): | | | |
| Silica/clay/20% sieve | 78 | 57 | 40 |
| Silica/clay/20% sieve + C | 84 | 69 | 58 |
| Matrix A (Table 6): | | | |
| Silica/clay/20% sieve[a] | 75 | – | 50 |
| 80% Matrix A/20% sieve[a] | 85 | – | 77 |
| Vanadium passivator: | | | |
| Super D | 70 | – | 19 |
| Super D + vanadium passivator[b] | 70 | – | 70 |

[a]Low (1%) Na$_2$O CREY.
[b]Proprietary Davison V Passivator.

comparisons with a standard commercial catalyst both for resid
and a vacuum gas oil. The data show noticeable improvements
achieved in liquid product yield after incorporation of matrix
component B.

Other preparations were tested for metals tolerance, with the
results shown in Table 13. In this case the use of selected matrix components A and C clearly improves resistance to vanadium
deactivation. Deactivation of a commercial catalyst by vanadium
can also be completely inhibited by the use of a proprietary vanadium passivator as illustrated in Table 13.

## REFERENCES

[1] A. G. Oblad, in Heterogeneous Catalysis (B. H. Davis
    and W. P. Hettinger, eds.), ACS Symposium Series 222,
    1983.
[2] C. E. Janig, H. Z. Martin, and D. L. Cambell, in
    Heterogeneous Catalysis (B. H. Davis and W. P.
    Hettinger, eds.), ACS Symposium Series 222, 1983.
[3] C. J. Plank, in Heterogeneous Catalysis (B. H. Davis
    and W. P. Hettinger, eds.), ACS Symposium Series 222,
    1983.
[4] P. B. Venuto and E. T. Habib, Fluid Catalytic Cracking,
    Dekker, New York, 1979.
[5] B. C. Gates, J. R. Katzer, and G. C. A. Schuit, Chem-
    istry of Catalytic Processes, McGraw-Hill, New York, 1979,
    Chap. 1.
[6] J. S. Magee and J. J. Blazek, Preparation and Perform-
    ance of Cracking Catalysts, Zeolite Chemistry and Cataly-
    sis (J. A. Rabo, ed.), ACS Monograph 171, 1976.
[7] J. S. Magee, Zeolite Cracking Catalysts—An Overview in
    Molecular Sieves—II (J. R. Katzer, ed.), ACS Symposium
    Series 40, 1977.
[8] J. J. Blazek, Update on Fluid Catalytic Cracking Trends,
    Davison Catalagram, No. 68, Davison Chemical Division,
    W. R. Grace & Co., Baltimore, 1981.
[9] L. R. Aalund, Oil Gas J., p. 71 (April 11, 1983).
[10] L. R. Aalund, Ibid., p. 67 (March 20, 1978); p. 66
    (March 30, 1981); p. 69 (September 14, 1981); J. B.
    Rush, Chem. Eng. Prog., 77(12), 29 (1981).
[11] O. J. Zandona, L. E. Busch, W. P. Hettinger Jr., and
    R. P. Kroch, Oil Gas J., p. 82 (March 22, 1982).
[12] R. Dean, J. L. Mauleon, and W. S. Letzsch, Ibid., p. 75
    (October 4, 1982); p. 168 (October 11, 1982).

[13]  W. Letzsch, J. L. Mauleon, and S. Long, Div. Pet. Chem.
      Prepr. Am. Chem. Soc., 28, 944 (1983).
[14]  P. G. Thiel, Davison 1982 Survey of Residuum Fluid Cata-
      lytic Cracking in the United States, Davison Catalagram,
      No. 66, 1983, Davison Chemical Division, W. R. Grace &
      Co., Baltimore, 1983.
[15]  Tests D 189 and D 524, respectively, 1980 Annual Book of
      ASTM Standards, Part 23, American Society for Testing
      and Materials, Philadelphia, 1980.
[16]  R. E. Ritter, L. Rheaume, W. A. Welsh, and J. S. Magee,
      Oil Gas J., p. 103 (July 6, 1981).
[17]  S. Jaras, Appl. Catal., 2, 207 (1982).
[18]  U.S. Patents 4,289,608 (1981) and 4,292,169 (1981).
[19]  U.S. Patent 4,228,036 (1980).
[20]  U.S. Patent 4,374,019 (1983).
[21]  G. H. Dale and D. L. McKay, Hydrocarbon Processing,
      p. 97 (September 1977), and references therein.
[22]  U.S. Patent 4,321,129 (1982).
[23]  U.S. Patent 4,377,494 (1983).
[24]  U.S. Patent 4,364,847 (1982).
[25]  U.S. Patent 4,290,919 (1981).
[26]  W. S. Letzsch and D. N. Wallace, Oil Gas J., p. 58
      (November 29, 1982).
[27]  M. C. Poutsma, in Zeolite Chemistry and Catalysis (J. A.
      Rabo, ed.), ACS Monograph 171, 1976.
[28]  R. Mone and L. Moscou, Proceedings of the International
      Conference on Molecular Sieves, 3rd, Zurich, 1973, Leuven
      University Press, p. 351.
[29]  D. Barthomeuf and R. Beaumont, J. Catal., 30, 288
      (1973).
[30]  L. Upson, S. Jaras, and I. Dalin, Oil Gas J., p. 135
      (September 20, 1982).
[31]  U.S. Patent 3,402,996 (1966).
[32]  C. V. McDaniel and P. K. Maher, "Zeolite Stability and
      Ultrastable Zeolites," in Zeolite Chemistry and Catalysis
      (J. A. Rabo, ed.), ACS Monograph 171, 1976.
[33]  J. G. Speight, Chemistry and Technology of Petroleum,
      Dekker, New York, 1980.
[34]  R. L. Richardson and S. K. Alley, Div. Pet Chem.,
      Prepr., Am. Chem. Soc., 20(2), 554 (1975).
[35]  D. M. Nace, Ind. Eng. Chem., Prod. Res. Dev., 9(2),
      203 (1970).
[36]  Ph. Courty and Ch. Marcilly, Preparation of Catalysts
      III, Elsevier, Amsterdam, 1983, p. 485; T. Ono, Y.
      Ohguchi, and O. Togari, Ibid., p. 631.

[37]  K. Rajagopalan and A. W. Peters, Eighth North American
      Meeting of the Catalysis Society, Philadelphia, 1983.

[38]  Program from Simulation Sciences, Inc., Fullerton, Cali-
      fornia.

[39]  C. N. Satterfield, Mass Transfer in Heterogeneous Cataly-
      sis, MIT Press, Cambridge, Massachusetts, 1970, p. 19.

[40]  P. B. Weisz and J. S. Hicks, Chem. Eng. Sci., 17, 265
      (1962).

[41]  A. Z. Dorogochinski, S. N. Khadzhiev, and S. M.
      Gairbekova, in Application of Zeolites in Catalysis (G. K.
      Boreskov and Kh. M. Minachev, eds.), Akademiai Kiado,
      Budapest, 1979.

# Monomers and Polymers

# Studies with High Activity Catalysts for Olefin Polymerization

FREDERICK J. KAROL
UNIPOL Systems Department
Union Carbide Corporation
Bound Brook, New Jersey

XI.  CATALYSIS AND THE POLYETHYLENE
     REVOLUTION

XII. CONCLUDING REMARKS

     REFERENCES


I.  INTRODUCTION

Catalysis continues to play a vital role in polymerization of
such olefins as ethylene and propylene.  A voluminous patent
and scientific literature describing transition metal catalysts for
olefin polymerization has emerged since the original discoveries
by Ziegler, Natta, and other workers [1-6].  Significant progress in polymerization catalysis has been made in the last 15
years, particularly with the development of methods to increase
the efficiency of transition metal catalysts in olefin polymerization.  Success in this area has provided the basis of simplified,
less costly plant operations which do not require removal of
residual catalyst from the polymer [3-9].

Although much attention has been directed toward improvements in catalyst productivity, there has been and continues to
be much active research on other important features of olefin
polymerization.  The specific nature of the catalyst has an important effect on polymer molecular weight, polymer molecular
weight distribution, copolymerization kinetics, and degree of
stereoregularity.  Moreover, the size, shape, and porosity
(morphology) of the catalyst particle play an important role in
regulating the morphology of the resultant polymer [6-9].  Many
modern day industrial catalysts show attractive behavior in most
or all of the areas described above.

While significant technological advances based on new and improved catalysts have been made since the original discoveries,
there is still a widely held view that understanding of these catalysts is meager and empirical.  It is true that olefin catalytic
processes are not generally understood in a detailed mechanistic
sense.  One cannot predict a priori the behavior of a particular
catalyst composition or the specific characteristics of the polymer produced by the catalyst.  Nevertheless, significant progress has been made through studies with a wide variety of high
activity catalysts.  Certain general requirements have been identified to be critical in polymerization.  This paper reviews segments of current catalyst research activities in olefin polymerization and discusses some current perspectives in these research
areas.

## II. TYPES OF HIGH ACTIVITY
## CATALYSTS

Developments toward higher activity ($\geq$200 kg polymer/g Ti versus 1-5 kg polymer/g Ti) Ziegler-Natta catalysts have been based in large part on reactions of specific magnesium, titanium, and aluminum compounds [2, 3, 6, 10]. Some of the early impetus in this area was provided by catalysts, chemically anchored on Mg(OH)Cl supports [11]. Other studies concentrated on the use of $MgCl_2$ as a substrate [12-14]. Grinding of $MgCl_2$, and treatment with $TiCl_4$, provided one route to a higher surface area substrate of magnesium and titanium. Some developments focused on reaction products of magnesium alkyls and titanium compounds [15]. Other workers described the advantages of preparing trimetallic sponges by the addition of certain aluminum compounds to a magnesium substrate which had been treated with a titanium compound [16]. Catalysts based on reaction of magnesium alkoxides with transition metal compounds have also received attention [17]. Chemical complexes consisting of magnesium and titanium compounds with electron donors also serve to catalyze olefin polymerization [18].

Attractive, high activity catalysts for propylene polymerization have been described. Patents to Montedison/Mitsui discuss catalysts comprising an aluminum alkyl compound with an electron donor such as ethylbenzoate, and a solid matrix containing the reaction products of halogenated magnesium compounds with a Ti(IV) compound and an electron donor [12, 13]. The specific surface area of the solid matrix after treatment can be in the range of 100-200 $m^2$/g. High catalyst productivities based on titanium have been reported for polymerization with these catalysts. Some activities have centered on treatment of the solid matrix with metal halides to improve the polymerization activity and/or stereoregularity in isotactic polypropylene polymerization [14].

Catalyst research with transition metal oxides supported on refractory metal oxides, e.g., $CrO_3/SiO_2$ for ethylene polymerization [19] continues to be quite active with attention devoted to factors involved in chemical anchoring of $CrO_3$ to the silica surface [20, 21] and to the effect of silica surface modifications on the nature and behavior of active sites [22, 23]. The direct use of organometallic compounds of transition metals for preparation of solid catalysts for ethylene polymerization represents a third type of high activity catalyst [2, 10, 24]. Research with this catalyst type has recently been well-documented [2] and active research continues in this area.

### III. PREPARATIVE METHODS TO HIGH
### ACTIVITY CATALYSTS

The efficiency of olefin polymerization catalysts can be improved by several methods. Usually the data available in patents do not provide sufficiently detailed information to develop a rigorous classification relating to catalytic efficiency. Nevertheless, certain features of these catalysts have been described to allow a classification to be made based on operational methods used to prepare the catalysts. These classes are distinguished by the mode of formation of a precursor composition containing a transition metal compound. These operational classes for compounds of transition metals include chemical anchoring to the surface of a substrate [2, 11, 19-25], formation of bimetallic complexes [18, 26-29], insertion into defects of a substrate [6, 12, 13, 30-33], formation of high surface area sponge [17, 34-37], and formation of solid solutions by cocrystallization [15, 38-40]. Illustrative examples of each of these classes are contained in Table 1.

The operational methods used to improve catalyst productivities appear to have a common basis for their effectiveness. Those techniques that provide the opportunity to expose and isolate the maximum amount of transition metal compound, either in a solid matrix or a bimetallic complex, have been highly effective. Methods that assist in stabilizing exposed and isolated transition metal centers can provide an added boost to the polymerization efficiency of many catalysts.

### A. Chemical Anchoring to Surface of Substrate

In chemical anchoring a chemical reaction takes place between surface groups on the substrate (silica, magnesium hydroxychloride) and the transition metal compound to form a new surface composition (Eqs. 1-3). Chemical attachment of the compound to the support provides an anchoring device preventing destruction of potentially active sites by mutual interaction. The temperature to which the support is heated can have a significant effect on the polymerization activity of the catalyst [19, 25, 41].

$$\text{(1)}$$

## TABLE 1

### Operational Methods to High Activity Catalysts

1. Chemical Anchoring to Surface of Substrate

   a) $TiCl_4/Mg(OH)Cl$     b) $CrO_3/SiO_2$     c) $(C_5H_5)_2Cr/SiO_2$

2. Formation of Bimetallic Complexes

   a) $MgCl_2 + 2TiCl_4 + 8POCl_3 \rightarrow [Ti_2Cl_{10}]^{2-}[Mg(POCl_3)_6]^{2+} \cdot 2POCl_3$

   b) $2MgCl_2 + TiCl_4 + 7THF \rightarrow [TiCl_5(THF)]^-[Mg_2Cl_3(THF)_6]^+$

   c) $MgCl_2 + TiCl_4 + 4CH_3CO_2C_2H_5 \rightarrow TiMgCl_6(CH_3CO_2C_2H_5)_4$

3. Insertion into Defects of Substrate

   a) $MgCl_2 + TiCl_4 +$ ethyl-p-toluate $\xrightarrow{\text{ball-milling}}$

   b) $MgCl_2 + TiCl_4 \cdot$ dioxane $\xrightarrow{\text{ball-milling}}$

4. Formation of High Surface Area Sponge

   a) $TiCl_4 + Et_2AlCl \longrightarrow \beta\text{-}TiCl_3 \cdot xEtAlCl_2$
   
   $[TiCl_3 \cdot (EtAlCl_2)_{0.03}(\text{ether})_{0.01}] \xleftarrow[\text{treated solid}]{TiCl_4} \Big\downarrow \text{isoamyl ether}$

   b) $Mg(OEt)_2 + TiCl_4 \rightarrow [MgCl_2 \cdot Mg(OEt)_2 \cdot Mg(TiCl_6)]$

   c) $Mg(OEt)_2 + Ti(O\text{-}nC_4H_9)_4 + EtAlCl_2 \rightarrow$ trimetallic sponge

5. Formation of Solid Solutions by Cocrystallization (Coprecipitation)

   a) $EtMgCl + TiCl_4 \rightarrow TiCl_3 \cdot MgCl_2 +$ organic fragments

High surface area (>100 $m^2$/g) silica supports have proven particularly attractive with chromium catalysts, probably because such
high surface areas provide an efficient means of isolating active
centers. At a 30-100 ppm chromium loading with the $CrO_3/SiO_2$
catalyst, chromium efficiency was at a maximum [19]. Each chromium atom at this loading was assumed to be an active site.

## B. Formation of Bimetallic Complexes

Bimetallic complexes containing at least one atom of magnesium,
manganese, or calcium, and at least one atom of titanium, vanadium,
or zirconium, when combined with organometallic compounds of the
metals belonging to one of Groups I-III of the Periodic Table, are
claimed as catalysts, highly active in the polymerization of olefins
[26]. The complexes have the general formula $M_mM'X_{2m}Y \cdot nE$,
wherein

M   = Mg, Mn, Ca
m   = a number from 0.5 to 2
M'  = Ti, V, Zr
X   = Cl, Br, I
Y   = one or more atoms or groups of atoms selected from atoms of
      halogen, $-NR_2$, $-OR$, $-SR$, and others of a similar nature
n   = a number from 0.5m to 20m
E   = electron donor compound selected from classes of compounds
      which include esters, alcohols, ethers, amines, nitriles, and
      certain phosphorus compounds

Similar bimetallic complexes of the general formula $MM'TY \cdot nE$,
wherein the symbols have the significance indicated above, and
T = oxygen, a carbonate anion, or a certain pair of monovalent
groups, have been shown to provide, on activation, highly active catalysts [27].

These bimetallic halides can be prepared by reacting the halide
$MX_2$ (or MT) with the compound M'Y at a temperature of from 25-
150°C in electron donor solvents E (see Table 1). Unlike the surface complexes formed by chemical anchoring to a substrate, bimetallic complexes are well-defined compounds with characteristic properties. The simplicity of these bimetallic complexes offers the opportunity of identifying basic requirements for reaching high polymerization activity.

The product of interaction between the components $TiCl_4$, $MgCl_2$,
and $CH_3COOC_2H_5$, namely $MgTiCl_6(CH_3COOC_2H_5)_4$, which is illustrated below, has been investigated by means of single-crystal,
x-ray diffraction methods [29].

The molecular structure of $MgTiCl_6(CH_3COOC_2H_5)_4$ may be described in terms of a slightly deformed bioctahedron. In formation of this bimetallic complex, the $TiCl_4$ acceptor molecule has changed its coordination from tetrahedral to octahedral. The $MgCl_2$ has retained its octahedral coordination.

In one study [18] bimetallic complexes were obtained by reacting $TiCl_4$, $VOCl_3$, $MoOCl_4$, $WOCl_4$, or $AlCl_3$ with Be, Mg, Ca, or Sr chlorides in the presence of electron donors such as $POCl_3$ (L) or $C_6H_5POCl_2$ (L'). These complexes showed well-defined stoichiometry, ionic character, and crystalline structure. Complexes $(TiCl_6)MgL_6$, $(TiCl_5L')_2MgL_6'$, and $(Ti_2Cl_{10})MgL_6$ treated with $(i-C_4H_9)_3Al$ were found very active in ethylene polymerization, but completely unable to polymerize propylene. Conclusions about the role of magnesium ions in these catalysts, and other high-activity catalysts, require discussion:

Titanium centers may be diluted by magnesium ions, which influence the number of active centers. This dilution process can take place in several ways, including bimetallic complex formation, formation of solid solutions by cocrystallization, and chemical anchoring to the surface of a substrate, which may not involve magnesium ions.

The presence of magnesium ions can stabilize active titanium centers from deactivation processes. The type of ligand on the magnesium ion influences the magnitude of such stabilization.

The presence of magnesium ions enhances chain transfer processes since $M_n$ of HDPE decreases when Mg/Ti ratio increases.

The presence of magnesium ions leads to catalysts which provide polyethylenes with a narrow molecular weight distribution ($\overline{M}_W / \overline{M}_n \sim 3\text{-}5$).

The similarity in size, coordination preference, electronic structure, and electronegativity of Ti(IV), Mg(II), and Al(III) ions is reflected in structural parameters and chemical properties (Table 2) [42]. All three metal ions have an inert gas electron number, but differ in their formal charge. The similarity in size between Mg(II) and Ti(IV) and the common electronegativity and size/ charge ratio for Al(III) and Ti(IV) permit an easy substitution between metal ions in a catalyst framework [43]. All three metal ions should fit equally well into the interstices between chloride anions in a crystalline lattice. The metal ions can react to form organometallic compounds which can be stabilized by formation of adducts with electron donors. Ligand exchange and substitution of one metal ion species for another are common.

Chloride ion plays an important role in many high-activity Ziegler-Natta catalysts. Chloride ion permits facile bridge formation, thus providing the "cement" between the metal centers of the catalyst framework. Alkoxide bridges probably introduce significant steric constraints and decrease catalyst crystallinity and uniformity. Replacement of a bulky alkoxide by a chloride ion may also lead to vacancies in the catalyst lattice and increase the effective surface area of the catalyst particles.

## C. Insertion into Defects of Substrate

Several studies have examined the nature and performance of high activity $MgCl_2/TiCl_4$ precursors prepared by a milling method [6, 44-47]. The primary effect of milling is to break the layered structure of the magnesium chloride crystal. More specifically, the structural variations introduced by ball milling are mainly associated with rotational disorder of the Cl-Mg-Cl triple layers. This breakage into crystallites around 10 nm occurs along the weakly bonded layers. With prolonged grinding times for $MgCl_2$, each peak in the x-ray diffraction pattern becomes progressively lower and broader, indicating crystallite size becomes smaller and smaller during grinding [45].

The rate of lowering and broadening of x-ray diffraction peaks of $MgCl_2$ milled with $TiCl_4$ was found to occur more rapidly as compared with milling of pure $MgCl_2$ [45]. The process of reduction in the crystallite dimensions of $MgCl_2$ during milling was greatly accelerated. The $TiCl_4$ diffuses to the inner layers of the $MgCl_2$ structure and through this diffusion process makes the cleavage of the $MgCl_2$ crystal easier. By this process the $TiCl_4$ adsorbed on the surface of $MgCl_2$ is in a state of high dispersion. When ethylene molecules enter into the internal voids of the $MgCl_2$ substrate and polymerize, the particle disintegrates into individual crystallites. The crystallite size can be directly correlated with the activity of the catalyst.

TABLE 2

Geometric and Electronic Properties of Ions of Catalyst Components

| Ion | Radius (nm) | Size/charge | Pauling electronegativity | Electronic structure | Coordination number | Geometry |
|---|---|---|---|---|---|---|
| Ti(IV) | 0.068 | 0.017 | 1.5 | $3s^2 3p^6 3d^0$ | 6 | Octahedral |
| | | | | | 4 | Tetrahedral |
| | | | | | 5 | Distorted trigonal bipyramidal |
| Mg(II) | 0.065 | 0.033 | 1.3 | $2s^2 2p^6$ | 4 | Tetrahedral |
| | | | | | 6 | Octahedral |
| | | | | | 5 | - |
| Al(III) | 0.050 | 0.017 | 1.6 | $2s^2 2p^6$ | 4 | Tetrahedral |
| | | | | | 6 | Octahedral |
| | | | | | 5 | - |
| Cl(-) | 0.18 | 0.18 | 3.2 | $3s^2 3p^6$ | 1 | - |
| | | | | | 2 | Bent-bridging group |

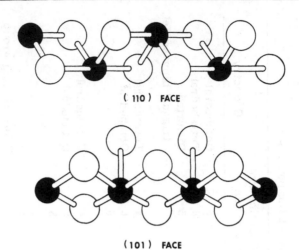

( 110 ) **FACE**

(101) **FACE**

FIG. 1. Faces of magnesium chloride after ball milling [46]:
O = Cl, ● = Mg.

Activated magnesium halides would appear to be ideal substrates
for titanium halides. The coordinatively unsaturated magnesium
ions located at the side surfaces and crystal edges can easily bond
with titanium halides. On the more probable cleavage surfaces of
the crystals, the (110) and (101) faces, the magnesium atoms are
coordinated with four and five chlorine atoms, respectively (Fig.
1). Strongly bonded surface complexes involving halogen bridges
between titanium and magnesium can form (Fig. 2). The presence
of electron donor molecules can enhance the stability of these surface complexes. It is possible that interaction of $TiCl_4$ with activated $MgCl_2$ to form surface complexes proceeds with formation of
a $Mg[TiCl_6]$-type ionic complex, similar to the $K_2[TiCl_6]$ complex
[44, 48].

## D. Formation of High Surface Area Sponges

Previously most catalyst systems based on $TiCl_3$, and used commercially, consisted of partially reduced $TiCl_3$ cocrystallized with
$AlCl_3$. Precise composition and extent of cocrystallization depend
upon the method of preparation and heat treatment. Usually the
formula $TiCl_3 \cdot 0.33AlCl_3$ is used for convenience. The effect of
dry milling the $TiCl_3$ to increase polymerization activity has been
an important development [49].

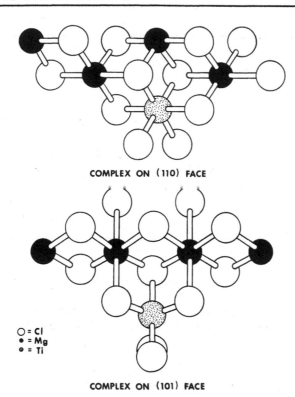

COMPLEX ON (110) FACE

O = Cl
• = Mg
• = Ti

COMPLEX ON (101) FACE

FIG. 2. Possible complexes of $TiCl_4$ on faces (110) and (101) of $MgCl_2$ [46].

The discovery of ether-treated $TiCl_3$-based catalysts of high activity and stereospecificity in propylene polymerization provided a novel route to monometallic $TiCl_3$ catalysts with surface areas >75 $m^2/g$ [35]. The catalysts can be prepared in a three-step process (Table 1). In a first step $TiCl_4$ is reduced by organoaluminum compounds such as $Et_2AlCl$. The reduced solid consists of a β-$TiCl_3$-based composition containing aluminum compounds which have halogen and/or hydrocarbon fragments. The specific surface area of this reduced solid is about 1 $m^2/g$, and the catalytic properties are not impressive.

In a second step, treatment of this β-$TiCl_3 \cdot 0.33AlCl_3$ with diisoamyl ether provides a means to remove the $AlCl_3$ from the solid matrix. The treated solid has a physical form similar to that of the reduced solid and a similar surface area. Catalytic properties

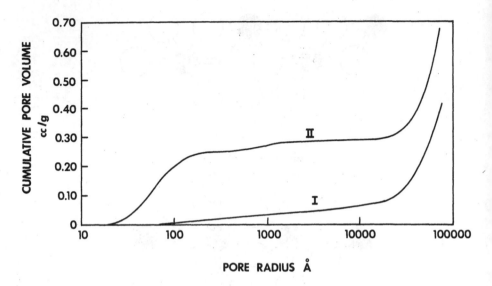

FIG. 3. Cumulative pore volume of TiCl₃ particles versus radius for treated solid (I) and catalytic complex (II).

of this treated solid, like those of $\beta$-TiCl₃·0.33AlCl₃ are not impressive.

In the third step the treated solid is reacted with TiCl₄ near 65°C to give a product of composition $\delta$-TiCl₃·(AlR$_n$X$_{3-n}$)$_x$·(C)$_y$, where C is the complexing agent, diisoamyl ether, and $0 \leq n \geq 2$, $x < 0.3$, and $0.11 > y > 0.009$. This catalytic complex of $\delta$-TiCl₃ has a high surface area which is usually >150 m²/g. In the presence of Et₂AlCl, this catalytic complex has about a fivefold higher activity than the conventional TiCl₃ catalysts, and produces only 2-5% of amorphous polypropylene.

Certain detailed studies have described the effect of the reaction of TiCl₄ on the treated solid to form the catalytic complex. In Fig. 3 [35] there are two main zones that can be distinguished. One zone corresponds to pores the radii of which are smaller than 15,000 Å. The other zone, corresponding to pores of larger radii, was related to the volume of spaces between spherical particles. For the treated solid the total pore volume of the spherical particles is small and comprises only the spaces between the microparticles constituting the spherical particles. In the catalytic complex the high value of the total volume of the spherical particles is due to a particular, cellular structure (sponge) forming a network.

Highly active, bimetallic sponge catalysts based on magnesium-titanium precursors can be obtained by reaction of magnesium alkoxides with tetravalent titanium halides (Table 1) [17, 34-36]. During the course of precursor preparation, the original structure is changed as the magnesium and titanium compounds are firmly linked as new crystalline species. The new complexes are probably a mixture of magnesium chloride, magnesium alkoxide, and magnesium-titanium complexes. The final precursor (catalyst) has a much higher surface area than the original magnesium alkoxide. The most interesting, activated solids have specific surface areas greater than 40 $m^2/g$ and preferably higher than 100 $m^2/g$. These high and unexpected surface areas are believed to be a major reason for the very high activity of catalysts prepared from the activated solids.

Another step in the development of high surface area catalysts relates to the preparation and use of trimetallic sponges based on magnesium-titanium-aluminum (or silicon) compounds (Table 1) [39]. These trimetallic sponges appear attractive in ethylene polymerization as a route to provide polyethylenes with superior impact resistance. The maximum catalytic activity in ethylene polymerization was observed using trimetallic sponges with the highest surface areas (273-339 $m^2/g$) and the lowest titanium concentration (0.34-0.41 $mg/m^2$). These results illustrate that it is advantageous to isolate titanium centers from each other in the preparation of these complexes.

### E. Formation of Solid Solutions by Cocrystallization

In traditional Ziegler-Natta catalysts, reduction of $TiCl_4$ by an organoaluminum compound ultimately leads to a cocrystallized product of $AlCl_3$ and $TiCl_3$,

$$3TiCl_4 + R_3Al \rightarrow 3TiCl_3 \cdot AlCl_3 + \text{organic fragments} \qquad (4)$$

with a typical content of one-third $AlCl_3$ for one $TiCl_3$. Likewise, the reduction of $TiCl_4$ by an organomagnesium compound leads to a solid solution of $TiCl_3$ in $MgCl_2$ [15, 16, 38],

$$TiCl_4 + RMgCl \rightarrow TiCl_3 \cdot MgCl_2 + \text{organic fragments} \qquad (5)$$

Cocrystallization of $MgCl_2$ with $TiCl_3$ leads to a disordered structure. This disordering can be verified using x-ray diffraction data by comparison of the $TiCl_3 \cdot MgCl_2$ solution with the $\beta$-$TiCl_3$ structure. These titanium-magnesium complexes, when treated with an organoaluminum compound, show high polymerization

activity. The high polymerization activity has been ascribed to
the disordered structure of $TiCl_3 \cdot MgCl_2$ [15].

In one study involving the preparation of cocrystallized $TiCl_3 \cdot$
$MgCl_2$, no identifiable x-ray lines could be assigned to either
$MgCl_2$ or to $TiCl_3$, which indicated a very small crystallite size
must be present. A transmission electron micrograph showed a
very fine nodular structure for the organomagnesium-reduced
titanium composition. Apparently the primary particle size was
<50 nm [15].

Solid solutions of binary chlorides have been obtained by reacting $TiCl_4$ with $V(CO)_6$, $Cr(CO)_6$, $Mn_2(CO)_{10}$, $Mn(CO)_5Cl$,
$Ni(CO)_4$, $CO_2(CO)_8$, $Fe(CO)_5$, or $Mn(CO)_6$ [39, 40]. The reactions yield mixed chlorides of the general formula $MCl_n \cdot nTiCl_3$
where n = 2 or 3 and M is a divalent or trivalent transition metal
cation:

$$nTiCl_4 + M(CO)_x \rightarrow [M(CO)_x \cdot nTiCl_4] \rightarrow MCl_n \cdot nTiCl_3 + xCO \quad (6)$$

The binary chlorides, prepared as indicated in Eq. (6), have been
characterized by elemental analysis, spectroscopic investigations,
x-ray analysis of the powders, and surface area measurements.

Crystallographic evidence, together with compositional data
and electronic spectra, showed that binary chlorides were present as solid solutions of the two constituent chlorides. The solid
solutions ($FeTi_2Cl_8$, $MnTiCl_5$), displaying lower crystallinity, exhibited high surface area values (113-163 $m^2/g$). The disorder
of crystal structure and surface area values seem related to the
ionic radius of the metals with respect to that of Ti(III). The
largest surfaces were found for the binary chlorides of Fe (ionic
radius 0.078 nm) and Mn (ionic radius 0.082 nm).

Catalyst efficiencies two or three times higher than that of
$AlCl_3 \cdot 3TiCl_3$ were observed in polymerization of ethylene when
some of the binary chlorides described above were activated with
$(i-C_4H_9)_3Al$ (Table 3). However, it was difficult to relate overall catalyst efficiencies to the surface area of the binary chlorides. This difficulty has been attributed to different kinetic profiles during polymerization, due to an instability of active centers,
which depend on the metal chloride associated with $TiCl_3$. Actually the $(MgCl_2)_{1.5} \cdot TiCl_3$-based catalyst is much more active
than all the other binary chlorides. This influence of $MgCl_2$ was
not adequately explained, although the relatively inert behavior
of $MgCl_2$ with aluminum alkyls was offered as a possibility for the
greater active site stability.

## TABLE 3

Catalytic Activity of Systems Based on Binary
Chlorides Containing $TiCl_3$ [39, 40][a]

| Binary chloride, composition | Catalyst efficiency, kg PE/g Ti | Specific activity, relative to $AlCl_3 \cdot 3TiCl_3$ |
|---|---|---|
| $MoCl_3 \cdot TiCl_3$ | 1.0 | 0.09 |
| $NiCl_2 \cdot 2TiCl_3$ | 4.2 | 0.38 |
| $CrCl_3 \cdot 3TiCl_3$ | 9.0 | 0.82 |
| $CoTi_{1.6}Cl_{6.6}$ | 19.1 | 1.74 |
| $MnCl_2 \cdot 2TiCl_3$ | 22.6 | 2.05 |
| $FeCl_2 \cdot 2TiCl_3$ | 24.0 | 2.18 |
| $MnCl_2 \cdot TiCl_3$ | 28.1 | 2.55 |
| $VTi_{1.1}Cl_{6.7}$ | 24.1 | 2.65 |
| $AlCl_3 \cdot 3TiCl_3$ | 11.0 | 1.00 |
| $(MgCl_2)_{1.5} \cdot TiCl_3$ | 132 | 12 |

[a]Experimental conditions: n-hexane = 1000 $cm^3$; $[(i-C_4H_9)_3Al]$ =
40 mmol/L; [Ti] = 0.5-2 mmol/L; T = 85°C; $C_2H_4$ = 5 atm; $H_2$ = 5
atm; time = 2 h.

## IV.  ACTIVE SITE CONCENTRATIONS

The determination of the concentration of active centers in olefin polymerization systems has received considerable attention [2,
5, 51-53]. Knowledge of active center concentrations provides the
basis for evaluating absolute values of propagation rate constants.
Approaches that have been used for measuring the concentration
of active sites have been based on (a) kinetic and molecular weight
data; (b) quenching reagents such as radioactive iodine, tritiated
or deuterated alcohols or water; and (c) radio-tagging methods
using [14]C-labeled aluminum alkyl compounds. Later, the use of
catalyst poisons such as [14]CO and [14]CO$_2$ has received considerable attention. For reaction with [14]CO a polymerization reaction

is quenched by the addition of labeled CO. The chemical reactions postulated to occur are

$$
-M\text{-}R + {}^{14}CO \longrightarrow -M\text{-}P \longrightarrow -M\text{-}{}^{14}CO\text{-}P \xrightarrow{ROH} H\text{-}{}^{14}CO\text{-}P \tag{7}
$$

where ☐ = vacant coordination site
     P   = growing polymer chain

    Results of experiments to determine the number of active centers (Cp) have indicated that for first generation Ziegler-Natta catalysts only a small fraction of the total amount of transition metal compound is catalytically active at any time. Generally for titanium catalysts of low productivity, Cp values range from $10^{-2}$ to $10^{-4}$ mol/mol titanium compound. Titanium catalysts of higher productivity (second/third generation catalysts) often show higher Cp values of $7 \times 10^{-1}$ to $10^{-2}$ mol/mol titanium compound (Tables 4 and 5) [51-53].

    The large changes in polymerization activity using high activity catalysts are mainly due to an increase in the number of active sites (Table 4). In propylene polymerization the presence of $MgCl_2$ in the catalyst has led to an increase in both the number of active sites and the propagation rate constant ($k_p$) (Table 5) [46, 52, 53]. Interestingly, the ratio of $Cp^i/Cp^a$ (isotactic active centers/atactic active centers) for propylene polymerization catalysts did not change very much with catalyst type. Likewise the $k_p$ values for the isotactic ($k_p^i$) and atactic active centers ($k_p^a$) were similar, with the exception of the catalyst containing ethylbenzoate (EB).

    Active centers for catalysts based on chromium and zirconium fall in the same concentration range as active centers in low and high activity titanium catalysts. Comparative data in Table 4 on the effect of ligands on the ethylene propagation rate constants of chromium and zirconium-based catalysts reveal the effect of changing the electron density of the active transition metal center on the propagation rate constants. The propagation rate constant for supported chromium catalysts increase by one order of magnitude by changing from an oxide to a cyclopentadienyl ligand. Even in the family of titanium catalysts the propagation rate constant for the $(\phi CH_2)_4 Ti/Al_2O_3$ catalyst is one order of magnitude lower than the titanium halide/aluminum alkyl catalysts.

TABLE 4

Active Centers and Propagation Rate Constants in Ethylene Polymerization [51]

| Catalyst | $C_p(max)^a$ X $10^3$ mol/mol·M | $k_p$ X $10^{-3}$ L/(mol·s) (80°C) | Maximum activity (g/mmol·M·h·atm) | Polymerization temperature (°C) |
|---|---|---|---|---|
| TiCl$_2$ | 0.15 | 12 | 11 | 90 |
| δTiCl$_3$·0.3AlCl$_3$ + Et$_3$Al | 1.2 | 12 | 80 | 80 |
| TiCl$_4$/MgCl$_2$ + Et$_3$Al | 400 | 12 | 25,000 | 80 |
| Ti(CH$_2$φ)$_4$/Al$_2$O$_3$ | 70 | 1.0 | 300 | 200 |
| CrO$_3$/SiO$_2$ | 120 | 0.7 | 300 | 80 |
| (C$_5$H$_5$)$_2$Cr/SiO$_2$ | 10 | 9.0 | 50 | 80 |
| (φ$_3$SiO)$_2$CrO$_2$/SiO$_2$/Et$_3$Al | - | - | 300 | 80 |
| Cr(C$_3$H$_5$)$_3$/SiO$_2$ | 100 | 2.9 | 1,500 | 80 |
| Zr(C$_3$H$_5$)$_4$/SiO$_2$ | 400 | 0.2 | 50 | - |
| Zr(C$_3$H$_5$)$_4$/Al$_2$O$_3$ | 20 | 2.2 | 80 | 200 |
| Zr(C$_3$H$_5$)$_3$X/SiO$_2$ (X = Cl, Br, I) | - | - | 350-1,000 | 80 |

$^a$$C_p$(max) designates the number of propagation centers corresponding to the maximum polymerization rate; M = transition metal compound.

TABLE 5

Active Centers and Propagation Rate Constants for Isotactic ($Cp^i$ and $kp^i$) and Atactic ($Cp^a$ and $kp^a$) [52, 53]

| Catalyst | Rate, g $C_3H_6$/g Ti·h | $Cp^i \times 10^3$, mol/mol Ti | $Cp^a \times 10^3$, mol/mol Ti | $kp^i \times 10^{-2}$, L/(mol·s) | $kp^a \times 10^{-2}$, L/(mol·s) |
|---|---|---|---|---|---|
| | | | Polymerization of propylene | | |
| $TiCl_4$ + $Et_3Al$ | 40 | 0.04 | 0.05 | 1.0 | 1.1 |
| $TiCl_4$ + $Bu_2Mg$ | 500 | 0.11 | 0.15 | 4.1 | 5.0 |
| $TiCl_4 \cdot EB$ /$MgCl_2$ + $Et_3Al$ | 10,000 | 2.7 | 1.3 | 8.7 | 3.3 |
| $TiCl_4$/$MgCl_2$ + $Et_3Al$ | 27,000 | 5.2 | 4.1 | 7.4 | 6.2 |
| Highly active supported catalyst ($MgCl_2$/$TiCl_4$/$R_3Al$/electron donor) [46] | 35,000 | 75 | - | 5.0 | - |

Measurements which show active centers as only a small fraction of the total amount of transition metal have led to a widely accepted view that these centers must have a special chemical or spatial environment. Such an outlook seems reasonable for heterogeneous catalysts of low activity. However, the fraction of active centers in many high activity polymerization catalysts may be quite large. A monolayer distribution of active components offers the potential for all transition metal centers to be active. This does not mean that all centers work simultaneously with the same efficiency. Growth of polymer chains at some active centers could prevent neighboring centers from participating in the polymerization reaction. However, the choice for each exposed center to act as an active center should be equal [45].

## V.  LIGAND EFFECTS ON CATALYTIC BEHAVIOR

Ligand environment at the transition metal center can exert a significant effect on polymerization behavior and the nature of the polymer produced [54-56]. Studies comparing titanium tetrachloride with titanium alkoxides have claimed that polyethylenes of narrower molecular weight distribution were obtained with titanium alkoxides were used [57]. For polymerization of isotactic polypropylene the preferred ligand is chlorine, particularly in combination with Ti, V, Cr, or Nb as in $TiCl_3$, $VCl_3$, $CrCl_3$, and $NbCl_5$ [1]. In catalysts used for the polymerization of conjugated dienes, the ligand that is attached to the transition metal plays an important role in determining to what degree a particular stereochemical structure is formed. The nature of the ligand may also determine if a particular catalyst will become soluble or heterogeneous.

Comprehensive studies have evaluated the effect of $\pi$-bonded organic ligands attached to chromium on polymerization parameters of several chromium alkyl-free catalysts [25, 56, 58]. In these studies (Table 6) differences in hydrogen response, comonomer incorporation, and polymerization activity with these catalysts suggested that the nature of the active sites was different due, at least in part, to changes in ligand environment. In these supported catalysts with $\pi$-bonded ligands it appears that at least one of these ligands (L) remains coordinated to the active chromium center during the polymerization process:

$$Cr-R + nCH_2=CH_2 \xrightarrow{k_p} Cr-(CH_2-CH_2)-_n R \qquad (8)$$

TABLE 6

Supported Chromium Catalysts with Different Ligands

| Catalyst | Ligand | Valence state of chromium[a] |
|---|---|---|
| $CrO_3/SiO_2$ | Oxide | 6 |
| $(C_3H_5)_2Cr/SiO_2$ | Allyl | 2 |
| $(C_5H_5)_2Cr/SiO_2$ | Cyclopentadienyl | 2 |
| $(C_9H_7)_2Cr/SiO_2$ | Indenyl | 2 |
| $(C_9H_{12})_2Cr/SiO_2 \cdot Al_2O_3$ | Cumene | 0 |
| $(C_{13}H_9)_2Cr/SiO_2$ | Fluorenyl | 2 |
| $(C_{14}H_{11})_2Cr/SiO_2$ | 9-Methylfluorenyl | 2 |

[a]Valence state of chromium refers to the initial chromium compound used to prepare the catalyst, and not necessarily to the active valence state.

$$\text{Cr-(CH}_2\text{-CH}_2)_n\text{R} + \text{H}_2 \xrightarrow{k_h} \text{Cr-H} + \text{H-(CH}_2\text{-CH}_2)-_n\text{R} \quad (9)$$

The decrease in the capability of cyclopentadienyl chromium-containing active centers to produce lower molecular weight polyethylenes in the presence of hydrogen was observed in the following series of ligands: cyclopentadienyl > indenyl > fluorenyl ≥ 9-methylfluorenyl. This decrease in hydrogen response is probably related to the decreased electron density on the chromium ion in the ligand series above.

Other studies with supported metal alkyl-free catalysts have shown that the molecular weight of polyethylene decreased with an increase in the electron density on the metal ion [51]. Polymerization studies with silica-supported $(C_3H_5)_2ZrX_2$ and $(C_3H_5)_3ZrX$, where X = Cl, Br, I, showed that polymer molecular weight decreased in the order Cl > Br > I. Furthermore, the introduction of halide into the composition of active centers led to a substantial increase in catalytic activity.

The $CrO_3/SiO_2$ polymerization catalysts become more active by
the incorporation of a small amount of titania either in or on the
support [23]. This presence also enhances the chain transfer
rate, thereby providing polyethylenes of lower molecular weight.
The promotional effect probably originates from formation of Ti-
O-Cr links which change the electronic environment on the chromium active center.

## VI.   FACTORS IN CONTROL OF MOLECULAR
## WEIGHT DISTRIBUTION

Considerable experimental data support the view that a dispersity of chemically distinct active species is primarily responsible for explaining the range of molecular weight distributions
($\overline{M}_W/\overline{M}_n = Q$) measured in ethylene ($Q = 2-50$) and propylene polymerizations ($Q = 5-12$) [59]. Among the factors that could affect
the dispersity of active species are the specific transition metal
compound including ligand environment and oxidation state, the
type of cocatalyst and electron donor used to generate the catalytically active species, the physical state of the catalyst, and
the nature of the catalyst substrate.  Changes in molecular
weight distribution brought about by diffusion limitations do not
provide an adequate explanation for the broad range of molecular weight distributions encountered with different catalyst systems.  Predictions of the diffusion theory of the effects of polymerization activity are not in accord with experimental data.
Table 7 contains a list of experimental observations in support
of the proposal of chemical factors being responsible for changes
in MWD.
The type of transition metal can have a pronounced influence
on polymer molecular weight distribution.  Numerous examples of
the effect of transition metal compound on molecular weight distribution may be found, particularly in the patent literature [60-
65].  Combinations of two or more transition metal compounds frequently provide a route to polymers of broad MWD, suggesting
the active sites from each transition metal compound gives rise to
its own specific MWD.
Ligands attached to the transition metal center can have a significant effect on polymerization behavior including MWD.  The
polydispersity of polyethylenes obtained with $TiCl_3-TiX_4-RAlX_2$
was found to narrow progressively as the X groups (X = chlorine
atom) were substituted with alkoxide groups [66].  With some
$MgCl_2$-supported compositions based on $TiX_4$ (X = $NR_2$, OR, Cl)
and i-$C_4H_9)_3Al$ as cocatalyst, the breath of the polyethylene MWD

TABLE 7

Experimental Support for Chemical Factors in Control of
Molecular Weight Distribution (MWD) [59]

---

Different transition metal catalysts can provide large changes in
polymer MWD

Chemical modifications of catalyst changes MWD

Heterogeneous catalyst can produce broad MWD, even when polymer
is in solution

Homogeneous, soluble catalysts provide narrow MWD, even when
polymer is insoluble in medium

Higher activity catalysts do not necessarily provide broader MWD

Electron donors, acting as poisons, can broaden MWD

---

increased with an increase in the electron acceptor power of the
ligand in the following manner [59]:

$$Ti(NEt_2)_4 < Ti(Ot\text{-}Bu)_4 < Ti(On\text{-}Bu)_4 \leq Ti(OEt)_4 < Ti(OPh)_4 < TiCl_4$$

With high activity catalysts, very little data have been published
on the effect of aluminum alkyls as cocatalysts on MWD. In many
cases changes in the specific aluminum alkyl cocatalyst did not
have any noticeable influence on the MWD. One study [67] did
find when using a $TiCl_4$-$MgCl_2$ catalyst that an increase in alkyl
chain length ($R_3Al$) from triethyl aluminum to trioctyl aluminum
led to a broader MWD (Q = 6 to Q = 12).

Use of $MgCl_2$ substrates has provided much of the impetus for
reaching high activity with Ziegler-Natta catalysts. Several research groups have shown that the presence of $MgCl_2$ in high activity, titanium-based catalysts leads to polyethylenes with a narrow molecular weight distribution [18, 62, 68]. Addition of third
components such as Lewis bases to high activity catalysts can have
different effects on polymer MWD, depending on the particular catalyst. There has been a hint that certain electron donors added
to high activity polypropylene catalysts can significantly alter
MWD [59].

Although research into the factors controlling molecular weight
distribution has received considerable attention, no comprehensive

theory has evolved that satisfactorily explains or predicts the molecular weight distribution of an olefin polymer with any level of
certainty. Analytical techniques for measuring the degree of het-
erogeneity in solid catalysts have not proven sufficiently reliable
for predicting the molecular weight distribution of the polymer
produced with any specific solid catalyst.

## VII. STEREOREGULARITY AND PROPYLENE POLYMERIZATION CATALYSTS

Extensive studies have been carried out on the stereochemistry
of the propagation step in propylene polymerization. Features of
isotactic and syndiotactic propagation include cis-addition of the
olefin to the active metal-carbon bond, with participation of the
primary $[M-CH_2-CH(CH_3) \sim\sim]$ and secondary $[M'--CH(CH_3)-$
$CH_2 \sim\sim]$ carbon atoms of the inserted monomer using isospecific
and syndiospecific catalysts, respectively [3, 69, 70]. Isotactic
placements originate from catalyst-monomer interactions; these
placements may not require the participation of metal alkyl in the
active site. Syndiotactic placements originate from nonbonded interactions between the monomer molecule undergoing insertion and
ligands (including the growing polymer chain) on the vanadium
atom [71]. Recent views of the origin of stereospecificity in the
synthesis of isotactic polymers connect this origin with the ability
of the catalyst-growing polymer chain to distinguish between the
two prochiral faces of the α-olefin [3], as illustrated below:

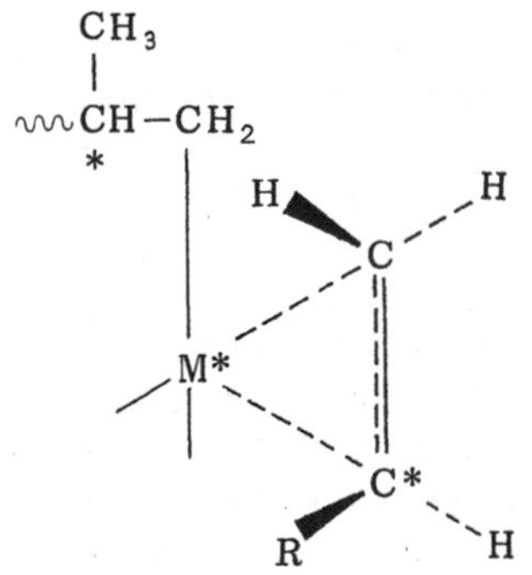

Schematic representation of monometallic catalytic
centers for isotactic polymerization. An asterisk
indicates a chirality center.

Depending on the type of catalyst used, regioselectivity in the polymerization of olefins can vary from 95-99% of "1-2" type and over 85% of "2-1" type.

Attractive, high activity catalysts for propylene polymerization have been described and characterized by many industrial and academic workers [3, 6, 8, 72-79]. Typically these catalysts comprise (a) an aluminum alkyl complexed with an electron donor such as ethylbenzoate, and (b) a solid matrix containing the reaction products of halogenated magnesium compounds with a Ti(IV) compound and an electron donor. High activity catalysts for propylene polymerization show an inverse relationship between catalyst productivity and isotactic index [6]. This effect has been ascribed to the existence of two types of active centers—one type being stereospecific and unstable with time, and the other type nonstereospecific and stable with time.

Several studies have been carried out to identify the role of Lewis bases in high activity catalysts. Investigations involving the stereoselective polymerization of racemic 4-methyl-1-hexene with $MgCl_2/TiCl_4/Al(C_4H_9)_3$ and (-) menthyl anisate as chiral Lewis base have shed some light on the role of Lewis bases in these types of catalysts [3, 80]. Results showed that the active centers had different stereospecificities. The highly stereospecific centers had different tendencies to complex the optically active Lewis base used. The complexation with the base decreased the activity of the catalytic centers. Centers having a low stereospecificity have a larger Lewis acidity and are not stereoselective.

In propylene polymerization using $MgCl_2/TiCl_4/Al(C_4H_9)_3$, ethyl benzoate catalysts, an increase in the ethyl benzoate/$Al(C_4H_9)_3$ ratio from 0.01 to 0.3 led to a significant lowering of the heptane-soluble fraction while the amount of isotactic polymer decreased by only 25% [3]. These results suggest that even among centers producing stereoirregular, linear polypropylene, large differences in Lewis acidity exist.

Other investigations [81] with first generation Ziegler-Natta catalysts in the presence of electron donors have suggested that active sites undergo exchange reactions or structural changes that affect stereospecificity. These changes involve metal alkyl participation. Higher metal alkyl concentrations and Lewis bases decrease the frequency of exchange reactions by increasing site stability. With high activity catalysts based on $MgCl_2/TiCl_4/$ ethyl benzoate in the presence of triethyl aluminum, the ethyl benzoate is reduced to an aluminum alkoxide ($Et_2AlOC-Et_2Ph$) [82]. Polymerization studies revealed that the interaction of ethyl benzoate with the catalyst was responsible for achieving high isotacticity. Ethyl benzoate is believed to improve isotacticity by inactivating the nonstereospecific polymerization sites

## TABLE 8

Effects of Electron Donor (ED) in High Activity Catalysts

Stabilize $MgCl_2$ crystallites by adsorption

Accelerate magnesium-titanium interaction

Reaction with aluminum alkyls with esters leads to formation of dialkylaluminum alkoxides

Effect of catalyst-cocatalyst-ED chemistry dependent on order of addition of components

Raises isotactic index

Increases hydrogen response of catalyst

to a greater degree than the stereospecific ones. Overall, many studies with high activity ball-milled catalysts in the presence of electron donors, particularly esters, show the effects outlined in Table 8.

## VIII. CATALYST CHARACTERIZATION STUDIES

Catalyst characterization studies using a wide range of modern analytical techniques have led to greater insight into the structure and composition of olefin polymerization catalysts. X-ray diffraction studies have been used extensively to monitor the effect of different degrees of activation of $MgCl_2$ on its crystal structure [6, 44-46]. $^{13}C$-NMR investigations have served to elucidate the molecular dynamics and equilibria in homogeneous Ziegler-Natta catalysts [83]. These NMR studies provided evidence in support of olefin insertion into a titanium-carbon bond. Studies using $^{13}C$-NMR for microstructure characterization of polypropylenes have proven of immense value in understanding the isospecific and syndiospecific tendencies of different $\alpha$-olefin catalysts [3].

Scanning electron microscopy has been widely used in the study of catalyst morphology and the replication of this morphology in the resultant polymer particles [6, 84-86].

Electron paramagnetic resonance (EPR) spectroscopy has served to probe the valence state and ligand environment in a highly active Ziegler-Natta catalyst ($MgCl_2$/$TiCl_4$/ethyl

benzoate/p-cresol/Et₃Al) [77]. This catalyst showed a single EPR
Ti(III) signal, from which analysis indicated the Ti(III) was
strongly attached to the catalyst surface with no other Ti(III)
ions on an adjacent site. About 20% of the total trivalent titanium
could be attributed to this EPR signal. After treatment of this
catalyst with AlEt₃/methyl-p-toluate complexes, a single Ti(III)
species, having the features for a stereospecific active site, was
observable by EPR. There appears to be a constant fraction of
1/4 to 1/5 of the titanium which is isolated while the remainder
is in bridged clusters, independent of the oxidation states of titanium. Studies using FTIR have monitored the reaction of $MgCl_2$
with ethyl benzoate [78]. Ethyl benzoate is incorporated into the
$MgCl_2$ matrix by Lewis acid-base complexation involving both oxygen atoms of the ester.

Infrared studies have also proven highly effective as a way to
measure the extent of reaction of silica surface silanol groups with
organotransition metal complexes or transition metal oxides [2, 24].
X-ray photoelectron spectroscopy (ESCA) coupled with chemical
methods has been used to probe the nature of the $CrO_3/SiO_2$ polymerization catalyst [87]. After heat treatment with oxygen at
600-800°C, only Cr(VI) was found. Exposure of the catalyst to
CO at 350°C led to a clean reduction to Cr(II). Reduction of
Cr(VI) with ethylene produced similar results. Contact of the
Cr(II), after CO treatment, with ethylene provided an active
catalyst. These results support the view that Cr(II) is the active state in $CrO_3/SiO_2$ catalysts. ESCA studies have also been
used to correlate the total amount of surface titanium of $TiCl_3$-
based catalysts with catalytic activity in propylene polymerization [88].

## IX.  EFFECTS OF CATALYST MORPHOLOGY
## ON POLYMER MORPHOLOGY

Heterogeneous olefin polymerization catalysts are capable of
replicating their morphology into the morphology of polymer particles [84]. Because of this observation, workers have been able
to regulate polymer particle morphology (size, shape, porosity)
by regulating the morphology of the catalyst particle. The catalyst particle acts as a template for growth of the polymer particle
(Fig. 4).

Under controlled conditions for growth of spherical polymer
particles, the behavior of the polymer particle size as a function
of time may be expressed by

$$R_t = \sqrt[3]{\frac{3}{4\pi \delta N} \int_0^t R_p \, dt}$$                               (10)

where $R_t$ = radius of polymer particle at time t
$\quad$ $R_p$ = total polymerization rate
$\quad$ N $\;$ = number of polymer particles per unit volume
$\quad$ $\delta$ $\;$ = polymer density under polymer conditions

For high activity catalysts in olefin polymerization, the average particle size of the polymer is 15-20 times larger than the catalyst particle size.

Although the polymer particle replicates the morphology of the catalyst particle, experimental data show that the catalyst particle breaks down or shatters during the polymerization process and the fragments become dispersed throughout the polymer particle (Fig. 5) [89].

The polymer initially formed on the catalyst particle may act as a cement for the smaller catalyst particles that bear the active sites for polymerization. This breakdown of the catalyst support has been measured for both the $CrO_3/SiO_2$ and the $Bu_2Mg/TiCl_4/SiO_2$ plus $R_3Al$ catalysts. Silica catalyst particles with initial particle sizes up to 250 μm were found to shatter very early in the polymerization process. However, even the most fragile silica seemed to stop fracturing when the median particle diameter was fractured to the 7-10 μm range.

Morphological studies with low and high activity Ziegler-Natta catalysts have suggested that microspheres provide the basic or elementary morphology of nascent polyolefins [90]. Globular and wormlike morphologies are generated by aggregation of microspheres. Intraparticle features such as void volume, globular and wormlike morphology, and aggregation of microspheres provide exciting new opportunities for understanding polymer growth mechanisms and the effect of catalyst and kinetic parameters on this internal structure.


## X. MECHANISM OF OLEFIN POLYMERIZATION

Olefin polymerization is believed to occur via a coordinated anionic mechanism involving a transition metal-carbon bond [1, 91]. Considerable data [1, 52] support the view for a two-step

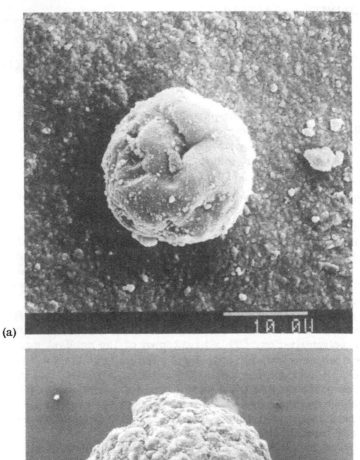

(a)

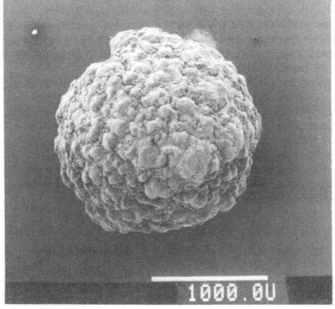

(b)

FIG. 4. Replication of catalyst particle morphology by polymer particle. a: SEM of catalyst particle ($\sim$20 µm). b: SEM of large polymer particle ($\sim$1800 µm).

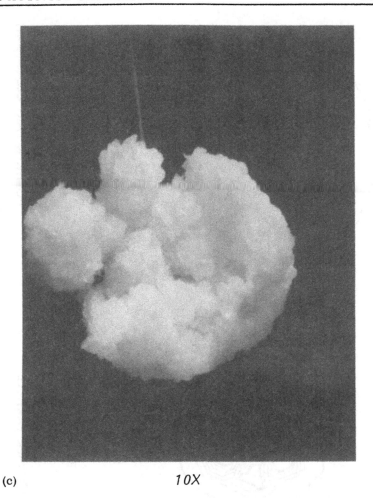

(c)                    *10X*

FIG. 4 (continued). c: Optical microscopy of polymer particle produced under conditions of poor morphological control.

mechanism in which the monomer coordinates to the active site in one step and then inserts into the transition metal-carbon bond:

$$M-R + CH_2=CH_2 \xrightleftharpoons{\text{coordination}} \begin{array}{l} M-R \\ CH_2=CH_2 \end{array}$$
$$\downarrow \text{ insertion}$$
$$M-CH_2-CH_2-R \quad \text{etc.} \qquad (11)$$

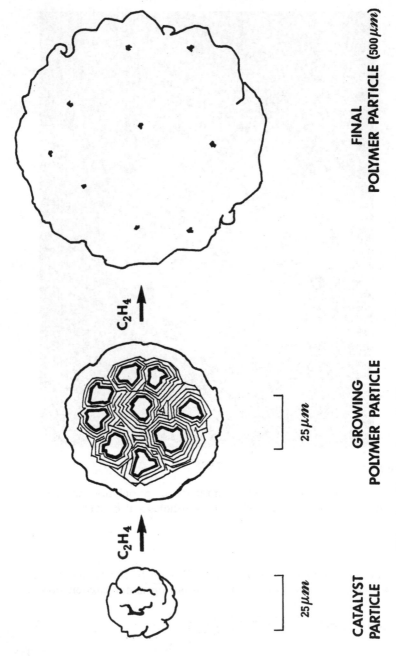

CATALYST
PARTICLE

25 $\mu m$

GROWING
POLYMER PARTICLE

25 $\mu m$

FINAL
POLYMER PARTICLE (500 $\mu m$)

C$_2$H$_4$

C$_2$H$_4$

FIG. 5.  Shattering of silica support during polymerization process.

With a two-step mechanism and a first-order propagation reaction
with respect to monomer, the propagation rate can be determined
by the rate of monomer coordination or by the rate of coordinated
olefin insertion and its surface concentration. Different workers
have supported one or the other of these views. In early work
on the mechanism of the polymerization, the overall activation energy of 10-15 kcal/mol was taken to be the activation energy of
the propagation step. This value was considered too high if it
corresponded to monomer coordination as the rate-determining
step. Later the real activation energy (3-6 kcal/mol) of the
propagation step was found to be much lower.

Some workers have proposed the participation of a carbene
complex in the propagation reaction [92, 93]. Although much experimental data are known confirming that olefin metathesis proceeds via carbene intermediates, no similar supporting data have
been generated to support a carbene mechanism in olefin polymerization. Studies to test the carbene mechanisms have not
provided any rigorous supporting evidence [94, 95].

The type of cocatalyst in Ziegler-Natta catalysts can exert a
significant effect on polymerization rate and polymer isotacticity
[97]. Such effects form the basis of the proposal for a bimetallic
mechanism for the propagation reaction

Comparative data (Table 5) of propagation rate constants ($k_p$) and
the stereoregularity of polymer fractions for one- and two-compo-
nent catalysts based on titanium chlorides indicate the cocatalyst
did not affect the reactivity and stereospecificity of propagation
centers [52]. The effect of the cocatalyst may be due to a change
in the total number of active centers and the ratio of stereospecific
and nonstereospecific centers. Furthermore, temporary deactivation of active sites can take place due to adsorption by the cocatalyst. Addition of $Et_2AlCl$ to $TiCl_2$ significantly lowered the number
of isospecific active centers. Similar results to the poisoning by
$Et_2AlCl$ were observed by addition of triphenyl phosphine [98].

The initiation reaction in metal alkyl-free catalysts such as
$CrO_3/SiO_2$ and $(C_5H_5)_2Cr/SiO_2$ still remains unclear [25, 96].
Generally, studies to test various hypotheses ($\pi$-coordinated
growth, carbene mechanism involving hydrogen transfer) have

not been supported by experimental tests. Formation of a metal
hydride with the hydrogen atom originating from monomer or hydrogen molecules still seems reasonable, but as yet unproven.


## XI. CATALYSIS AND THE POLYETHYLENE REVOLUTION

Production of low-density polyethylene, the world's largest
volume plastic, is undergoing the kind of revolution in the polyolefin field that has not been seen since the discoveries by Ziegler
and Natta [7, 9, 99]. The excitement over polymerization catalysis, which has already had enormous impact for high-density polyethylene and polypropylene production, now encompasses the
whole low-density polyethylene field. Distinctly different reaction processes operate at high pressures to produce LDPE and at
low pressures to produce linear low density polyethylene (LLDPE).
The conventional high-pressure process operates by a free-radical mechanism. For production of LLDPE, ethylene is copolymer-
ized with α-olefins such as butene-1. Coordinated anionic catalysts of high activity and good comonomer response are used to
produce polymers with different levels of short chain branching.
With the high pressure process, reactor control of the branching
reactions is a key to providing a range of product properties.
For LLDPE it is the coordination and insertion processes that
make the role of the catalyst so significant in influencing process
and product behavior.

Union Carbide has developed a unique and versatile low-pressure, fluid-bed process called UNIPOL for the production of
LLDPE (Fig. 6) [7, 9]. Key to the success of the UNIPOL process is the proprietary catalysts that operate at low pressure
and low temperatures and which are suitable for use in a gas-
phase, fluid-bed reactor. With these catalysts molecular weight
may be controlled by reaction temperature and the concentration
of chain transfer agent in the system. Hydrogen is an effective
chain transfer agent with many catalysts. Catalyst type significantly affects molecular weight distribution. Some investigations
continue to exploit solution [100], slurry [101], and even high
pressure routes using transition metal-based catalysts [102].
The most thoroughly described transition metal compounds are
based on titanium, zirconium, vanadium, and chromium.

The molecular architectures of free-radical and transition-
metal catalyzed low density products show distinct differences.
Polymerization of ethylene with free radical initiators leads to
polyethylenes with both short- and long-chain branches. In

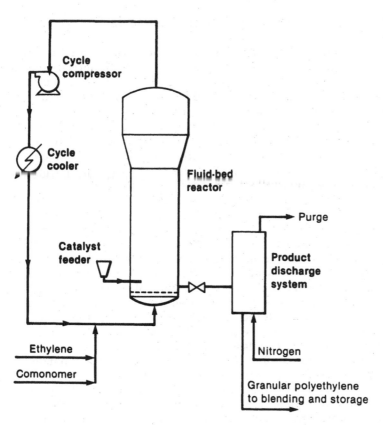

FIG. 6. Low-pressure fluid-bed process for linear low-density polyethylene.

transition-metal catalyzed copolymerizations of ethylene with $\alpha$-olefins (butene-1, hexene-1), polymers containing only short branches are produced. Fortunately, for transition-metal catalyzed polymerizations these differences translate into better products in numerous applications. Product properties in the high pressure, free-radical process are controlled by regulating a number of critical operating parameters. With transition-metal catalysts, product properties are controlled to a large extent by the specific catalyst used. The potential of catalyst technology to provide dramatic new process and product opportunities remains high.

## XII. CONCLUDING REMARKS

Olefin polymerization catalysis continues to be a fertile area of research, with worldwide participation in both industrial and academic laboratories. The intensity of research, documented in patents and publications, has shed light on important features in catalysis. Research to provide new catalyst compositions will focus on methods for generating stabilized, exposed, and isolated active centers. Methods for counting active sites need to be developed more fully. More importantly, attention needs to be devoted to the significance of active site measurements to kinetic analysis of the polymerization process.

Ligand effects in olefin polymerization have provided a better understanding of the steric and electronic factors of importance during the polymerization process. Changes in ligand environment offer the opportunity for altering and fine-tuning particular catalysts. Catalyst characterization studies and use of model reactions appear necessary in studies relating catalyst structure to polymer molecular weight distribution. More structural and stereochemical analyses of synthesized polymers will be necessary if the origin of stereospecificity and the structure of catalytic centers for propylene polymerization are to be understood. Microscopy studies, coupled with kinetic profile studies of catalyst behavior, are needed if more progress in polymer particle growth mechanisms is to be made. Polymerization by a coordinated anionic route still seems most reasonable for olefin polymerization. Alternate proposals still need to be tested through critical experiments which will allow a distinction to be made. The polyethylene revolution has instilled even greater vitality to studies in olefin polymerization catalysis. This renewed vitality should provide an important stimulus for catalyst research in the 1980s and beyond.

## REFERENCES

[1]  J. Boor, Jr., Ziegler-Natta Catalysts and Polymerizations, Academic, New York, 1979.
[2]  Yu. I. Yermakov, B. N. Kuznetsov, and V. A. Zakharov, Catalysis by Supported Complexes (Vol. 8 of Studies in Surface Science and Catalysis), Elsevier, New York, 1981.
[3]  P. Pino and R. Mülhaupt, Angew. Chem., Int. Ed. Engl., 19, 857-875 (1980).
[4]  H. R. Sailors and J. P. Hogan, J. Macromol. Sci.—Chem., A15(7), 1377-1402 (1981).

[5]   P. J. T. Tait, in Developments in Polymerization, Vol. 2
      (R. N. Haward, ed.), Applied Science Publishers, London,
      1979, Chap. 3.
[6]   P. Galli, L. Luciani, and G. Cecchin, Angew. Makromol.
      Chem., 94, 63-89 (1981).
[7]   R. B. Staub, Paper Presented at Golden Jubilee Confer-
      ence for Polyethylene 1933-1983, London, June 8-10, 1983,
      Paper No. B 5.4.
[8]   G. DiDrusco and L. Luciani, J. Appl. Polym. Sci., Appl.
      Polym. Symp., 36, 95-100 (1981).
[9]   F. J. Karol, Chemtech., pp. 222-228 (April 1983).
[10]  F. J. Karol, in Encyclopedia of Polymer Science and Tech-
      nology, Supplement I (H. F. Mark and N. M. Bikales,
      eds.), Wiley-Interscience, New York, 1976, pp. 120-
      146.
[11]  P. Dassesse and R. Dechenne, U.S. Patent 3,400,110
      (1968).
[12]  U. Giannini, E. Albizzati, and S. Parodi, U.S. Patent
      4,149,990 (1979).
[13]  L. Luciani, N. Kashiwa, P. C. Barbé, and A. Toyota,
      U.S. Patent 4,226,741 (1980).
[14]  B. L. Goodall and J. C. van der Sar, U.S. Patent
      4,343,721 (1982).
[15]  R. N. Haward, A. N. Roper, and K. L. Fletcher, Polymer,
      14, 365-372 (1973).
[16]  D. G. Boucher, J. W. Parsons, and R. N. Haward,
      Makromol. Chem., 175, 3461-3473 (1974).
[17]  K. Weissermel, H. Cherdron, J. Berthhold, B. Diedrich,
      K. D. Keil, K. Rust, H. Strametz, and T. Toth, J.
      Polym. Sci., Polym. Symp., 51, 187-196 (1975).
[18]  A. Greco, G. Bertolini, and S. Cesca, J. Appl. Polym.
      Sci., 25, 2045-2061 (1980).
[19]  J. P. Hogan, J. Polym. Sci., Part A-1, 8, 2637-2652
      (1970).
[20]  M. P. McDaniel, J. Catal., 67, 71-76 (1981).
[21]  M. P. McDaniel, Ibid., 76, 37-47 (1982).
[22]  M. P. McDaniel and M. B. Welch, Ibid., 82, 98-109 (1983).
[23]  M. P. McDaniel, M. B. Welch, and M. J. Dreiling, Ibid.,
      82, 118-126 (1983).
[24]  D. G. H. Ballard, Adv. Catal., 23, 263-325 (1973).
[25]  F. J. Karol, G. L. Karapinka, C. Wu, A. W. Dow, R. N.
      Johnson, and W. L. Carrick, J. Polym. Sci., Part A-1,
      10, 2621-2637 (1972).
[26]  U. Giannini, E. Albizzati, S. Parodi, and F. Pirinoli,
      U.S. Patents 4,124,532 (1978) and 4,174,429 (1979).

[27] U. Giannini, E. Albizzati, S. Parodi, and F. Pirinoli,
U.S. Patent 4,174,229 (1979).

[28] K. Yamaguchi, N. Kanoh, T. Tanaka, N. Enokido, A.
Murakami, and S. Yoshida, U.S. Patent 3,989,881 (1976).

[29] E. Albizzati, J. C. J. Bart, U. Giannini, and S. Parodi,
Preprints of IUPAC International Symposium on Macromolecules, Florence, Italy, September 7-12, 1980, Vol. 2,
pp. 40-43.

[30] U. Giannini, A. Cassata, P. Longi, and R. Mazzocchi,
British Patent 1,387,890 (1975).

[31] A. Mayr, P. Galli, E. Susa, G. DiDrusco, and E. Giachetti,
U.S. Patent 4,298,718 (1981).

[32] A. Mayr, A. Susa, A. Leccese, V. Davoli, and E. Giachetti,
British Patent 1,310,547 (1973).

[33] N. Kashiwa, T. Tokuzumi, and H. Fujimura, U.S. Patent
3,642,746 (1972).

[34] B. Diedrich, J. Appl. Polym. Sci., Appl. Polym. Symp.,
26, 1-11 (1975).

[35] J. P. Hermans and P. Henrioulle, British Patents
1,391,067 and 1,391,068 (1975) and U.S. Patents
4,210,735 and 4,210,738 (1980).

[36] A. Delbouille, J. Derroitte, E. Berger, and P. Gerard,
British Patent 1,275,641 (1972).

[37] E. Berger and J. Derroitte, U.S. Patent 3,901,863 (1975).

[38] D. Lassalle, J. L. Vidal, J. C. Roustant, and P. Mangin,
5th Conf. European Plastics Caoutch. Soc. Chim. Ind.,
Paris, France, 1978, Paper A-7.

[39] A. Greco, G. Perego, M. Cesari, and S. Cesca, J. Appl.
Polym. Sci., 23, 1319-1332 (1979).

[40] A. Greco, G. Bertolini, M. Bruzzone, and S. Cesca, Ibid.,
23, 1333-1344 (1979).

[41] See Ref. 2, pp. 135-137.

[42] F. A. Cotton and G. Wilkinson, Advanced Inorganic Chemistry, 4th ed., Wiley-Interscience, New York, 1980.

[43] F. J. Karol, Polym. Prepr., 24(1), 107-108; Paper Presented at American Chemical Society Meeting in Seattle,
Washington, March 1983.

[44] S. I. Makhtarulin, E. M. Moroz, E. E. Vermel, and V. A.
Zakharov, React. Kinet. Catal. Lett., 9(3), 269-274 (1978).

[45] X. Youchang, G. Linlin, L. Wanqi, B. Naiyu, and T. Youqi,
Sci. Sin., 22(9), 1045-1055 (1979).

[46] U. Giannini, Makromol. Chem. Suppl., 5, 216-229 (1981).

[47] P. Galli, P. Barbé, G. Guidetti, R. Zannetti, A. Martorana,
A. Marigo, M. Bergozza, and A. Fichera, Eur. Polym. J.,
19, 19-24 (1983).

[48] A. G. Sharpe, J. Chem. Soc., 117, 2907-2908 (1950).

[49]  Z. W. Wilchinsky, R. W. Looney, and E. G. M. Tornqvist, J. Catal., 28, 351-367 (1973).
[50]  See Ref. 2, pp. 144-148.
[51]  V. A. Zakharov and Yu. I. Yermakov, Catal. Rev.—Sci. Eng., 19(1), 67-103 (1979).
[52]  V. A. Zakharov, G. D. Bukatov, and Yu. I. Yermakov, Adv. Polym. Sci., 5, 101-153 (1983).
[53]  G. D. Bukatov, S. H. Shepelev, V. A. Zakharov, S. A. Sergeev, and Yu. I. Yermakov, Makromol. Chem., 183, 2657-2665 (1982).
[54]  See Ref. 1, pp. 89-92 and 136-137.
[55]  See Ref. 2, pp. 130-133.
[56]  F. J. Karol, W. L. Munn, G. L. Goeke, B. E. Wagner, and N. J. Maraschin, J. Polym. Sci., Polym. Chem. Ed., 16, 771-778 (1978).
[57]  R. C. Schreyer, U.S. Patent 2,986,531 (1961).
[58]  F. J. Karol and C. Wu, J. Polym. Sci., Polym. Chem. Ed., 12, 1549-1558 (1974).
[59]  U. Zucchini and G. Cecchin, Adv. Polym. Sci., 5, 101-153 (1983), and references therein.
[60]  See Ref. 59, Tables X-XIII.
[61]  I. J. Levine and F. J. Karol, U.S. Patent 4,011,382 (1977).
[62]  F. J. Karol, G. L. Goeke, B. E. Wagner, W. A. Fraser, R. J. Jorgensen, and N. Friis, U.S. Patent 4,302,566 (1981).
[63]  F. Baxmann, A. Frese, J. Dietrich, and O. Hahmann, U.S. Patent 3,920,621 (1975).
[64]  E. Albizzati and E. Giannetti, European Patent Appl., 0 032 734 (1981).
[65]  W. L. Carrick, R. J. Turbett, F. J. Karol, G. L. Karapinka, A. S. Fox, and R. N. Johnson, J. Polym. Sci., Polym. Chem. Ed., 10, 2609-2620 (1972).
[66]  H. Wesslau, Makromol. Chem., 26, 102 (1958).
[67]  See Ref. 59, p. 125.
[68]  See Ref. 2, pp. 225-227.
[69]  See Ref. 1, Chap. 15.
[70]  Yu. I. Yermakov, in Structural Order in Polymers (F. Ciardelli and P. Giusti, eds.), Pergamon, New York, 1981, pp. 37-50.
[71]  A. Zambelli and G. Allegra, Macromolecules, 13, 42 (1980).
[72]  U. Giannini, E. Albizzati, and S. Parodi, U.S. Patent 4,149,990 (1979).
[73]  L. Luciani, N. Kashiwa, P. C. Barbé, and A. Toyota, U.S. Patent 4,226,741 (1980).
[74]  G. DiDrusco and R. Rivaldi, Hydrocarbon Process., pp. 153-155 (May 1981).

[75]  J. C. W. Chien, J. C. Wu, and C. I. Kuo, J. Polym. Sci.,
      Polym. Chem. Ed., 20, 2019-2032 (1982).
[76]  J. C. W. Chien and J. C. Wu, Ibid., 20, 2445-2460 (1982).
[77]  J. C. W. Chien and J. C. Wu, Ibid., 20, 2461-2476 (1982).
[78]  J. C. W. Chien, J. C. Wu, and C. I. Kuo, Ibid., 21,
      725-736 (1983).
[79]  J. C. W. Chien, J. C. Wu, and C. I. Kuo, Ibid., 21,
      737-750 (1983).
[80]  P. Pino, G. Fochi, A. Oschwald, G. Piccolo, R. Mülhaupt
      and U. Giannini, in Coordination Polymerization (Poly-
      mer Science and Technology Series, Vol. 19) (C. C.
      Price and E. J. Vandenberg, eds.), Plenum, 1983, pp.
      207-223.
[81]  A. W. Langer, Jr., Ann. N.Y. Acad. Sci., 295, 110-126
      (1977).
[82]  A. W. Langer, T. J. Burkhardt, and J. J. Steger, Ab-
      stracts of Papers, 181st National Meeting of the American
      Chemical Society, Atlanta, Georgia, April 1981; American
      Chemical Society, Washington, D.C., 1981, Poly 031.
[83]  G. Fink, Proceedings of 28th Macromolecular Symposium of
      IUPAC, Amherst, Mass., July 12-16, 1982, p. 264.
[84]  See Ref. 1, Chap. 8.
[85]  J. Wristers, J. Polym. Sci., Polym. Phys. Ed., 11, 1601-
      1617 (1973).
[86]  J. F. Revol, W. Luk, and R. H. Marchessault, J. Cryst.
      Growth, 48, 240-249 (1980).
[87]  R. Merryfield, M. McDaniel, and G. Parks, J. Catal., 77,
      348-359 (1982).
[88]  M. Furuta, J. Polym. Sci., Polym. Phys. Ed., 19, 135-141
      (1981).
[89]  M. P. McDaniel, J. Polym. Sci., Polym. Chem. Ed., 19,
      1967-1976 (1981).
[90]  A. Muñoz-Escalona, Polym. Prepr., 24(1), 112-113, Paper
      Presented at American Chemical Society Meeting in Seattle,
      Washington, March 1983.
[91]  F. J. Karol and W. L. Carrick, J. Am. Chem. Soc., 83,
      2654-2658 (1960).
[92]  K. J. Irvin, J. J. Rooney, C. D. Stewart, M. L. H.
      Green, and R. Mahtab, J. Chem. Soc., Chem. Commun.,
      323, 604-606 (1978).
[93]  G. Ghiotti, E. Garrone, S. Coluccia, C. Morterra, and
      A. Zecchina, Ibid., 801, 1032-1033 (1979).
[94]  A. Zambelli, P. Locatelli, M. C. Sacchi, and E. Rigamonti,
      Macromolecules, 13, 798-800 (1980).

[95] P. Locatelli, M. C. Sacchi, E. Rigamonti, and A. Zambelli, Preprints of IUPAC International Symposium on Macromolecules, Vol. 2, Florence, Italy, September 7-12 (1980), pp. 28-31.

[96] M. P. McDaniel and D. M. Cantor, J. Polym. Sci., Polym. Chem. Ed., 21, 1217-1221 (1983).

[97] See Ref. 1, Chap. 13 and 15.

[98] V. A. Zakharov, G. D. Bukatov, and Yu. I. Yermakov, Makromol. Chem., 176, 1959-1968 (1975).

[99] Chem. Eng., 86, 80-85 (December 3, 1979) (1979 Kirkpatrick Chemical Engineering Achievement Award [Union Carbide Corporation]).

[100] T. K. Moynihan, Paper Presented at Golden Jubilee Conference for Polyethylene, 1933-1983, London, June 8-10, 1983, Paper B 5.3.

[101] J. P. Hogan, Ibid., Paper B.3.

[102] J. P. Machon, Ibid., Paper B 2.5.

# Mobil Zeolite Catalysts for Monomers

WARREN W. KAEDING, GEORGE C. BARILE, AND
MARGARDT M. WU
Mobil Chemical Company
Princteon, New Jersey

## I.  INTRODUCTION

It has been about 20 years since Plank, Rosinski, and Hawthorne reported their spectacular results with metal-modified zeolite crack- ing catalysts for more efficient production of gasoline [1].  This discovery has saved an estimated 200 million barrels of crude oil each year in the United States alone [2].  In 1972, a patent by Argauer and Landolt described the preparation of a member of a generation of new synthetic zeolites, called ZSM-5.  It was unique because of its high silica/alumina ratio and greatly reduced coking rates for reactions with hydrocarbons by comparison with known low silica zeolites [3].  This material was an early member of a series of over 50 synthetic zeolitic substances prepared in Mobil laboratories.

The formation of Mobil Chemical Company in 1960, as a subsid- iary, focused work and attention on the expansion of chemicals production and applications.  In this report, some of the new chem- istry and processes based on Mobil synthetic zeolite catalysts for the production of monomers are reviewed.

## II.  p-XYLENE/TEREPHTHALIC ACID/ POLYESTERS

Fiber grade p-xylene (99+%) for oxidation to terephthalic acid, and subsequent reaction with ethylene glycol to make polyester resins comes from the $C_8$ aromatic fraction produced in catalytic reformers of petroleum refineries.  Two processes are used com- mercially to separate the para isomer from o- and m-xylene and ethylbenzene (EB) [4].  In order to utilize this remaining mix- ture, the o- and m-xylene isomers are isomerized to the thermo- dynamic equilibrium mixture (about 20% para including EB) for recycle (Eq. 1, $R = CH_3$).  In addition, the EB concentration is reduced by dealkylation and disproportionation to prevent its ac- cumulation in the recycle stream.

ISOMERIZATION

$$(1)$$

## TABLE 1

### Mobil Xylene Isomerization Processes for p-Xylene Production

1. MVPI: Mobil vapor phase isomerization (with hydrogen)

| | Material balance (wt%) | |
|---|---|---|
| | Feed | Product |
| $C_1$-$C_5$ | 0 | 0.2 |
| Benzene | 0 | 1.6 |
| Toluene | 0 | 1.0 |
| Ethylbenzene | 14.0 | 10.0 |
| Xylene | | |
| Para | 10.0 | 20.1 |
| Meta | 64.0 | 45.0 |
| Ortho | 12.0 | 18.4 |
| $C_9$-$C_{10}$ aromatic | 0 | 3.7 |

2. MLPI: Mobil low pressure isomerization (without hydrogen)

3. MHTI: Mobil high temperature isomerization

4. MHSI: Mobil high severity isomerization

## A. Xylene Isomerization

Mobil ZSM-5 catalyst is tailor-made to be used in four distinctive process schemes to accommodate available raw material and recycle streams, existing process equipment, cost, and hydrogen availability, etc. A listing of the process variations and a weight material balance are shown in Table 1 [5]. The presence of benzene and $C_9$+ aromatics indicates products from EB removal by deethylation (Eq. 2) or disproportionation.

DEALKYLATION

$$R \quad \longrightarrow \quad + \quad R'-CH=CH_2 \tag{2}$$

## TABLE 2

Unselective and Selective Toluene Disproportionation
to Produce Xylenes and Benzene

| Catalyst | HZSM-5 (unselective) | Mg-ZSM-5 (selective) | |
|---|---|---|---|
| Conditions: | | | |
| Temperature, °C | 550 | 550 | 550 |
| Pressure | atm | atm | atm |
| WHSV | 3.5 | 3.5 | 10 |
| Conversion, % | 40 | 16 | 5 |
| Products: | | | |
| Benzene/Xylene mole ratio | 56/44 | 53/47 | 52/48 |
| Xylenes: | | | |
| Para | 24 | 59 | 83 |
| Meta | 52 | 32 | 10 |
| Ortho | 24 | 9 | 2 |

The concentration of p-xylene is enriched at the expense of the
ortho/meta isomers, which are recycled to extinction. Although it
is difficult to obtain precise production figures, about 50% of the
Free World's p-xylene is made with the aid of ZSM-5 catalysts.

### B. Toluene Disproportionation—Unselective

Toluene is the most abundant and lowest cost aromatic starting
material. With ZSM-5 catalyst it undergoes a disproportionation
reaction to near equimolar amounts of benzene and xylene (thermodynamic equilibrium mixture) (Eq. 3, R = CH$_3$; Table 2) [6].

DISPROPORTIONATION

(3)

Both products have higher values than the starting material. In addition, ethylbenzene is absent to simplify the purification and recycle steps. The o/m-xylene raffinate remaining after the removal of p-xylene may be recycled to extinction by the xylene isomerization process described in Section II-A, above.

## C.  Selective Toluene Disproportionation

When ZSM-5 was modified with phosphorus, magnesium [6], or a considerable variety of other metal salts [7] by ion exchange and/or impregnation with an aqueous or organic solution and calcined in air to produce the corresponding oxides, a significant change in the catalyst occurred. When toluene was disproportionated, p-xylene and benzene were the major products, in sharp contrast to the equilibrium mixture of xylenes obtained with the unmodified zeolite (Table 2).

Advantages of this process are production of the desired para-isomer directly and utilization of a starting material free from ethylbenzene. Roughly equimolar amounts of more valuable benzene and p-xylene are the major products.

## D.  Alkylation of Toluene with Methanol

Xylene (and water) may also be prepared by alkylation of toluene with methanol over ZSM-5 catalysts [8] (Eq. 4, R = R" = $CH_3$):

ALKYLATION

$$(4)$$

In this case, xylene is the major organic product. In a manner similar to toluene disproportionation (above), unmodified zeolite gives an equilibrium ratio of isomers while a para-selective, metal oxide modified ZSM-5 gives predominantly the desired para-isomer (Table 3). This provides another new route for the production of p-xylene where it is the major product of reaction.

TABLE 3

Unselective and Selective Alkylation of Toluene with
Methanol to Produce Xylenes and Water

| Catalyst | HZSM-5 (unselective) | Mg-ZSM-5 (selective) | P-ZSM-5 (selective) |
|---|---|---|---|
| Conditions: | | | |
| Temperature, °C | 495 | 600 | 600 |
| Pressure | atm | atm | atm |
| Toluene/MeOH, mole ratio | 1.5/1 | 2/1 | 1/1 |
| Conversion, wt%: | | | |
| Toluene | 40 | 29 | 40 |
| MeOH | 99 | 99 | 96 |
| Liquid effluent, wt%: | | | |
| Toluene | 56 | 67 | 57 |
| Xylene | 34 | 28 | 36 |
| Other | 10 | 5 | 7 |
| | 100 | 100 | 100 |
| Xylene isomers, %: | | | |
| Para | 24 | 86 | 90 |
| Meta | 53 | 10 | 7 |
| Ortho | 23 | 4 | 3 |

## III. ETHYLBENZENE/STYRENE/POLYSTYRENE

Although substantial amounts of ethylbenzene (EB) are present
in the $C_8$ aromatic fraction from reformers, the escalating cost of
energy for separation and purification by distillation has eliminated
this source for economic reasons. The alkylation of benzene with
ethylene in the liquid phase with an $AlCl_3 \cdot HCl$ catalyst to produce
EB has been the established process [9]. More recently, a ZSM-5
catalyst and process have been developed for this reaction in a
heterogeneous gas/solid phase reaction. An interesting account
of this effort has been reported by Dwyer, Lewis, and Schneider
[10, 11].

After joint development of the process with the Badger Company
of Cambridge, Massachusetts, this alkylation step was incorporated
in a world-scale styrene plant for American Hoechst Corporation at

TABLE 4

Ethylbenzene Production History and Advantages
Resulting from Use of HZSM-5 Catalyst

| | |
|---|---|
| 1943-1976 | Liquid phase process, $AlCl_3 \cdot HCl$ catalyst, >90% EB production |
| 1976 | Mobil/Badger vapor phase, fixed-bed process. ZSM-5 catalyst announced |
| 1980 | First world scale Mobil/Badger plant announced. American Hoechst Corp., Bayport, Texas |
| 1983 | Eight EB plants licensed. One-quarter of world capacity |
| Advantages: | Process simplification<br>More energy efficient<br>Corrosion problems eliminated<br>Nonpolluting effluents<br>Low heavies by-product |

Bayport, Texas. Demonstrated performance has resulted in the
licensing of seven additional plants, accounting for about one-
quarter of the world's capacity (Table 4) [5]. The conditions
of reaction and performance characteristics for Mobil ZSM-5 EB
catalyst [10, 11] are summarized in Table 5. By recycling the di-
and triethylbenzene by-products, a transalkylation reaction occurs with the benzene feed to produce the desired EB.

## IV.  p-ETHYLTOLUENE/p-METHYLSTYRENE/ POLY-p-METHYLSTYRENE

In a manner similar to styrene production (Section III, above),
toluene may be alkylated with ethylene to produce ethyltoluene
(ET). Unmodified catalyst gives a near equilibrium mixture of
isomers, similar to solution reactions with $AlCl_3 \cdot HCl$ catalyst
[12].

TABLE 5

Alkylation of Benzene with Ethylene to Produce
Ethylbenzene over ZSM-5 Catalyst

| Catalyst | HZSM-5 |
|---|---|
| Operating conditions: | |
|   Reactor configuration | 4 beds in series |
|   Temperature, maximum | 800°F, 425°C |
|   Pressure, outlet | 200-300 psig |
|   WHSV, Ethylene | 3-5 |
|       Total | 300-400 |
| | |
| Material balance—18 days—pilot operation: | |
|   Ethylbenzene, lb | 1,146,000 |
|   Residue, lb | 3,275 |
|   Light paraffins, lb | 1,150 |
|   Overall ethylbenzene yield, % | 99.6 |
| | |
| Equilibrium reactor effluent—pilot plant, wt%: | |
|   Benzene | 71.86 |
|   Ethylbenzene | 21.40 |
|   Cumene | 0.34 |
|   n-Propylbenzene | 0.66 |
|   Ethyltoluene | 0.39 |
|   Butylbenzene | 0.05 |
|   Diethylbenzene | 3.70 |
|   Triethylbenzene | 1.60 |
| | ——— |
| | 100.00 |

Physical treatment of the ZSM-5 catalyst with steam or coking
or modification with metal salts produces catalysts which give a
whole spectrum of isomeric mixtures containing up to 99% p-ethyltoluene in the product (Table 6). Subsequent dehydrogenation
of the ET mixture gives the corresponding p-methylstyrene isomers. Of the many options available, a unique 97% para, 3% meta
mixture was chosen for development and conversion to a variety
of homo- and copolymers [13].

A multimillion-pound-per-month demonstration plant has been
in operation to provide material for product, process, and market

TABLE 6

Products Formed with Different Catalysts for Alkylation
of Toluene with Ethylene to Give Ethyltoluene

| Catalyst | $AlCl_3$-HCl | Unmodi-fied zeolite | Treated zeolite | para-selective zeolite |
|---|---|---|---|---|
| Composition (wt%): | | | | |
| Gas and benzene | 0.2 | 1.0 | 2.8 | 0.9 |
| Toluene | 48.3 | 74.4 | 83.2 | 86.2 |
| EB and xylenes | 1.2 | 1.2 | 0 | .0 |
| | | | | |
| p-Ethyltoluene | 11.9 | 5.9 | 5.8 | 11.9 |
| m-Ethyltoluene | 19.3 | 13.3 | 8.2 | 0.4 |
| o-Ethyltoluene | 3.8 | 2.8 | 0 | 0 |
| | | | | |
| Aromatic $C_{10}$+ | 14.4 | 1.4 | 0 | 0.1 |
| Tar | 0.9 | 0 | 0 | 0 |
| | | | | |
| Total | 100.0 | 100.0 | 100.0 | 100.0 |
| | | | | |
| Ethyltoluene isomers (%): | | | | |
| Para | 34.0 | 26.8 | 41.4 | 96.7 |
| Meta | 55.1 | 60.6 | 58.6 | 3.3 |
| Ortho | 10.9 | 12.6 | 0 | 0 |

development. The unique and desirable properties of poly-PMS
starting with toluene and savings in aromatic raw material costs
(35-45¢/gal in 1983), compared to benzene, the starting material
for styrene, provide the potential for establishment of a major
new monomer in the marketplace [13, 14].

## V. DIETHYLBENZENE/DIVINYLBENZENE

In the period following the early development of styrene, ethylbenzene was alkylated to give diethylbenzene (DEB) in a liquid-
phase reaction with an $AlCl_3$·HCl catalyst (Eq. 3, R = $C_2H_5$) [9].
Approximate thermodynamic equilibrium mixtures of isomers were
produced. The ortho-isomer was separated by distillation and

TABLE 7

Alkylation of Ethylbenzene with Ethylene
to Produce Diethylbenzene

| Catalyst | Unmodified HZSM-5 | Modified zeolite catalyst | |
|---|---|---|---|
| Conditions: | | | |
| Temperature, °C | 250 | 400 | 525 |
| Pressure, psig | 0 | 100 | 100 |
| WHSV, EB | 7.15 | 29.7 | 30.2 |
| $C_2H_4$ | 0.48 | 1.2 | 1.2 |
| EB/$C_2H_4$, mole | 4.0/1 | 6.8/1 | 6.9/1 |
| Conversion, EB | 19.2 | 11.4 | 16.6 |
| $C_2H_4$ | 85.7 | 51.1 | 28.8 |
| Selectivity, wt%: | | | |
| Benzene | 3.5 | 8.0 | 58.0 |
| Toluene | 1.7 | 1.0 | 1.6 |
| Xylene | 2.1 | 0.4 | 0.4 |
| Ethyltoluene | 1.3 | 1.1 | 3.8 |
| Diethylbenzene | 77.4 | 88.4 | 32.7 |
| Other aromatics | 9.4 | 0.6 | 0.4 |
| Gas | 4.6 | 0.5 | 3.1 |
| Total | 100.0 | 100.0 | 100.0 |
| Diethylbenzene: | | | |
| Para | 35.5 | 99.2 | 99.6 |
| Meta | 60.2 | 0.8 | 0.4 |
| Ortho | 4.3 | 0 | 0 |

recycled. However, separation of the meta/para-isomers was im-
practical because of their very close boiling points. The mixture
obtained, roughly 35% para/65% meta isomers, was catalytically dehydrogenated to give the corresponding isomeric mixtures of di-
vinylbenzene (DVB). Some of the corresponding ethylstyrene precursors were also present. Because of the higher reactivity and
boiling points, separation to pure isomers was not achieved. However, the DVB mixture found use as a cross-linking agent for styrene. When ethylbenzene was alkylated with ethylene over ZSM-5,
a 2/1 meta/para ratio of diethylbenzene was produced with only 4%
ortho-isomer (Table 7). However, when this catalyst was modified
to induce para-selectivity, over 99% p-diethylbenzene was produced
(Table 7).

Some DEB was also made by disproportionation of ethylbenzene as indicated by the presence of benzene in the product (Eq. 3, R = $C_2H_5$). This reaction occurs more easily than starting with toluene to produce xylene and benzene (see Section II-C). At high temperature, disproportionation becomes the major reaction path (column 3, Table 7). Subsequent dehydrogenation of the diethylbenzene gives the corresponding p-DVB (Eq. 5). Some of the intermediate p-ethylstyrene is also present and may not be separated from the DVB product mixture:

DEHYDROGENATION

$$(5)$$

## VI. 3,4-DIMETHYLETHYLBENZENE/
## 3,4-DIMETHYLETHYLSTYRENE

The remarkable ability of the ZSM-5 class of zeolite catalysts to discriminate between isomers of different sizes is further demonstrated by the alkylation of an equilibrium mixture of xylenes with ethylene to produce ethylxylenes. Six dimethylethylbenzene (DMEB) isomers are possible (Eq. 6). At 350°C, approximately two-thirds of the products are DMEB. The major compound is 3,4-DMEB, the isomer with the smallest cross-sectional diameter. By steaming the catalyst, a remarkable 94% of the 3,4-DMEB is produced with even higher overall selectivity (Table 8). The latter may be dehydrogenated to give the corresponding styrene analogue (Eq. 7):

$$(6)$$

2.3          3.4          2.6          2.4          2.5          3.5

TABLE 8

Ethylation of an Equilibrium Mixture of Xylenes with Ethylene
over ZSM-5 Catalyst (reaction conditions: temperature
350°C, atmospheric pressure, xylene composition 23%
para/55% meta/22% ortho, xylene/ethylene feed
WHSV 8.6/.65, mole ratio 3.5/1)

| Catalyst | HZSM-5 | Steamed HZSM-5 |
|---|---|---|
| Product effluent, wt%: | | |
| Benzene, toluene | 1.1 | 0.2 |
| Xylenes: | | |
| Para | 18.9 | 19.2 |
| Meta | 46.6 | 47.3 |
| Ortho | 19.7 | 21.1 |
| Higher aromatics | 2.2 | 1.0 |

Dimethylethylbenzenes:

| | Equilibrium %[a] | % | % | % | % |
|---|---|---|---|---|---|
| 2,3- | 3.2 | 0.2 | 2.3 | 0 | 0 |
| 3,4- | 20.9 | 6.5 | 73.0 | 9.4 | 94.0 |
| 2,5- | 1.9 | 0 | 0 | 0 | 0 |
| 2,4- | 16.4 | 1.0 | 11.2 | 0.4 | 4.0 |
| 2,5- | 23.9 | 0.7 | 7.9 | 0.2 | 2.0 |
| 3,5- | 33.7 | 0.5 | 5.6 | 0 | 0 |
| | | | 100.0 | | 100.0 |
| Gas | | 1.4 | | 0.9 | |
| Other | | 1.4 | | 0.3 | |
| Total | | 100.0 | | 100.0 | |

[a]At 315°C.

$$\text{(ethyltoluene, } CH_2CH_3, CH_3, CH_3) \longrightarrow \text{(vinyltoluene, } CH=CH_2, CH_3, CH_3) + H_2 \qquad (7)$$

It is interesting to note that the unreacted xylene has an isomer composition similar to the starting material. Perhaps isomerization to o-xylene and alkylation occur simultaneously or in close
sequence. The net result is the selective production of the isomer with the smallest minimum dimension and the fastest diffusion
rate out of the catalyst.

## VII. m-METHYLSTYRENE/POLY-m-METHYLSTYRENE

The versatility of para-selective catalysts is further demonstrated by a selective hydrogenation reaction to produce m-methylstyrene (MMS). The vinyltoluene (VT) of commerce is a mixture
of approximately 35% p/65% m-methylstyrene fixed by the ratio of
ethyltoluene isomers produced during the alkylation step (see Section IV). Furthermore, the m/p-VT isomers also have very close
boiling points. As a result, practical methods have not been found
to separate the isomers.

When ZSM-5 catalyst was impregnated with soluble platinum reagents to deposit the metal within the zeolite pores and the acidity
was neutralized by exchange with alkali cations, i.e., sodium, a
selective reduction catalyst was formed. These principles were
reported by Dessau with the selective reduction of certain linear
olefins mixed with branched isomers and for reduction of styrene
in a mixture with 2-methylstyrene [15]. When the VT mixture and
hydrogen was passed over this catalyst, the smaller para-isomer
easily diffused into the zeolite pores along with hydrogen and was
catalytically reduced with hydrogen, at platinum catalyst sites, to
the corresponding p-ethyltoluene. The larger meta-isomer could
not easily enter the pores and passed through the catalyst bed
unchanged. The resulting MMS and p-ethyltoluene had boiling
points sufficiently different to separate by distillation to give 99%
MMS as a product (Table 9).

## VIII. SUMMARY AND DISCUSSION

During the past 10 years, Mobil zeolites have replaced older
catalysts for xylene isomerization to produce p-xylene (about 50%

TABLE 9

Selective Reduction of Meta/Para-Methylstyrene
(commercial VT) to m-Methylstyrene
and p-Ethyltoluene

Conditions:

| | |
|---|---|
| Temperature, °C | 370 |
| Pressure | atm |
| | |
| WHSV, Methyl styrene (VT) | 1.4 |
| $H_2$ | 0.1 |
| $H_2$/methyl styrene (VT) | 11.4 |
| | |
| Time on stream, h | 63 |

Conversion, %:

| | |
|---|---|
| Para-methyl styrene | 97.8 |
| Meta-methyl styrene | -3.8 |

Product composition, wt%

| Ethyltoluene, | | |
|---|---|---|
| Para | | 30.6 |
| Meta | | 2.1 |
| | | |
| Methylstyrene, | (feed) | |
| Para | 36.6 | 0.8 |
| Meta | 62.7 | 65.1 |
| Other | 0.7 | 1.4 |
| Total | 100.0 | 100.0 |

Methylstyrene, %:

| | |
|---|---|
| Meta | 98.8 |
| Para | 1.2 |

of world capacity) and benzene alkylation to give ethylbenzene (25%
of world capacity). This is a result of improved performance and
economics, better heat recovery, reduction of corrosion, and elimination of spent catalyst disposal.

Discovery and development of new para-selective catalysts for p-ethyltoluene and dehydrogenation catalysts for p-methylstyrene (PMS) have resulted in the construction and operation of a demonstration plant in the summer of 1982 to produce multimillion-pound quantities of PMS for commercial scale testing and market development. We believe that PMS will prove to be the first large-scale monomer to appear in the marketplace in the last 25 years.

Reactions to give other monomers or precursors which illustrate the unique properties which shape-selective zeolite catalysts offer are also shown. A key factor is the dimension of the desired product molecule in comparison with other isomers or related products. p Diethylbenzene can be prepared in 99+% isomeric selectivity. A closer fit of the p-diethylbenzene and its corresponding activated complex is realized than with a dimethyl or methylethyl combination of substituents within the zeolite pores where the reaction occurs. This permits a closer discrimination between the para- and ortho/meta-isomers. A similar situation was observed with the selective production of 3,4-DMEB (94%), the isomer with the smallest dimension.

The preparation of 99% MMS is an example of a selective reaction where the smallest isomer, p-methylstyrene, selectively diffused into the pores to react with hydrogen, while the larger MMS could not penetrate the catalyst and passed through the bed unreacted.

Many more unique reactions and applications have and will be found in the future to produce new products more efficiently and economically with zeolite catalysts.

## REFERENCES

[1] C. J. Plank, E. J. Rosinski, and W. P. Hawthorne, Ind. Eng. Chem., Prod. Res. Dev., 3, 165 (1964).
[2] C. J. Plank, Heterogeneous Catalysis, Selected American Historics (B. H. Davis and W. P. Hettinger, eds.), ACS Symposium Series 222, 1983.
[3] R. J. Argauer and R. G. Landolt, U.S. Patent 3,702,886 (1972).
[4] H. C. Ries, Stanford Research Institute Handbook No. 25A, 1970.
[5] F. N. Fagan and P. B. Weisz, Japanese Petroleum Institute Conference, Tokyo, May 9, 1983.
[6] W. W. Kaeding, C. Chu, L. B. Young, and S. A. Butter, J. Catal., 69, 392 (1981).

[7]   W. W. Kaeding, L. B. Young, and C. Chu, Ibid., In Press.
[8]   W. W. Kaeding, C. Chu, L. B. Young, B. Weinstein, and
      S. A. Butter, Ibid., 67, 159 (1981).
[9]   R. H. Boundy and R. F. Boyer (eds.), Styrene—Its Poly-
      mers, Copolymers and Derivatives, Reinhold, New York,
      1952.
[10]  F. G. Dwyer, P. J. Lewis, and F. M. Schneiter, "Effi-
      cient, Non-Polluting Ethylbenzene Producers," Chem. Eng.,
      p. 90 (January 5, 1976).
[11]  P. J. Lewis and F. G. Dwyer, Prepr., Am. Chem. Soc.
      Div. Pet. Chem., 22(3), 1077 (1977).
[12]  J. L. Amos, K. E. Coulter, A. C. Wilcox, and F. J.
      Soderquist, in Styrene—Its Polymers, Copolymers and
      Derivatives (R. H. Boundy and R. F. Boyer, eds.),
      Reinhold, New York, 1952, p. 1232.
[13]  W. W. Kaeding, L. B. Young, and A. G. Prapas, Chem-
      tech., p. 556 (1982).
[14]  W. W. Kaeding and G. C. Barile, in New Monomers and
      Polymers (W. M. Culbertson and C. U. Pitman, eds.),
      Plenum, New York, 1983.
[15]  R. M. Dessau, J. Catal., 77, 304 (1982).

# A Fundamental Study of High Activity Catalyst for Olefin Polymerization

JAMES C. W. CHIEN
Department of Chemistry
Department of Polymer Science and Engineering
Materials Research Laboratories
University of Massachusetts
Amherst, Massachusetts

# I. INTRODUCTION

Soon after the commercialization of the Ziegler-Natta catalyst for polyolefin production, efforts were begun to improve the productivity of the catalyst. It was realized that the α- and δ-forms of TiCl₃ have layered crystal structures having only chlorine atoms in the basal planes. Exposed and coordinatively unsaturated Ti sites are only found along c-axis edges of the crystallites. Therefore, if a suitable support is found and Ti atoms can be anchored to its surface with stereoelectronic characteristics resembling those in the AA-TiCl₃, then a high activity catalyst would result. This search progressed from metal oxides to metal hydroxides, to metal hydroxychlorides, and finally to metal chlorides, in particular magnesium chloride. The historical developments have been discussed in detail by Karol and Hsieh at this conference.

In the numerous patents on the high activity $MgCl_2$ supported catalysts, advantageous modifications with alcohol and weak Lewis base such as organic ester have been claimed. Yet there is a lack of any understanding about how these additives modify the physical and/or chemical structures of the $MgCl_2$. In this work we have distilled a protocol for the preparation of a $MgCl_2$ supported high activity catalysts from various patents which include all reasonable modifications. The central purpose is to study the structure of the support following each step of catalyst preparation by using a number of physical, chemical and analytical techniques. The results show that each modification achieves certain desirable physiocochemical changes in the support.

# II. EXPERIMENTAL SECTION

## A. Catalyst Preparation

The entire catalyst preparation was carried out in an oxygen- and moisture-free environment using the Schlenck apparatus. Transfers were made in a dry box. Anhydrous $MgCl_2$, obtained from Toho Titanium Corp., was large silvery flakes. It was heated for 30 min at 380°C under a stream of HCl; 8 mL of $H_2O$ was collected from 73 g of $MgCl_2$. The temperature was increased to 430°C under an argon atmosphere, followed by vacuum drying for 5 h to give Support I.

To 33.5 g of Support I was added 7.8 mL of ethyl benzoate (EB) in a molar ratio of 1 $MgCl_2$:0.155 EB and ball-milled with cylindrical porcelain elements. The resulting product, after thorough washing with n-heptane and drying, is referred to as Support II.

2.7 g of Support II was suspended in n-heptane, and 1.5 mL of p-cresol (PC) ($MgCl_2$/EB:PC = 1:0.5) was added dropwise over a 1-h period with stirring at ambient temperature. The reaction mixture was heated to 50°C for 1 h and subsequently washed and dried to give Support III.

To 21 g of Support III was added 5.2 mL of $AlEt_3$ (A) which corresponds to a PC:A molar ratio of 2:1. The mixture was stirred for 1 h at room temperature, filtered, washed, and dried under vacuum to give Support IV. During the course of reaction, the evolved gas was quantitatively collected and determined to be ethane by mass spectrometry. One mole of $C_2H_6$ was liberated from 1 mol of PC in the support.

The final process is the reaction of 1.55 g of Support IV in 5 mL of heptane with 10 mL of $TiCl_4$ at 100°C followed by thorough washing to form the "catalyst."

## B.  Determination of Ti Oxidation States

The oxidation states of Ti were determined by redox titration and elemental analysis. Because the samples were air-sensitive and a dry box is unsuited for quantitative titrations, a double-glove-bag technique was developed. Two large polyethylene glove bags were used, one inside the other. Purified nitrogen flowed into the inner bag and exited through the outer bag, thus constantly sweeping the space between the two bags. All solvents, solutions, and chemicals were freed of air before being placed in the inner bag.

Two redox titrations were carried out for a given sample.

### 1. Titration A

A precisely weighed quantity of catalyst (about 30-50 mg containing about 0.1-0.5 mmol of titanium) was dissolved in 10 mL of 0.1 $\underline{M}$ ferric sulfate solution. To the stirred mixture was added 1 mL of 85% phosphoric acid to improve the endpoint; sodium diphenylamine sulfonate in a 2% concentration was used as the indicator. A standard 2 $m\underline{N}$ dichromate solution was used for titration in this procedure. Ferric ion oxidizes the titanium according to

$$Ti^{2+} + 2Fe^{3+} \rightarrow Ti^{4+} + 2Fe^{2+} \qquad (1)$$

$$Ti^{3+} + Fe^{3+} \rightarrow Ti^{4+} + Fe^{2+} \qquad (2)$$

and the $Fe^{2+}$ ions produced are titrated by the dichromate. Therefore, $Ti^{2+}$ and $Ti^{3+}$ act effectively as two-electron and one-electron reductants, respectively. Ferric sulfate solution was titrated for the blank; Stauffer AA $TiCl_3$ served as the standard.

## 2. Titration B

Similar-sized samples were dissolved in 2 $\underline{N}$ $H_2SO_4$ and titrated as in Titration A. $Ti^{2+}$ species in the catalyst were oxidized by the protons

$$Ti^{2+} + H^+ \rightarrow Ti^{3+} + \frac{1}{2}H_2 \tag{3}$$

and titrated directly by dichromate. Therefore both $Ti^{2+}$ and $Ti^{3+}$ were titrated as one-electron reductants. Reaction (3) does not take place in the presence of $Fe^{3+}$ ion. The sulfuric acid solution was titrated for the blank; dicyclopentadienyldicarbonyltitanium, prepared from the reduction of dicyclopentadienyltitanium chloride, was the standard in this case.

The concentration of $Ti^{2+}$ was obtained from the difference of titers in A and B. Subtracting this quantity from the total of B gave the concentration of $Ti^{3+}$ ions. The concentration of $Ti^{4+}$ species is obtained from the elemental analysis of total titanium by subtracting from it the $Ti^{2+}$ and $Ti^{3+}$ contents.

All these analyses were performed on triplicates with errors of $\sim\pm19$, $\sim\pm6$, and $\sim\pm3\%$ for $Ti^{2+}$, $Ti^{3+}$, and $Ti^{4+}$ ions, respectively. The low accuracy for $Ti^{2+}$ is due to its low concentration in the catalyst, and because it was obtained from the difference of two titers.

## C. Elemental Analysis

All elemental analyses were performed by Galbraith Laboratories, Inc. of Knoxville, Tennessee, in duplicate or triplicate.

## D. EPR

A Varian E-9 spectrometer was used in this work. A sample of diphenylpicrylhydrazyl (DPPH) in the dual cavity provides precise field calibration for the calculation of the g value. The EPR spectral intensity of $Ti^{3+}$ species was measured quantitatively by comparing the double integrated spectra of the catalyst and that of a standard heptane solution of 2,2,6,6-tetramethylpiperidinoxyl.

## E. FTIR

Following each step the solid support or catalyst was washed repeatedly, dried, and transferred to a dry box, mixed with

predried KBr powder, and pulverized. The dry box was also
equipped with a hydraulic press for the preparation of KBr pellets, which were kept in glass jars filled with argon. The specimen was quickly transferred to the FTIR chamber and flushed
with nitrogen. A Nicolet 7199 spectrometer was used to obtain
the FTIR spectra.

## F. BET

The apparatus for surface-area measurement was constructed
according to a well-established design. Samples were evacuated
at 10⁻⁴ torr for 12 h at 25°C to remove physisorbed molecules.
The sample container was filled with helium to measure the volume of the apparatus that contained the samples, assuming no
absorption of He under these conditions. Helium was then
pumped out for 1 h at 25°C. After reimmersing the sample in
liquid nitrogen, nitrogen gas was admitted and the pressure
drop was measured. Sufficient time must be allowed for the
absorption equilibrium to be established for each successive
measurement. Four or more measurements were made for $P/P_0$
from 0.02 to 0.3, where P is the pressure and $P_0$ is the sat-
urating vapor pressure of the gas. The results were plotted
according to

$$\frac{P}{V(P_o - P)} = \frac{1}{V_m C} + \frac{C - 1}{V_m C}\frac{P}{P_o} \tag{4}$$

where V is the volume at pressure P, $V_m$ is the volume of gas
corresponding to the formation of a monolayer, and C is a constant.

## G. Pore Size Analysis

A Quantachrome scanning porosimeter, used to determine the
pore spectrum of the catalyst materials, was coupled to a micro-
computer which places the continuous pressure-volume scan in
the computer memory. The pressure is increased from 0 to
60,000 psi as mercury is forced into the pores. The intrusion
measurement was followed by extrusion.

## H. X-Ray Diffraction

Powder x-ray diffraction was obtained with a Phillips instru-
ment operating at 35 keV and 15 mA. The sample was loaded
into a thin-walled quartz x-ray capillary in a dry box. The

large-diameter portion of the sample tube was removed with a
triangular metal file and sealed with molten black Apiezon wax.
Three-hour exposures were taken and the diffraction intensities
were measured with a microdensitometer.


## III.  RESULTS AND DISCUSSION

### A.  Chemical Composition

After ball milling the EB-modified $MgCl_2$, Support I was washed
six times with heptane (total volume 300 mL), vacuum dried, and
pulverized with a mortar and pestle.  The results of elemental
analysis are (in wt%) Mg 19.61, Cl 16.21, C 13.04, H 1.45, and
O 4.67.  This analysis corresponds to a composition of $MgCl_{2.14}$-
$EB_{0.146}$.  After correction for the small amount of carbon and hydrogen in the original $MgCl_2$, the calculated elemental contents
for this empirical formula are Mg 19.82, Cl 61.86, C 13.10, H
1.27, and O 3.94.  The molar ratio of $MgCl_2$ to EB before ball
milling was 1:0.154.  Therefore, 94.8% of the EB was complexed
with $MgCl_2$ and was not dissociated by washing with inert solvent.

Table 1 summarizes the other compositions.  Support III retains almost quantitatively all the p-cresol in the reaction mixture.  Treatment with $AlEt_3$ liberated two equivalents of ethane
as determined by volumetry and mass spectrometry.  The catalyst contains only $MgCl_2$, $TiCl_{3.5}$, $AlCl_3$, and EB.  All the p-
cresol moiety was removed by the $TiCl_4$ treatment.

Hydrolysis-GC was used to analyze the organic components in
the support and the catalyst (Table 2, Columns 2 and 4); the cor-
rected values obtained, assuming that the extraction efficiencies
were 77% for EB and 75% for PC, are given in Columns 3 and 5.
Furthermore, the estimates from elemental analysis are shown in
Columns 4 and 7 for comparison.  The results showed remarkably
good agreement between hydrolysis-GC results and the compositions deduced from elemental analysis.  However, it should be
noted that GC gave low amounts of EB and PC after activation;
for instance, after Catalyst IV was activated with $AlEt_3$/MPT, hydrolysis-GC found 0.07 mmol/g of EB and 0.19 mmol/g of PC.

The low amount of PC is in agreement with FTIR, but the low
content of EB suggests that the composition of activated catalyst
is sensitive to the experimental procedure (vide infra).  Another
possibility is that the reaction products of EB with the coactivator
have principal IR bands similar to EB, but on hydrolysis the products differed from the original EB.

TABLE 1

Chemical Composition of Typical Support

| | Support III | | Support IV | | Catalyst | | |
|---|---|---|---|---|---|---|---|
| | Found | Calcd.[a] | Found | Calcd.[b] | Found (wt%) | Found (mmol/g) | Calcd.[c] (mmol/g) |
| Mg | 12.84 | 13.19 | 12.98 | 12.8 | 15.62 | 6.05 | 6.25 |
| Ti | | | | | 4.48 | 0.93 | 0.94 |
| Al | | | 2.79 | 2.58 | 1.73 | 0.64 | 0.67 |
| Cl | 38.55 | 38.50 | 39.44 | 35.2 | 64.18 | 18.1 | 17.6 |
| C | 32.30 | 36.87 | 35.68 | 38.3 | 11.13 | 9.27 | 9.0 |
| H | 4.0 | 3.49 | 3.62 | 3.9 | 1.22 | 1.21 | 1.6 |
| O | 12.25 | 7.93 | 5.49 | 7.24 | 1.64 | 1.03 | 1.9 |

[a]Calculated for $(MgCl_2)_1 \cdot (EB)_{0.146} \cdot (PC)_{0.621}$.

[b]Calculated for $(MgCl)_1 \cdot (EB)_{0.146} \cdot (EtAl(OR)_2)_{0.193} \cdot (PC)_{0.235}$, where R is p-methylphenoxy.

[c]Calculated for $(MgCl_2)_1(TiCl_{3.5})_{0.15}(AlCl_3)_{0.1}(EB)_{0.16}$.

TABLE 2

Organic Components in Support and Catalyst

| | EB $(mmol/g^1)$ | | | PC $(mmol/g^1)$ | | |
|---|---|---|---|---|---|---|
| | Hydrolysis-GC | | | Hydrolysis-GC | | |
| Sample | Found | Cor-rected[a] | Elemental analysis | Found | Cor-rected[b] | Elemental analysis |
| Support (I) | 0.89 | 1.16 | 1.29 | | | |
| Support (II) | 0.68 | 0.88 | 0.88 | 2.11 | 2.81 | 2.90 |
| Support (III) | 0.63 | 0.82 | 0.92 | 1.78 | 2.37 | 2.16 |
| Catalyst (IV) | 0.46 | 0.73 | 0.88 | 0.40 | 0.53 | 0.63 |

[a]Corrected for extraction efficiency of 77%.
[b]Corrected for extraction efficiency of 75%.

TABLE 3

BET Surface Areas of $MgCl_2$-Based
Supports and Catalyst

| Materials | | Sample | BET results | |
|---|---|---|---|---|
| Type | Promoters | no. | S ($m_2/g$) | C |
| II | EB /PC | VI-70 | 31 | 13.2 |
| II | EB /PC | VI-188 | 34 | 6.35 |
| III | EB /PC | X-127 | 17 | 7.82 |
| III | EB /PC /TEA | VI-76 | 53 | 50.5 |
| III | EB /PC /TEA | VII-7 | 73 | 22.3 |
| III | EB /PC /TEA | X-130 | 58 | 42.8 |
| IV | EB /PC /TEA /TiCl$_4$ | VI-116 | 110 | 32.2 |
| IV | EB /PC /TEA /TiCl$_4$ | X-142 | 149 | 27.9 |

## B.  Surface Area

The surface areas of $MgCl_2$ and Support I were too small to be
measured by BET.  However, the specific surface area

$$S = \frac{V_m As}{w(22,400)} \ m^2/g \qquad (5)$$

where A is Avogadro's number and w is the sample weight in grams,
increase steadily with each step of catalyst preparation (Table 3).
The intercept of the BET plot also gives

$$C = \exp - \frac{(\Delta G_1^0 - \Delta G_2^0)}{RT} \qquad (6)$$

## C.  Pore Size

Mercury porosimetry was performed on all the materials.  The
results on pore volumes and pore surfaces are summarized in
Table 4.  During the preparation, V and $S_p$ showed steady increases.

TABLE 4

Pore Size of MgCl$_2$-Based Supports and Catalysts
Determined by Mercury Porosimetry

| Materials | Promoters | Sample no. | V, pore volume (cm$^3$/g) | S$_p$, pore surface (m$^2$/g) |
|-----------|-----------|------------|---------------------------|-------------------------------|
| MgCl$_2$  | HCl                    | X-116 | 0.41 | 15.9 |
| I         | HCl/EB                 | X-117 | 0.41 | 30.5 |
| II        | HCl/EB/PC              | X-127 | 0.93 | 81.5 |
| III       | HCl/EB/PC/TEA          | X-130 | 1.21 | 89.0 |
| IV        | HCl/EB/PC/TEA/TiCl$_4$ | X-142 | 1.31 | 104  |

### D. X-Ray Diffraction (Fig. 1)

X-ray diffraction is a useful technique for measuring maximum crystallite sizes of ~300 Å at large angles; the limit is 100 Å with medium angles. The Toho MgCl$_2$ is a crystalline substance with large crystallites. Ball milling with EB significantly reduces its crystallite size, as shown in Table 5.

TABLE 5

Change of MgCl$_2$ Crystallite Sizes with Ball Milling
in the Presence of EB

| Reflection ball milling (h) | $d_{h,k,l}$ (Å) | | | |
|------------------------------|---------|---------|---------|---------|
|                              | (003) | (104) | ($1\bar{1}0$) | (021) |
| 1  | 161 | 123 | 305 | 309 |
| 2  | 89  | 73  | 162 | 309 |
| 4  | 89  | 42  | 85  | 170 |
| 8  | 59  | 38  | 85  | 113 |
| 16 | 19  | 19  | 36  | 31  |

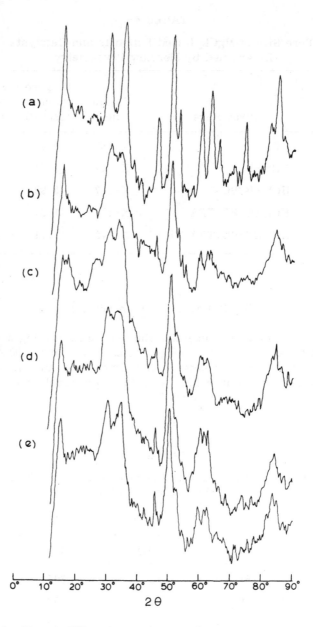

FIG. 1.  X-ray diffraction patterns of (a) Support I ball milled
for 11 h; (b) Support I ball milled for 72 h with EB; (c) Support
III; (d) Support IV; (e) catalyst.

Most of the x-ray linewidth decreases occurred during the first 16 h of ball milling, in which time the crystallite thicknesses were reduced to less than 36-20 Å, depending on the lattice plane. If we assume that the crystallites are approximately spherical with an average diameter of 30 Å, Material I should have a much higher surface area (of the order of 210 m$^2$/g). Consequently, it must be in the form of the large aggregates. The small MgCl$_2$ crystallites are held together by complexation with EB. Subsequent treatments reduce the sizes of the aggregates: Support III 760-1500 Å, Support IV 350-480 Å, and the catalyst 170-230 Å.

## E. FTIR

The infrared (IR) spectrum of neat EB is characterized by a cluster of bands centered about 2980 cm$^{-1}$; they are assigned to $\nu$C—H of the aromatic system, $\nu_{as}$ and $\nu_s$ of CH$_3$, and $\nu_s$ of —CH$_2$—. The other intense and informative bands in EB are $\nu$C=O at 1719 cm$^{-1}$, $\nu$C—O at 1279 cm$^{-1}$ and 1109 cm$^{-1}$, and $\delta$CH$_2$ rock at 712 cm$^{-1}$.

Upon ball milling of MgCl$_2$ with EB there are intense and broad $\nu$OH bands with a maximum at 3440 cm$^{-1}$ and a sharp $\delta$H$_2$O band at 1620 cm$^{-1}$, even though the Toho MgCl$_2$ is said to be anhydrous and was thoroughly dried. Therefore, this support cannot be said to be free of Mg(OH)Cl or even of Mg(OH)$_2$. Superimposed on the OH band is the cluster of bands of EB at 2980 cm$^{-1}$. The other characteristic absorptions of EB at 1601, 1584, 1543, 1364, 1315, 1183, 1073, 1027, and 712 cm$^{-1}$ were clearly observable.

The most important changes in the IR spectrum of EB as a result of complexation with MgCl$_2$ occurred in the bands associated with the ester group. The $\nu$C=O band was shifted to a lower frequency at 1683 cm$^{-1}$. The two $\nu$C—O bands in EB were shifted to higher frequencies of 1329 and 1306 cm$^{-1}$ in I. These changes are well known when the ester forms complexes. The shift in the C=O band is $\Delta\nu = -36$ cm$^{-1}$, and $\Delta\nu = +50$ and $+197$ cm$^{-1}$ in the C—O bands. Two structures for the complex are proposed:

A neat IR spectrum of PC is characterized by a broad $\nu$OH at 3340 cm$^{-1}$, $\nu$CH of the phenyl group at 3010 cm$^{-1}$, and $\nu_{as}$ and $\nu_s$ of CH$_3$ at 2920 and 2860 cm$^{-1}$, respectively. Other phenyl group bands were found at 1640, 1610, and 1514 cm$^{-1}$ and a broad 1218 cm$^{-1}$ band appeared for $\nu$C$-$OH. The other useful fingerprint bands are at 815 and 742 cm$^{-1}$.

The spectrum of Support II showed that the OH band had greatly increased in intensity and bandwidth and the band shape was distorted because of the absorption by the OH groups in MgCl$_2$ and PC. All the bands of EB remained in Support II, albeit with some broadening, and the $\nu$C=O band shifted to a slightly higher frequency. All the characteristic bands of PC are present and accounted for. In agreement with expectations, the FTIR spectrum of Support III showed a large decrease in the intensity of the OH absorption. In fact, the band resembled that of support I. Some of the strong PC bands, that is, 1514 and 1218 cm$^{-1}$, were little affected by the reaction with TEA. However, the intense 815 cm$^{-1}$ band in Support II was apparently split into three bands in Support III at 864, 826, and 778 cm$^{-1}$. In this frequency region the vibrations are due to the $\delta$CH out-of-plane bending mode which is sensitive to the number of adjacent hydrogen atoms among other effects. Possibly these bands arise from the different reaction products between PC and TEA.

The reaction of TEA with Support II has apparently little or no effect on the EB in the support, except that the low frequency complexed $\nu$C=O band has broadened significantly. These results show that no appreciable reaction occurs between TEA and EB because the reaction between TEA and PC dominates.

The most prominent band in PC at 1511 cm$^{-1}$ disappeared after TiCl$_4$ reaction. The 1228 cm$^{-1}$ band became very weak. The three $\delta$CH out-of-plane bending absorptions also disappeared. A new band at 821 cm$^{-1}$ could be due to $\delta$CH for EB complexed with Ti or Al chlorides. Thus the p-cresol moieties were largely removed by the TiCl$_4$ reaction.

When mixed, the bands of free ester disappeared immediately. The complexed ester bands of MT are found at 1669 and 1618 cm$^{-1}$ for $\nu$C=O and 1335 and 1315 cm$^{-1}$. In addition, there is a new band at 650 cm$^{-1}$ of some reaction product.

The activated catalyst was obtained by adding the coactivators to the catalyst. The FTIR spectrum showed that the OH band is like the one in Support I and much narrower in width than those in the other intermediate supports and the catalyst. The bands at 2923 and 2865 cm$^{-1}$ may be attributable to the $\nu_{as}$ CH$_3$, $\nu_s$ CH$_3$, and $\nu_s$ CH$_2$ of MT and EB. Both the 715 and 820 cm$^{-1}$

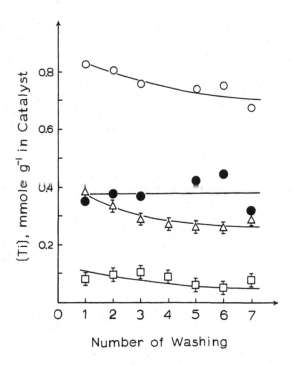

FIG. 2. The effect of n-hexane washing on the titanium contents of the catalyst: (O) total Ti, (●) $Ti^{4+}$, (△) $Ti^{3+}$, (□) $Ti^{2+}$.

bands of EB in the catalyst remained after activation. The $\delta$CH band of MT was found at 758 cm$^{-1}$. The 1228 cm$^{-1}$ band in the catalyst, which may be a remnant of p-cresol moiety, is absent after activation. The $\nu$C—O bands broadened, as might be expected, because they contain absorptions from both EB and MT. Qualitatively, the activation resulted in the incorporation of MT, further removal of p-cresol moiety, and retention of most of the complexed EB.

### F. Oxidation States of Titanium in the Catalyst

The concentrations of $Ti^{2+}$ and $Ti^{3+}$ in the catalyst were determined by titration, and the total titanium was obtained by elemental analysis. The difference gave the $Ti^{4+}$ content of the catalysts. The changes in the titanium concentrations of various oxidation states due to washing was monitored (Fig. 2). There was a loss

of about 0.1 mmol/g of total titanium after seven washings, virtually all of which may be attributed to $Ti^{3+}$ species. It may be concluded that some soluble $Ti^{3+}$ species, which would most likely become nonstereospecific active sites, were removed by the washing.
The concentrations of titanium after the first washing were 0.38,
0.38, and 0.1 mmol/g for $Ti^{4+}$, $Ti^{8+}$, and $Ti^{2+}$, respectively.

The amounts of reduced titanium must depend on the number
of Al-Et groups in the $AlEt_3$ promoted support. This support
before reaction with $TiCl_4$ contains $\sim 0.74$ mmol of Al-Et/g of
catalyst. If two equivalents of Al-Et are needed to reduce a
$Ti^{4+}$ to $Ti^{2+}$ and one equivalent of Al-Et can reduce one $Ti^{4+}$ to
$Ti^{3+}$, then about 0.56 mmol of Al-Et/g of catalyst is required to
obtain the amount of reduced titanium species. This represents
a rather efficient 78% use of the Al-Et groups.

The formation of $Ti^{3+}$ species is the result of alkylation. The
$Ti^{4+}$ alkyls are unstable at moderate temperatures. The usual
mechanism for reduction is

$$2Ti^{4+}Et \rightarrow 2Ti^{3+} + C_2H_4 + C_2H_6 \tag{7}$$

If applied to the present example, this would require the presence of $Ti^{4+}$ ions in proximity on the support (vide infra). In addition, because the $TiCl_4$ reaction was carried out at 100°C, at
which condition homolysis is also possible,

$$Ti^{4+}Et \xrightarrow{\Delta} Ti^{3+} + Et \tag{8}$$

It is not known whether the support has a destabilizing influence
on the $Ti^{4+}$-Et bond. The formation of divalent titanium is well
known among scientists working with Ziegler-Natta catalysts to
be due to over alkylation

$$Ti^{4+}Et_2 \longrightarrow Ti^{2+} + C_2H_6 + C_2H_4$$

or

$$\xrightarrow{\Delta} Ti^{2+} + 2Et \cdot \tag{9}$$

The catalyst has a characteristic EPR spectrum as shown in
Fig. 3 with values of 1.945 and 1.913. Cooling to -195°C had
virtually no effect on the EPR lineshape or linewidth; there was
a fivefold increase in intensity. A fourfold increase would be
expected by simple Curie law dependence. It is important to
note that this $Ti^{3+}$ signal represents only $\sim 20\%$ of the concentration of trivalent titanium species determined by redox titration

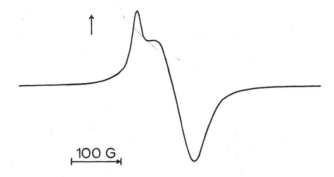

FIG. 3.  EPR of MgCl$_2$ supported Catalyst (IV); temperature = 25°C, rg = 6.2 X 10$^1$.

TABLE 6

Oxidation States of Ti in Catalyst after Activation

|  | Ti$^{2+}$ | Ti$^{3+}$ | | Ti$^{4+}$ |
|---|---|---|---|---|
|  |  | Titration | EPR |  |
| mmol/g | 0.113 | 0.474 | 0.12 | 0.05 |
| $\Sigma[Ti^i]$ (%) | 17.7 | 74.4 | 18.8 | 7.9 |

and amounted to less than 6% of the total titanium complexes present in all oxidation states.  Therefore, there must be a Ti$^{3+}$ species present in the catalyst that is EPR-silent.

The process of activation is clearly one of alkylation of Ti and the reduction of those Ti$^{4+}$ species in the catalyst.  Table 6 showed that more than 90% of the Ti$^{4+}$ complexes were reduced to Ti$^{3+}$ and Ti$^{2+}$ states on reaction with TEA-MPT complexes.  Activation by TEA alone probably will result in the formation of more divalent titanium sites.  EPR studies showed that activation by the TEA-MT complex produced a single chiral Ti$^{3+}$ species (Fig. 4).  The results in Table 6 also show that only 25% of the Ti$^{3+}$ in the activated catalyst is EPR-observable.  This compares with about 20% of the Ti$^{3+}$ in the catalyst before activation, which gave EPR signals.  It appears that

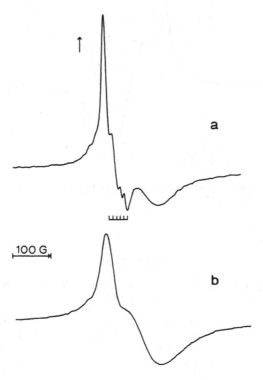

FIG. 4. EPR spectra of Catalyst (IV) activated with TEA/MT complex: (a) 20 min after mixing temperaturel = 25°C, rg = 2 X $10^3$; (b) temperature = -195°C, rg = 4 X $10^2$.

a constant fraction of titanium in the catalyst is isolated and the remaining is chlorine bridged, regardless of the oxidation states.

In conclusion, the morphology changes during the preparation of a $MgCl_2$-supported catalyst are shown in Fig. 5. The ultimate crystallite of $MgCl_2$ was produced by ball-milling with EB. Treatments with PC and TEA reduce the sizes of the aggregates. The reaction of TEA with PC resulted in a weak alkylating complex. $TiCl_4$ treatment led to a very high surface area catalyst with isolated and clustered Ti species of various oxidation states. Upon activation there were about 75% of the Ti in the 3+ oxidation state.

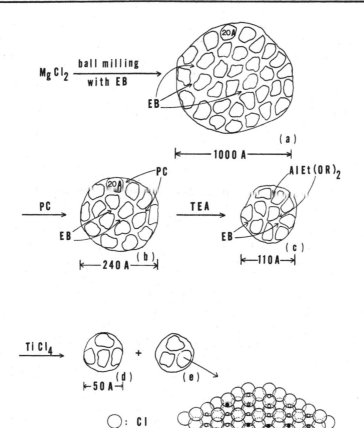

FIG. 5.  Schematic representation of the physicochemical changes of $MgCl_2$ during catalyst preparation.

# Olefin Polymerization Catalysis Technology

H. L. HSIEH
Research and Development
Phillips Petroleum Company
Bartlesville, Oklahoma

I. INTRODUCTION

II. CHROMIUM CATALYSTS FOR HDPE
PRODUCTION

III. Z-N CATALYSTS FOR HDPE PRODUCTION

IV. CATALYSTS FOR POLYPROPYLENE
PRODUCTION

V. CONCLUSIONS

REFERENCES


## I. INTRODUCTION

In the mid-1950s a series of patents on a new ethylene polymer
was issued. These patents were all similar in that solid catalysts
were used to produce polyethylene at relatively low ethylene pressures (Table 1). In contrast to the high pressure process which
produced highly branched polymers with densities of 0.910 to
0.930 g/cm$^3$, the new resins were linear and had densities of
0.940 to 0.970. Thus, the material made by the early ICI process became known as high pressure or low density (LDPE) polyethylene, and the new materials were called low pressure or high
density (HDPE) polyethylene. This early technology for HDPE
may be broadly grouped in terms of three different processes:
Phillips, Standard Oil of Indiana, and Ziegler.

TABLE 1

Early Discoveries of Low Pressure Linear
Polyethylene Processes

| Research group | Catalysts | Time of initial discovery | Early polymer density $(g/cm^2)$ |
|---|---|---|---|
| Zletz [1], Standard Oil (Indiana) | a) Nickel oxide on activated carbon | Last half of 1950 | 0.96 |
| | b) Molybdenum oxide on alumina | | |
| Hogan and Banks [2], Phillips Petroleum Co. | Chromium oxide on silica alumina | Last half of 1951 | 0.95- 0.97 |
| Ziegler et al. [3] | $TiCl_4 + R_3Al$ | November 1953 | 0.945 |

The first catalysts developed by Standard Oil of Indiana were
nickel oxide on activated carbon supports, but this was soon
changed to molybdenum oxide on alumina with alkali or alkaline
earth metals or hydrides as promoters. While tremendous amounts
of research money have gone into developing these catalysts, today
they are of no commercial significance.

As with the discovery of LDPE, Phillips HDPE process was discovered by curious researchers while they were investigating
other systems. In the late 1940s, J. P. Hogan and R. L. Banks
were assigned to study the catalytic dimerization and trimeriza-
tion of olefins, especially ethylene and propylene. The interest
in this stemmed from the large reserves of feedstocks which
Phillips possessed at that time and which could easily be thermally cracked into olefins. If high octane motor fuels could be
produced from these light olefins, an attractive means of utilizing these feedstocks would be available. A catalyst, nickel
oxide on silica-alumina, was developed which accomplished this,
but the life of the catalyst was restricted to 8-10 turnovers before regeneration was necessary. In order to help solve this
problem, a chromium salt was added to the catalyst. After activation and exposure to ethylene this catalyst rapidly solidi-
fied, with a solid white material being formed. The material was

analyzed and found to be polyethylene. From this discovery in 1951, Phillips management recognized the potential of this new polymer, HDPE, and proceeded rapidly to develop it commercially.

The process as commercialized used a catalyst which consisted of chromium oxide supported on silica or silica-alumina support. The process utilized a volatile hydrocarbon solvent in which the ethylene dissolved. As the polymer formed, it dissolved in the solvent and was recovered by precipitation after filtration to remove catalyst. Since this initial process, catalyst productivities have been greatly increased, which has permitted the use of a new process in which the polymer precipitates as it is formed. This is Phillips particle form process. Because of the higher productivities, no catalyst removal is necessary.

The chief competition of the Phillips process is the Ziegler process. The catalyst, titanium tetrachloride activated by an aluminum alkyl, is the basic difference between the two processes. The Ziegler catalysts are well suited for use in Phillips reactor technology. Another difference between the two polymerization methods is the manner in which molecular weight is controlled. Chromium oxide on silica is activated in air at various temperatures to control molecular weight while hydrogen is used in polymerization to control the molecular weight of Ziegler resins. Hydrogen also has some effect on the chromium oxide/silica systems, but to a much lesser extent than on the Ziegler systems. The Ziegler system not only polymerizes ethylene but also propylene. Phillips catalyst does both as well, but the polymerization of propylene with it is of little commercial value.

## II. CHROMIUM CATALYSTS FOR HDPE PRODUCTION

The Phillips catalysts are generally prepared from an amorphous silica gel and $CrO_3$. $CrO_3$ is dissolved in water and the resulting orange solution is used to impregnate a spray-dried silica, already containing a small amount of free water ($H_2O$) and bound water ($-OH$ groups). After mixing in enough of the aqueous $CrO_3$ solution to slightly more than fill the pores and to provide a Cr concentration of about 1%, the catalyst is heated while being mixed until it is again free-flowing. The catalyst is then activated with dry fluidizing air at 500-1000°C for a few hours. At the end of the hold period, the catalyst is allowed to cool with dry air passing through, and the catalyst is then stored under a positive dry air or dry nitrogen pressure.

More than one type of silica is used in the preparation of commercially used catalysts.  Of particular importance are surface
area, pore size, and pore size distribution.  Variations in these
properties bring about differences in catalyst activity, molecular weight, and molecular weight distribution of the polymer produced.  For example, as pore size is increased, the molecular
weight of the polyethylene decreases (melt index increases) at
constant polymerization conditions.  Special techniques for the
control of silica morphology have been developed by Phillips
(and commercialized) for the production of large pore silicas.

Most of the commercial catalysts contain about 1% total Cr.
This concentration is by no means critical but gives a good balance between catalyst activity and resistance to deactivation by
feed impurities.  The particle size is normally in the range of
30 to 150 μm.

The major weakness of Phillips chromium oxide catalyst in the
late 1960s was its limited melt index potential.  It was possible to
produce high melt indices in the solution form process and low
melt indices in the particle form process.  The desire to produce high melt index polymers in the particle form process
stems from the cost advantage of it versus the solution process.  The discovery which put Phillips on the track to higher
melt index potentials in the particle form process was that the
presence of a small amount of titania on the Phillips Cr/silica
polymerization catalyst can be a powerful promoter, enhancing
both activity and especially melt index response.  The promotional effect probably derives from the creation of Ti—O—Cr
links which change the electronic environment on the Cr active
center.  The proportion of Ti—O—Cr links formed, relative to
Si—O—Cr links, is extremely sensitive to subtle variations in the
preparation of catalyst, and determines the magnitude of the promotional effect [4].

Two ways of incorporating titania onto Cr/silica catalysts have
been described in the patent literature.  In the earliest method
[5-7] the surface of the silica was coated with a layer of titania by
treating it with a titanium ester capable of reacting with silanols.

Unreacted ester groups were then burned away during calcination. How the Cr becomes attached to Ti is not always clear in some procedures.

The second method of incorporating titania onto the catalyst consists of coprecipitating hydrous titania along with the silica gel [8]. This can be accomplished by adding a water-soluble titanium salt to the silicate solution before gellation. Although the degree of dispersion is unknown, presumably some of the titania is exposed on the surface and later, during calcination, chromium may attach to it. Several recent publications describe the effects of titanium on Phillips Cr/silica catalysts and discuss the chemistry of the modified catalysts [1, 0 11].

Following the discovery and commercial development of Cr/silica-titania catalyst, another major breakthrough occurred in the middle 1970s. McDaniel and Welch found the control of the molecular weight (MW) of polyethylene from Cr/silica and Cr/silica-titania catalysts was greatly extended by utilizing two principles. First, both MW and activity were found to be inversely related to the hydroxyl population on the support. And this, they discovered, could be additionally lowered by calcining the support in reducing agents like carbon monoxide. Second, two types of active polymerization centers were proposed to exist, one producing low MW polymer, the other high. The low MW center was favored by impregnating the Cr anhydrously onto a highly dehydrated support so that the Cr reacted with oxide bridges rather than hydroxyls [12]. Taking advantage of these two principles, McDaniel and Welch developed a simplified procedure which requires only one activation step and does not resort to anhydrous impregnation. They call this procedure "R/R activation" after the two processes it entails—first a high temperature reduction of the Cr, followed by its reoxidation at a lower temperature [13].

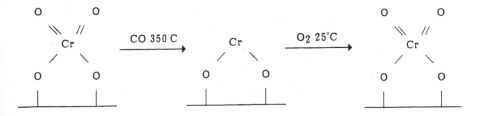

During reduction, not only is the support chemically dehydroxylated, but the bonding of the Cr to the support is also disrupted, destroying the normal high MW centers. With this breakthrough, melt indices of up to 30 g/10 min could be achieved.

Pullukat et al. [9-11, 14] developed another activation procedure which also involves a reduction followed by a reoxidation.
In this case, Cr(VI) was reduced by a titanium ester, $Ti(OR)_4$,
to form what was believed to be a chromium(III) titanate, Cr-
$(OTi(OR)_3)_3$. This material was then heated in a nonoxidizing
atmosphere and later reoxidized, during which some Ti—O—Cr
links were thought to remain intact.

These results make the chromium oxide catalyst developed at
Phillips the most versatile catalyst on the market today. One
can achieve broad molecular weight distributions, high melt indices in the particle form process, and a very wide range of
physical properties by simple modification of the catalyst or activation conditions.

The single most disputed point of the Phillips Cr/silica polymerization catalyst since its development in the early 1950s is
that every valence state from Cr(II) to Cr(VI) has been proposed as the active species, either alone or in combination with
another valence [15-24]. The preparations studied usually contained several valence states, and conclusions were often based
on an indirect correlation between polymerization activity and
some other parameter, such as an ESR signal or lack of it, often
without knowledge of the average oxidation state or consideration of the dispersion of the Cr.

The second most controversial issue has been whether the Cr
is isolated on the surface (chromate) or exists in pairs (dichromate). Hogan [25] reported that two hydroxyls were removed
for each $CrO_3$ attached to the surface at 150°C and, therefore,
that the Cr must bond as chromate. Others, working at 500-
600°C, reported one OH lost per Cr, implying dichromate [15,
16]:

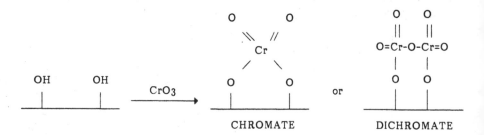

The most recent data obtained by McDaniel and his co-workers
at Phillips [26-28] suggest the following view of the Phillips catalyst: That Cr(VI) is stabilized on silica through the formation of
mainly chromate and that the active site is formed by reduction
to the divalent form, thus creating a high state of coordinative

unsaturation. Adsorption of ethylene decreases the electron density on the Cr.

Other chromium catalysts for ethylene polymerization have been developed and described. For example, Ballard obtained highly active catalysts by supporting $\pi$-cyclopentadienyl and $\pi$-allyl chromium [29]. Union Carbide developed chromocene, $(C_5H_5)_2Cr/SiO_2$, catalyst [30]. Several recent U.S. patents assigned to Phillips describe chromium catalysts supported on aluminum phosphate-containing base as well as chromium catalyst supported on a silica-containing base having mixed valent states. The quest to develop new catalysts to produce resins with superior properties is a challenging and never-ending job.

## III. Z-N CATALYSTS FOR HDPE PRODUCTION

The use of Ziegler catalysts to produce polyethylene initially grew at a much slower pace than the chromium oxide on silica catalysts. The main difficulty with the original catalysts was that yield of polymer per unit of catalyst was so low that the catalyst had to be extracted from the polymer. A typical yield for these early catalysts was in the range of 1.5-3.0 kg of polyethylene/g of transition metal, which meant a polymer metal content of 300-1700 ppm. This much metal, especially titanium, left in the polymer resulted in color and stability problems, and the associated chloride caused severe corrosion problems. The only way to avoid these problems was to wash the polymer to remove these entities. This greatly increased the cost of using Ziegler catalysts as compared to chromium oxide-silica catalysts where yields based on chromium were sufficient so that washing was not necessary. This led to a great deal of research on Ziegler catalyst to find means of improving productivities.

This research effort has paid off and resulted in so-called "second generation" catalysts. The term "Ziegler catalyst," which is conventionally used to mean a titanium tetrachloride used in conjunction with an aluminum alkyl to polymerize olefins, is no longer appropriate. The "second generation" catalysts are supported on a variety of carriers. Table 2 shows the initial work done on supported transition metal catalysts. Although the productivity of these early catalysts was not sufficient to eliminate catalyst removal techniques, they did point the way to what can be truly called "second generation" catalysts where no catalyst removal is necessary. Notice how early, 1955, that work on supported systems was begun. These first catalysts were supported as oxides, halides, and carbonates.

## TABLE 2

### Early Supported Catalysts

| Patent no. | Date | Company | Catalyst preparation |
|---|---|---|---|
| British 841,822 | 1955 | British Petro-chemicals | $TiCl_4$ reduced by aluminum alkyls in the presence of solids such as $MgCO_3$ |
| U.S. 3,153,634 | 1956 | Sun Oil | $TiCl_4$ adsorbed on $SiO_2$/$Al_2O_3$ was reduced |
| U.S. 2,980,662 | 1956 | Sun Oil | Ti compounds were reduced in the presence of solid NaCl, $FeCl_3$, $AlCl_3$, $GaCl_3$, etc. |
| British 877,457 | 1957 | Sun Oil | $TiCl_4$ reduced in the presence of inorganic compounds |
| British 969,761 | 1960 | Cabot | Transition metal compounds were reduced with $SiO_2$, $Al_2O_3$, $SiO_2$/$Al_2O_3$, $ZrO_2$, $TiO_2$, ThO, MgO, etc. |
| British 969,767 | 1963 | Cabot | Transition metal compounds were reduced with $SiO_2$, $Al_2O_3$, $SiO_2$/$Al_2O_3$, $ZrO_2$, $TiO_2$, ThO, MgO, etc. |

The mid-1960s were really when the work on these transition metal supported systems began to come to fruition. Possibly stimulated by Cabot's patent, British Patent 969,761 (1960), Solvay, Hoechst, Mitsui, Montecatini, and B. F. Goodrich came out with a series of patents dealing with supporting halotitanium species on hydroxy magnesium halides (Table 3). The precise structural requirements for the support as well as its role in determining the polymer yield and properties have not yet been clearly defined. The degree of dehydration of the support seems to play a significant role in determining the activity of the catalyst. It also appears that an average particle diameter of 10-20 mμ and specific surface area of 20-30 $m^2$/g would yield optimum results.

## TABLE 3

### Magnesium Hydroxychloride Supported Catalysts

| Patent no. | Date | Company | Catalyst preparation |
|---|---|---|---|
| British 1,024,336 | 1963 | Solvay | Reaction of a transition metal compound with a hydroxychloride of a bivalent metal preferably Mg(OH)Cl |
| Japanese 45-40295 | 1967 | Mitsui | $TiCl_4$ reacted with Mg-$(OH)_2$ and reduced |
| U.S. 3,634,384 | 1968 | B.F. Goodrich | $TiCl_4$ + Al alkyl + Mg(OH) Cl |
| British 726,839 | 1968 | Solvay | Reaction of magnesium hydroxide with a transition metal halide |
| British 728,002 | 1968 | Montecatini | |
| British 735,291 | 1968 | Hoechst | |

The accepted reaction of the transition metal halide with support is

$$MgOH + TiCl_4 \rightarrow MgOTiCl_3 + HCl$$

At about the same time as the work on the magnesium hydroxy-chlorides was going on, companies started to explore the use of magnesium chloride as a support (Table 4). It was soon found that magnesium halides served as even better supports if Lewis bases such as alcohols were added to the catalysts. Much of this work was done utilizing ball mills where the Lewis base and/or the titanium halide were milled into the solid magnesium chloride. The other variation of this procedure is to mill the magnesium chloride with either Lewis base or titanium tetrahalide and wash the catalyst with the other reagent.

Another group of supported catalysts is formed by the reaction of magnesium alkoxides and transition metal halides (Table 5). During the preparation the original structure of the alkoxide is completely destroyed and a new crystalline species formed. This is also accompanied by a high increase in surface area. Modification

## TABLE 4

### Magnesium Chloride Support

| Patent no. | Date | Company | Catalyst preparation |
|---|---|---|---|
| British 1,286,867 | 1968 | Montecatini | $TiCl_4$ ball milled with $MgCl_2$ |
| Belgian 744,221 | 1969 | Montecatini | Reaction of $MgCl_2$ with halo- |
| Belgian 747,846 | 1969 | Montecatini | genated titanium compounds |
| Japanese 46-34092 | 1968 | Mitsui | $TiCl_4$ reacted with $MgCl_2$· |
| U.S. 3,642,746 | 1968 | Mitsui | $nROH$ or $MgCl_2 \cdot nH_2O$ or |
| Japanese 46-34093 | 1968 | Mitsui | $MgCl_2 \cdot n$Lewis base |
| Belgian 755,185 | 1969 | Hoechst | Reaction of a magnesium di-halide electron donor adduct, e.g., $MgCl_2 \cdot 6C_2H_5OH$, with a titanium compound |

## TABLE 5

### Magnesium Alkoxide Support

| Patent no. | Date | Company | Catalyst preparation |
|---|---|---|---|
| U.S. 3,644,318 | 1968 | Hoechst | $TiCl_4$ reacted with $Mg(OR)_2$ |
| Belgian 758,994 | 1969 | Hoechst | or complex magnesium alkox- |
| Belgian 743,325 | 1969 | Solvay | ide |
| Belgian 780,530 | 1971 | Hoechst | Reaction of a magnesium alkox- |
| Dutch 7,216,195 | 1971 | Solvay | ide and an acid halide with a tetravalent titanium compound |

of this catalyst type by introduction of additional elements in the catalytic complexes is possible by use of complex magnesium alkoxides, e.g., $Li_2[Mg(OC_2H_5)_4]$, or by carrying out the reaction in the presence of an acid halide, e.g., $SiCl_4$ or $BCl_3$. In fact, magnesium alkoxide reacted with titanium tetrachloride may have been one of the first, if not the first, "second generation" catalysts used.

## TABLE 6

### Catalysts Made with Magnesium Alkyls

| Patent no. | Date | Company | Catalyst preparation |
|---|---|---|---|
| Netherlands 7,100,931 | 1971 | Shell | $Bu_2Mg + MgBr_2 + TiCl_4$ |
| German 2,211,486 | 1972 | Stamicarbon | $Hexyl_2Mg + EtAlCl_2 + TiCl_4$ |
| German 2,519,071 | 1975 | Asahi | $Bu_2Mg/Et_3Al + TiCl_4$ |
| German 2,517,567 | 1975 | Dow Chemical Co. | $Bu_2Mg + Et_3Al + Ti(OiPr)_4$ |
| Japanese 76-64,586 | 1974 | Mitsui | $Bu_2Mg + EtOH + Et_2AlCl + TiCl_4$ |
| Japanese 76-111,281 | 1976 | Asahi | $Bu_2Mg + Et_3Al + BuOH + TiCl_4$ |

As early as 1968, Hoechst was able to perform HDPE production in several large-scale plants employing a high yield catalyst (possibly $Mg(OMe)_2$ treated with $TiCl_4$) system to avoid catalyst removal steps.

Catalysts prepared by reacting magnesium alkyls and titanium compounds started appearing in the late 1960s and have continued into the 1970s (Table 6). This is done using a variety of techniques. The development of these catalysts has been closely associated with the discovery and availability of soluble dialkyl magnesium compounds.

In the 1970s the primary interest has not been in catalyst productivity, but more in the area of polymer properties. As the polyolefin business becomes more competitive and the cost of production escalates, it becomes more and more important to control precisely such factors as molecular weight, polydispersity, and chain branching. If these physical parameters cannot be controlled during polymerization, then they must be controlled by physically blending of the polymer which is expensive and gives rise to quality control problems.

In the early 1970s at Phillips, we realized the importance of obtaining a good balance of physical properties. This goal, coupled with the compatibility of the "second generation" transition metal in our process technology, led to our involvement in this research in the early 1970s [31].

The phrase "high activity catalyst," in regard to transition metal olefin polymerization catalysts, has been abused and often times misused in recent years. The patent literature is full of examples of high activity catalysts when all that is meant is a high yield of polymer per unit weight of titanium. In many of these instances other catalyst components such as chloride ions are exceedingly high in the polymer, or the polymerization times and conditions are not commercially feasible. Thus, one must be cautious when dealing with published productivity values.

By a highly active catalyst we mean one which leaves no more than 10 ppm titanium and 30 ppm chloride in the polymer formed. Titanium levels higher than this lead to color problems, and higher chloride content results in excessive corrosion problems. In addition, other metals must be present in less than 10 ppm in the polymer, with the actual permissible level controlled on an individual basis. Any residues in the polymer must meet FDA regulations.

The advantages of highly active catalysts have been greatly overstated in recent years. In fact, only through activities high enough to eliminate catalyst removal is it possible for transition metal catalysts to compete with chromium oxide catalysts where catalyst removal is not used. In addition, the chromium oxide system contains no chloride. Therefore, the use of high activity transition metal catalysts to replace chromium oxide systems cannot be justified solely in terms of productivity. A similar situation is found for polypropylene. Here the new catalyst must be so high in productivity and stereospecificity that all post-polymerization purification steps are eliminated and the catalyst must be relatively inexpensive because "installed" purification systems are not exceedingly expensive to operate. The main cost savings would not be in existing plants but in newly constructed facilities.

High activity transition metal catalysts are generally prepared by one of two techniques. Either the catalyst is prepared with intensive ball milling or no milling is used at all. In general, milled catalysts are somewhat more difficult to prepare than their unmilled counterparts. Milling must be done under anhydrous conditions. This is difficult and expensive. The preferred method would be one in which the catalysts require no milling. It is much easier to handle a slurry in a closed system than it is to handle a solid material in a ball mill. Since 1980, some grades of commercial resins are being produced by Phillips with our proprietary catalysts. Active R&D work on new ethylene polymerization catalysts

TABLE 7

Desirable Features of Polyethylene Catalysts

| Catalyst feature | Benefits |
|---|---|
| High activity | Eliminate de-ashing, improve color and stability |
| High copolymerization capability | Broad product density |
| In-situ branching | Eliminate comonomer |
| MW and MWD control | Broad product capability |
| High product bulk density | Increase reactor throughput |
| Product particle size control | Direct sale of reactor product, energy reduction |
| Low cost | Economical |

is continuing.  In Table 7 are listed the desirable features researchers are seeking.

## IV.  CATALYSTS FOR POLYPROPYLENE PRODUCTION

The commercial catalysts used for the production of polypropylene from the late 1950s on are listed in Table 8.  Note the dramatic increase in relative activity based on titanium as well as the improvements in stereospecificity.

The high surface $\delta$-TiCl$_3$ is generally recognized as one of the best catalysts of the TiCl$_3$ type and has been used commercially to manufacture polypropylene in several locations.

The three-step TiCl$_3$ catalyst can be prepared in view of Solvay's original German application, 2,213,086, in 1972.

Step 1:  A.  Dilute 1 mol TiCl$_4$ with hexane and cool to 0°C
         B.  Dilute 1.1 mol Et$_2$AlCl with hexane, add to TiCl$_4$ slowly with stirring
         C.  Stir with graduate warming to 65°C, and hold 1 h
         D.  Wash the resulting brown solid thoroughly with hexane.

TABLE 8

Polypropylene Catalyst Development Periods

| Development stage | Commercial use | Relative activity | Isotactic index |
|---|---|---|---|
| In-situ production catalyst | 1957 | 1 | 75-85 |
| External production catalyst $\delta$-TiCl$_3$ | 1960 | 10 | 85-87 |
| Promoted catalyst $\delta$-TiCl$_3$· 1/3 AlCl$_3$ | 1970 | 13 | 88-94 |
| High surface catalyst Solvay $\delta$-TiCl$_3$ | 1976 | 40 | 94-96 |
| High activity catalyst | 1979 | 300-2000 | 94-98 |
| Super high activity catalyst | 1982 | 6000+ | 94-97 |

Step 2:  Add 1 mol diisoamyl ether to the slurry brown solid of Step 1 in hexane. Warm to 35°C and stir 1 h. Decant the brown solid complex and wash thoroughly with hexane.

Step 3:  Add 3 mol TiCl$_4$ to the slurry brown solid complex of Step 2 in hexane. Heat to 65°C and stir 2 h. Cool, decant the purple solid, and wash thoroughly with hexane.

This $\delta$-TiCl$_3$ catalyst has a high surface area of around 140 m$^2$/g compared to $\delta$-TiCl$_3$·1/3 AlCl$_3$ catalyst (4 m$^2$/g). It is generally viewed that the removal of AlCl$_3$ by extraction with diamyl ether from the cocrystalline TiCl$_3$-AlCl$_3$ particle results in the formation of a Ti with a significantly larger population of single and (especially) double vacancy sites and induces a microporosity that allows access to these sites. This may largely explain the higher activity of catalyst components of this composition in comparison with pure violet TiCl$_3$. The epitactic adsorption of TiCl$_4$ on single and double vacancy sites creates the proper site structure for subsequent conversion into highly stereospecific active centers upon reaction with Et$_2$AlCl cocatalyst [32].

TABLE 9

$MgCl_2$/Ethyl Benzoate/Hexene-1/$TiCl_4$ Catalyst

$$MgCl_2 + EtOBz + hexene\text{-}1 + TiCl_4 \xrightarrow[\text{24 h}]{\substack{\text{vibratory}\\ \text{ball mill}}} \text{catalyst}$$

Cocatalyst: $Et_3Al$/$Et_2AlCl$/ethyl anisate

Activity:  150 kg PP/g Ti
or         6 kg PP/g catalyst

Isotactic index = 96%

Polymer residuals = Ti 7 ppm, $\overline{Cl}$ 120 ppm

The supported catalysts, often referred to as high activity cat-
alyst or high millage catalyst, are mostly based on $MgCl_2$ support.
An early example of this type is the catalyst made in view of a
Montedison's German patent application, 2,347,577, in 1973.

1.  Complex 1 mol $TiCl_4$ with 1 mol ethyl benzoate. This gives
a bright yellow solid.
2.  Ball mill anhydrous $MgCl_2$ 4 days in a jar mill.
3.  Charge the yellow complex, milled $MgCl_2$, and durene in a
jar mill. Milling for 5 days gives a brown caked solid which must
be broken loose from the jar walls after 24 h. The final product is
broken up to pass 500 mesh. The catalyst contains 3% Ti.

The above catalyst, combined with a cocatalyst containing $Et_3Al$
and ethyl anisate or ethyl benzoate, will give an activity of 80-100
kg PP/g Ti in liquid propylene with hydrogen at 65°C. The propylene and xylene solubles are comparable to a commercially promoted catalyst.

Another early example of supported catalyst is shown in Table 9.
We prepared the catalyst by milling together magnesium chloride,
ethyl benzoate, hexene-1, and titanium tetrachloride [33]. Perhaps the most important feature of this catalyst is the cocatalyst.
We have had best results using a cocatalyst which contains equal
moles of $Et_3Al$ and $Et_2AlCl$ with an ester such as ethyl anisate or
ethyl benzoate. An activity of 150 kg PP/g Ti was obtained in
our pilot plant. This means that the titanium and chloride residuals are 7 and 120 ppm, respectively. The latter is much too high

TABLE 10

Examples of Some High Activity Catalysts for Propylene Polymerization in Slurry Phase [34]

| Catalyst | Productivity (kg/g Ti) | Isotactic index (%) | Company |
|---|---|---|---|
| [(MgCl$_2$ + EtOBz)milled + TiCl$_4$ + (ICl$_3$ in ClC$_2$H$_4$Cl)] / Et$_3$Al | 330 | 97 | Mitsubishi Petrochemical |
| [(TiCl$_3$ in ClC$_2$H$_4$Cl + (ICl$_3$ in ClC$_2$H$_4$Cl) + ((MgCl$_2$ + EtOBz)milled + ClC$_2$H$_4$Cl)]/Et$_3$Al | 274 | - | Mitsubishi Petrochemical |
| [(MgCl$_2$ + Phborate)milled + EtOBz + TiCl$_4$]milled + CH$_2$ClCH$_2$Cl + ICl$_3$/Et$_3$Al | 335 | 97 | Mitsubishi Petrochemical |
| [MgCl$_2$ + EtOBz + EtOH + SiCl$_4$] /Et$_3$Al | 203 | 97 | Mitsubishi Petrochemical |
| [MgCl$_2$·6EtOH + EtOBz] in SiCl$_4$ + TiCl$_4$/(i-Bu)$_3$Al | 372 | 90 | Mitsui Petroleum |
| [(MgCl$_2$ + EtOBz + EtOH)milled + TiCl$_4$] /Et$_3$Al + Me-4-toluene | 210 | 96 | Mitsui Petroleum |

| | | | |
|---|---|---|---|
| [(MgCl$_2$ + decane + 2-ethylhexylalcohol) + TiCl$_4$ + EtOBz] / EtOBz + TiCl$_4$/(i-Bu)$_3$Al + Et$_3$Al$_2$Cl$_3$ + Me-4-toluene | 438 | 98 | Mitsui Petroleum |
| [(Mg + (EtO)$_4$Si + (I$_2$ in MeI) + BuCl) + BzCl + TiCl$_4$] / Et$_3$Al + Me-4-toluene | 292 | 90 | Montedison |
| [(MgCl$_2$ + Phenol)milled + TiCl$_4$]100°C /Et$_3$Al | 875 | 94 | Phillips Petroleum [35] |
| [((MgCl$_2$ + (BuO)$_4$Ti + EtOBz + Phenol) + Et$_3$Al$_2$Cl$_3$) + TiCl$_4$]100°C /Et$_3$Al + Et-anisate + Et$_2$AlCl | 300 | 96 | Phillips Petroleum |
| [MgCl$_2$ + BzCl)milled + PhME + TiCl$_4$ + P(PhO)$_3$] / Et$_3$Al + EtOBz | 507 | 96 | Showa Denko |
| [(MgCl$_2$ + phenyl phosphoric dichloride)milled] + TiCl$_4$ + P(PhO)$_3$/Et$_3$Al | 519 | 99 | Showa Denko |
| [(MgCl$_2$ + decane + 2-ethylhexylalcohol + EtOBz) + TiCl$_4$ + EtOBz]/(i-Bu)$_3$Al + Et$_3$Al$_2$Cl$_3$ + Me-p-toluene | 438 | 98 | Montedison/Mitsui Petroleum |
| [[(MgCl$_2$ + EtOBz)milled + EtOBz + TiCl$_4$]milled] refluxed with CH$_2$ClCH$_2$Cl/Et$_3$Al | 546 | 95 | Mitsubishi Petrochemical |

TABLE 11

Examples of Some Super High Activity Catalysts for Polypropylene
Polymerization in Slurry Phase [34]

| Catalyst | Productivity (kg/g Ti) | Isotactic index (%) | Company |
|---|---|---|---|
| [(MgCl$_2$ + EtOBz)milled + TiCl$_4$] + CCl$_4$/Et$_3$Al + p-MeOC$_6$H$_4$CO$_2$Et | 2000 | 96 | Shell International Research |
| [(MgCl$_2$ + EtOH + YF + kerosene) + EtOBz + TiCl$_4$]/Et$_3$Al + MeC$_6$H$_4$CO$_2$Me | 4800 | 97 | Mitsui Petroleum |
| [(MgCl$_2$ + EtOH + kerosene + Emasol 320 + Et$_3$Al + EtOBz + TiCl$_4$]/Et$_3$Al + MeC$_6$H$_4$CO$_2$Me | 8577 | 96 | Mitsui Petroleum |

and is not acceptable. One common difficulty which has been found for most of the high activity catalysts is that it has not been possible to maintain high activities and at the same time reduce solubles (high isotactic indices) to the desired level. Some examples of the high activity catalysts are listed in Table 10. The list is by no means complete and serves only as an illustration. No less than 32 companies have filed patent applications describing supported catalysts for propylene polymerization. In the 1977-1978 period, over 120 applications were filed. In contrast, only about 10 were filed in the 1973-1974 period.

In Table 11, examples of some super high activity catalysts for propylene polymerization are listed. Super high activity catalysts which will give less than 10 ppm residual titanium and 30 ppm residual chloride, combined with 97%+ isotactic index, are the current goals of catalyst researchers. Ease of preparation is another important consideration.

## V. CONCLUSIONS

The new generations of olefin polymerization catalysts generated a lot of excitement initially. However, to be a viable commercial alternative to the established catalysts, they will have to demonstrate better polymer properties as well as cost savings.

## Acknowledgments

The author wishes to thank Dr J. P. Hogan, Dr M. P. McDaniel, and Dr M. B. Welch of Phillips Petroleum Company for their input and useful discussion in preparing this manuscript.

## REFERENCES

[1]  A. Zletz (to Sandard Oil, Indiana), U.S. Patent 2,692,257 (1954).

[2]  J. P. Hogan and R. L. Banks (to Phillips Petroleum Co.), U.S. Patents 2,827,721 and 2,846,425 (1958).

[3]  K. Ziegler, H. Breil, H. Martin, and E. Holzkamp (to Karl Ziegler), German Patent 973,626 (1960).

[4]  M. P. McDaniel, M. B. Welch, and M. J. Dreiling, J. Catal., 82, 118 (1983).

[5]  J. P. Hogan and D. R. Witt (to Phillips Petroleum Co.), U.S. Patent 3,622,521 (1971).

[6]  B. Horvath (to Phillips Petroleum Co.), U.S. Patent 3,625,864 (1971).

[7]   T. J. Pullukat and M. Shida (to Chemplex Co.), U.S.
      Patent 3,780,011 (1973).
[8]   R. E. Dietz (to Phillips Petroleum Co.), U.S. Patents
      3,887,494 (1975) and 4,119,569 (1978).
[9]   T. J. Pullukat, R. E. Hoff, and M. Shida, J. Polym. Sci.,
      Polym. Chem. Ed., 18, 2857 (1980).
[10]  T. J. Pullukat, R. E. Hoff, and M. Shida, J. Appl.
      Polym. Sci., 26, 2927 (1981).
[11]  T. J. Pullukat, M. Shida, and R. E. Hoff, in Transition
      Metal Catalyzed Polymerization, Part B (MMI Press Sympo-
      sium Series, Vol. 4, R. P. Quirk, ed.), Hardwood, 1983,
      p. 697.
[12]  M. P. McDaniel and M. B. Welch, J. Catal., 82, 98 (1983).
[13]  M. P. McDaniel and M. B. Welch (to Phillips Petroleum
      Co.), U.S. Patents 4,151,122 (1979), 4,177,162 (1979),
      4,182,815 (1980), 4,247,421 (1981), and 4,277,587 (1981).
[14]  R. E. Hoff, T. J. Pullukat, and M. Shida (to Chemplex
      Co.), U.S. Patent 4,041,224 (1977).
[15]  H. L. Krauss, in Proceedings, 5th International Confer-
      ence on Catalysis, Palm Beach, 1972, Vol. 1 (J. W.
      Hightower, ed.), North Holland, Amsterdam, 1973, p. 7.
[16]  A. Zecchina, E. Garrone, G. Ghiotti, C. Morterra, and
      E. Borello, J. Phys. Chem., 79(10), 966 (1975) and suc-
      ceeding papers to p. 988.
[17]  L. M. Baker and W. L. Carrick, J. Org. Chem., 33(2),
      616 (1968).
[18]  D. D. Beck and J. H. Lunsford, J. Catal., 68, 121 (1981).
[19]  L. K. Przhevalskaya, V. A. Shvets, and V. B. Kazanski,
      J. Catal., 39, 363 (1975).
[20]  C. Eden, H. Feilchenfeld, and Y. Haas, Ibid., 9, 367
      (1967).
[21]  D. D. Eley, C. H. Rochester, and M. S. Scurrell, Proc.
      R. Soc. London, A, 329, 361-390 (1972); J. Catal., 29,
      20 (1973).
[22]  L. L. Reijen and P. Cossee, Discuss. Faraday Soc., 41,
      277 (1966).
[23]  K. G. Miesserov, J. Polym. Sci., Part A-1, 4, 3047 (1966);
      J. Catal., 22, 340 (1971).
[24]  G. Villaume, R. Spitz, A. Revillon, and A. Guyot, J.
      Macromol. Sci.—Chem., A8(6), 1117 (1974).
[25]  J. P. Hogan, J. Polym. Sci., Part A-1, 8, 2637 (1970).
[26]  R. Merryfield, M. P. McDaniel, and G. Parks, J. Catal.,
      77, 348 (1982).
[27]  M. P. McDaniel, in Transition Metal Catalyzed Polymeriza-
      tion, Part B (MMI Press Symposium Series, Vol. 4, R. P.
      Quirk, ed.), Hardwood, Academic Publishers, 1983, p. 713.

[28]   M. P. McDaniel, J. Catal., 76, 17, 29, 37 (1982).
[29]   D. G. H. Ballard, Adv. Catal., 23, 263 (1973).
[30]   F. J. Karol, G. L. Karapinka, C. Wu, A. W. Dow, R. N.
       Johnson, and W. L. Carrick, J. Polym. Sci., Polym.
       Chem. Ed., 10, 2621 (1972).
[31]   H. L. Hsieh, 5th International Symposium on Cationic and
       Other Ionic Polymerization, Kyoto, Japan, April 15-18,
       1980. Also, Polym. J., 12(9), 597 (1980).
[32]   R. P. Nielsen, in Transition Metal Catalyzed Polymeriza-
       tion, Part A (MMI Press Symposium Series, Vol. 4, R. P.
       Quirk, ed.), Hardwood, 1983, p. 47.
[00]   M. B. Welch (to Phillips Petroleum Co.), U.S. Patent
       4,242,480 (1981).
[34]   For individual patent references, see p. 194-197 in Transi-
       tion Metal Catalyzed Polymerization, Part A (MMI Press
       Symposium Series, Vol. 4, R. P. Quirk, ed.), Hardwood,
       1983.
[35]   M. B. Welch and R. E. Dietz (to Phillips Petroleum Co.),
       U.S. Patent 4,243,552 (1981).

# Photocatalysis and Photovoltaics

# Hydrogen Evolving Solar Cells*

ADAM HELLER
AT&T Bell Laboratories
Murray Hill, New Jersey

Optimal conversion of sunlight by a single-threshold converter, whether semiconductor-based or molecular, requires an energy gap near 1.4 eV [1-3]. Tandem systems, based on two semiconductors or on two light-harvesting molecules, require materials with gaps near 1.8 and 1.0 eV [4]. At normal solar irradiance and at 27°C, the thermodynamic limit to the solar conversion efficiency is 27% for a single converter and 36% for tandem cells [4]. For nonconcentrated sunlight the actual efficiency that has been attained is 21.9% [5]. Although the thermodynamic efficiency limits for semiconductor and molecular systems are the same, all efficient systems today are semiconductor-based.

The foundations of the science of semiconductor-based photoelectrodes were laid between 1955 and 1970 [6-16]. In 1972 it was shown that when an oxygen-evolving metal anode is replaced by the stable semiconductor anode n-$TiO_2$ [17], a substantial part of the electrical energy required for the electrolysis of water is conserved [18]. Unfortunately, the band gap ($E_{BG}$) of n-$TiO_2$ is 3.0 eV, so the excitation spectrum of this material is confined to the UV. Consequently, sunlight could not be efficiently converted in the early cells based on n-$TiO_2$ or n-$SrTiO_3$ ($E_{BG}$ = 3.2 eV) photoanodes [19-24].

Photoanodes with band gaps appropriate for efficient solar conversion were first reported in 1960 [9], before the introduction of the concept of power-producing ("regenerative") cells [25]. However, these were quite unstable. Ideas for their stabilization emerged in 1966 [26], but the introduction of relatively stable photoanodes took nearly a decade [27-50].

Simultaneously, the second key problem of photoelectrodes, the reduction of quantum yield from radiationless recombination of electron-hole pairs at the semiconductor-solution interface, was addressed. It was shown that recombination can be reduced by properly controlling the interfacial chemistry [37, 51-66]. These advances led to the first power-producing photoelectrochemical cells of greater than 10% solar conversion efficiency [51, 52, 63, 64, 67]. Stabilization and reduction of surface recombination also opened the way to efficient and direct photoelectrochemical cells for producing hydrogen. These cells are the subject of this article.

## I.  PHOTOELECTRODES

Hydrogen can be evolved in either photocathode- or photoanode-based cells. Photocathodes and photoanodes are based on electrical

contacts between semiconductors and electrolytes, between semiconductors and metallic electrocatalysts, or between semiconductors and both electrolytes and catalysts. Associated with these
contacts is a barrier, $\psi_B$, which separates the photogenerated
electrons and holes (Fig. 1). Hydrogen evolution at photocathodes requires the presence of a catalyst on or near the surface.
When the catalyst is on the surface of the semiconductor, photogenerated electrons diffuse or drift to the interface of the semiconductor and the catalyst (Fig. 1a) where they reduce adsorbed protons to form hydrogen, according to $2e^- + 2H^+ \rightarrow H_2$.
Electrical neutrality is maintained by transport of holes through
both the bulk of the semiconductor and its back electrical contact
to the anode, where they oxidize either dissolved anions, such as
chloride, via the reaction $2h^+ + 2Cl^- \rightarrow Cl_2$, or water itself, via
the reaction $4h^+ + 2H_2O \rightarrow 4H^+ + O_2$. As in conventional solar
cells, the light generates a potential difference between the
front and the back of the semiconductor. If this photogenerated
potential is inadequate to produce electrolysis, it must be aug-
mented by an external dc power source.

In illuminated photoanodes, the photogenerated holes drift to
the contacting solution interface where they oxidize the electrolyte or water. The electrons traverse the semiconductor to the
back contact and reach the cathode via the external circuit,
where they reduce protons to hydrogen (Fig. 1b).

If the Gibbs free energy change in a thermodynamically uphill
reaction, such as $H_2O \rightarrow H_2 + 1/2O_2$, is $\Delta G$ kcal/mol, and if the
same reaction can be run under sunlight by investing only $\Delta G'$
kcal/mol, the Gibbs free energy efficiency of solar-to-chemical
conversion is $\eta = (\Delta G - \Delta G')/\int_0^\infty I\,d\lambda$, where the denominator
is the solar irradiance integrated over wavelengths from $\lambda = 0$
to $\lambda = \infty$. In an electrochemical cell a gain in free energy trans-
lates to a reduction in the potential bias needed for electrolysis.
If electrolysis requires in a thermodynamically reversible cell, a
potential bias of $V = \Delta G/nF$ volts (where n is the number of electrons transferred and F is Faraday's constant), and if under sunlight electrolysis results at a bias $V' = \Delta G'/nF$, a bias reduction
$\Delta V = V - V'$ is realized. The Gibbs free energy efficiency of solar-to-chemical conversion ($\eta$, Fig. 2) is now defined by

$$\eta = \frac{\Delta V_{max} i_{max}}{\int_0^\infty I\,d\lambda} \tag{1}$$

where $\Delta V_{max}$ and $i_{max}$ are the reduction in potential bias and the
photocurrent density at the point where their product reaches its

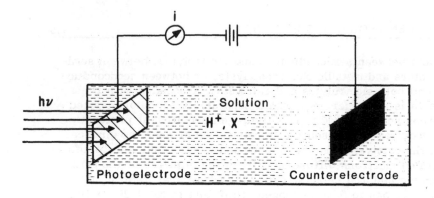

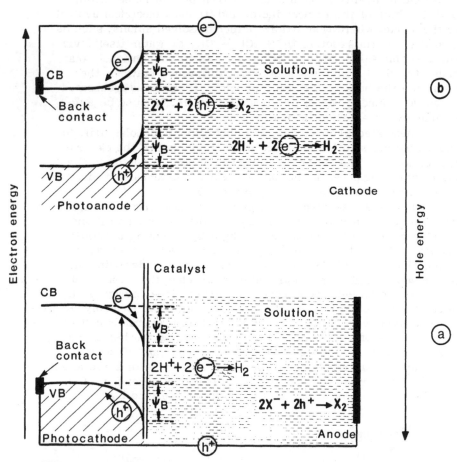

FIG. 1. (a) Separation, transport, and reaction of photogenerated electron-hole pairs in photocathode-based cells. Under intense irradiance the maximum bias reduction, $\Delta V_{OC}$ approaches the
height of the barrier that separates the charge carriers, $\psi_B$. (b)
Same, in photoanode based cells. Abbreviations: CB, conduction
band; VB, valence band.

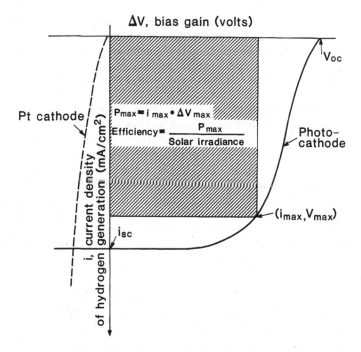

FIG. 2. Definition of the maximum bias reduction $\Delta V_{oc}$, the short circuit current $i_{sc}$, and the maximum solar-to-hydrogen conversion point $(\Delta V_{max}, i_{max})$. The Gibbs free energy efficiency is the ratio of the product $\Delta V_{max} \cdot i_{max}$ and the solar irradiance integrated over all wavelengths.

maximum value. Thus $\eta$ is 10% if, at pH 0, under a solar irradiance of 100 mW/cm$^2$ hydrogen is evolved at $\Delta V_{max}$ = +0.5 V vs SHE (standard hydrogen electrode) potential at a photocurrent density $i_{max}$ = 20 mA/cm$^2$. $\Delta V$ at the threshold of hydrogen evolution ($i = 0$) is $\Delta V_{oc}$. At $\Delta V = 0$ the photocurrent density is $i_{sc}$ (Fig. 2).

## II. QUANTUM EFFICIENCY

A photoelectrode reaches its theoretical integrated quantum efficiency of unity when all incident photons with energies exceeding $E_{BG}$ produce electron-hole pairs and when all of the charge carriers react with the electrolytic solution to produce the desired cathodic and anodic products. Losses in current efficiency are caused by radiationless recombination of photogenerated

electron-hole pairs, leakage of the "wrong" charge carriers across
the barrier, absorption and reflection of photons by the metallic
catalyst needed to accelerate the desired electrochemical reaction,
reflection of photons at the semiconductor-solution interface, absorption of light by the solution, and undesired electrode reaction
products.

Recombination, leakage, and reflection losses are common to
all solar cells [1]. In hydrogen-generating cells the quantum efficiency is determined by the ratio of the desired rate of hydrogen evolution to the sum of this rate and the undesired recombination and leakage rates [68]. All of these rates depend on
$\Delta V$ and on the surface chemistry of the semiconductor. Hydrogen evolution at high $\Delta V$ is much slower on most semiconductor
surfaces than are recombination and leakage, even if the latter
are suppressed by a properly controlled interfacial chemistry.
Hydrogen production is greatly enhanced when a Group VIII
metal catalyst, such as Pt, is incorporated in the semiconductor
surface [69].

An example of a catalyst-caused change from recombination
and leakage to efficient hydrogen production is seen in p-InP
photocathodes [69-72] which have ideal band gaps for solar
conversion ($E_{BG} = 1.34$ eV). In contact with aqueous acids,
their surfaces become covered with a 6-10 Å pore-free layer of
hydrated indium oxide, through which electrons freely tunnel
[73]. The thin indium oxide layer, together with cathodic protection by the photogenerated electrons [69-72], provides stability against oxidative corrosion [74, 75]. This layer also reduces
("passivates") the radiationless recombination of the photogenerated electrons with holes [62, 64, 65, 71, 72]. The presence
of a Group VIII metal at the surface catalyzes the reaction $H^+ +$
$e^- \rightarrow 1/2H_2$ and thereby increases the hydrogen evolution rate
and the quantum efficiency. For example, at pH = 0 and $\Delta V =$
0.5 V, incorporation of platinum yields an increase in quantum
efficiency of $10^4$ to 0.8 (Fig. 3) [72].

A uniform, 50 Å thick layer of platinum will, however, reflect
and absorb about one-half the useful photons in the solar spectrum. To avoid much of this loss, the catalyst must be deposited
in the form of islands with diameters smaller than the useful wavelengths of sunlight. Coverage of a semiconductor surface by such
catalyst islands results in greater transmission of light than expected from simple geometrical reasoning because of the screening of the optical electric field by the mobile electronic charge of
the metal. This screening channels the light into the regions between the islands. This phenomenon, well-known in the study of
the optical response of heterogeneous materials [76], is described
by the effective medium theory [77].

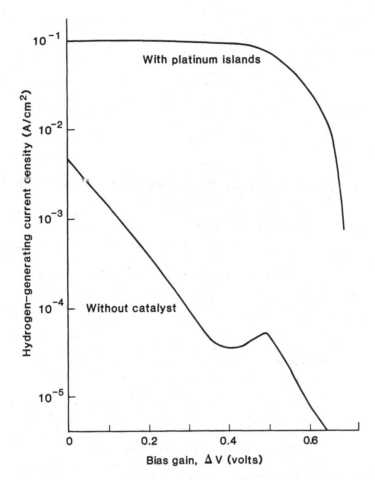

FIG. 3. Increase of the current density of hydrogen evolution upon incorporation of platinum in the surface of a p-InP photocathode, as a function of $\Delta V$. (3 $\underline{M}$ HClO$_4$, tungsten-halogen source.)

While increasing the island spacing improves the optical efficiency, it also reduces the collection efficiency of photogenerated electrons. The crossover point occurs where nearly all the electrons reach the islands before recombining with holes or leaking through the barrier. The maximum acceptable spacing between the islands can be increased and the transmission of photons to the semiconductor improved by increasing the lateral diffusivity/recombination-leakage ratio of the electrons. The most efficient

photoelectrodes for conversion of solar energy to hydrogen are
made, therefore, from high quality, defect-free semiconductors
with nonleaking barriers and with surface recombination passi-
vated by a controlled semiconductor surface chemistry. If the
surface diffusivity of electrons is small, one must choose between a severe light absorption loss caused by a high density
of catalyst islands, or a severe loss in collection efficiency from
a low density of catalyst islands. Thus, high quantum (current)
efficiency can be achieved only when three conditions are simultaneously met: the catalyst islands cover only a small fraction
of the surface, their diameter is much smaller than the wavelength of the exciting photons, and their spacing is not larger
than the minority carrier surface diffusion length.

Two examples which demonstrate that catalyst islands smaller
than the wavelength of exciting light do not reduce the quantum
efficiency in proportion to their surface coverage are given in
Fig. 4. In one (Fig. 4a) the Pt catalyst islands cause a loss of
less than 8% in the quantum efficiency of the p-InP photocathode,
even though the islands, with an average diameter of 700 Å, cover
about 40% of the surface and are several hundred Å thick. The
islands in Fig. 4(b), with an average diameter of 1400 Å, are
∿1000 Å high and cover about 60% of the p-InP surface. However, the light absorption loss is less than 30% [78]. Because
the surface diffusion length of electrons in p-InP is much greater
than the ∿1000 Å spacing between the islands, all of the photogenerated electrons are collected efficiently.

The surface chemistry, which determines the density and distribution of surface states in which photogenerated carriers recombine, is dependent on potential. Consequently, an island
spacing that ensures efficient hydrogen generation at one potential may not be suitable at another. This is illustrated for catalyzed p-InP photocathodes. Their quantum efficiency drops at
excessively reducing potentials or upon operation at low $\Delta V$.
The drop is caused by a partial loss of the indium oxide layer,
which introduces recombination centers [79]. Full efficiency is
readily restored upon reoxidation of the surface [69, 70, 72,
80].

The photocurrent efficiency is a function of the surface composition also in photoanodes. Layered-compound photoanodes,
such as n-MoSe$_2$ and n-WSe$_2$, have appropriate band gaps for
solar conversion. Relative to other photoanodes having such
band gaps, they are more stable against oxidative photocorrosion [35, 40-43, 57, 59-61, 63, 67, 81, 82], but stability and
high current efficiency are realized only for the cleavage (van
der Waals) planes of the crystals [43, 57, 59-61, 63, 67, 82].
Because electron-hole recombination is rapid on the other

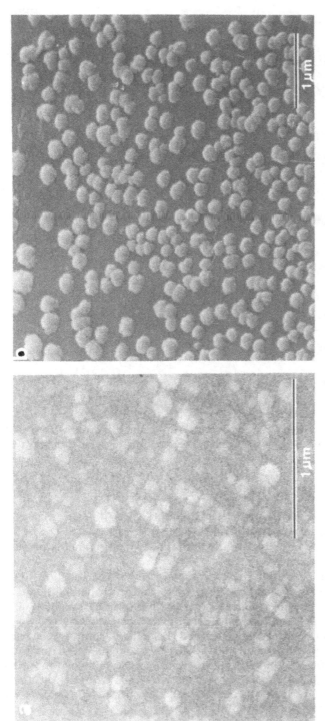

FIG. 4. Platinum islands on a hydrogen-evolving p-InP photocathode. The dimensions of the islands are small relative to the wavelengths of the solar spectrum exciting the semiconductor and are spaced at center-to-center distances smaller than the surface diffusion length of the photogenerated electrons. (a): About 40% of the surface is covered with an optically thick catalyst layer, but fewer than 8% of the solar photons are lost. (b): The optically thick catalyst covers 60% of surface, but less than 30% of the photons are lost [78].

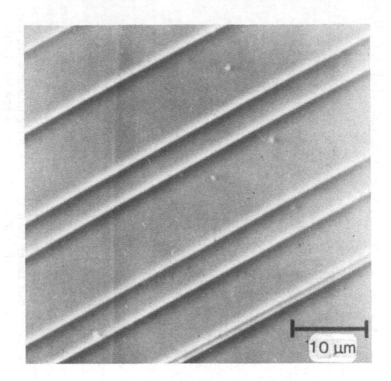

FIG. 5. Electron beam-induced current (EBIC) micrograph of a p-WSe$_2$/Al contact. Light areas, representing sites where electrons and holes recombine, are observed at steps in the van der Waals planes [60].

crystallographic planes, the solar conversion efficiency decreases with the density and height of steps in the surface (Fig. 5) [57, 59, 60]. With perfectly smooth photoanodes, efficiencies of 10% have been reached [63, 67].

Other than recombination, leakage is the most important loss process. Leakage of majority carriers (holes in photocathodes and electrons in photoanodes) from the bulk of the semiconductor to the catalyst or to the solution interface reverses the desired photoelectrochemical reaction. Leakage of minority carriers (electrons in photocathodes and holes in photoanodes) from the catalyst, or from the solution interface, to the bulk produces the same effect. If interfacial hydrogen atoms are oxidized either by bulk holes or by loss of electrons to the bulk, the quantum efficiency of hydrogen evolution drops.

Leakage is caused by thermionic emission across, or by chemical imperfections in, the barrier, $\psi_B$, that separates the photogenerated minority and majority carriers (Fig. 1). It is avoided if the interfacial potential barrier is made uniformly high. For example, metallic Ru forms a leaky barrier with n-GaAs. The leakage is reduced when the metal is saturated with hydrogen that reacts with the leakage-causing sites [83].

## III. BIAS REDUCTION ($\Delta V$)

In a hydrogen-evolving solar cell $\Delta V$ is the equivalent of the photovoltage in a power-generating cell. Like the open-circuit photovoltage $V_{oc}$ of a power-generating cell, $\Delta V_{oc}$, the bias reduction at the threshold for hydrogen generation, is determined by the barrier height, $\psi_B$, and by the irradiance. In photocathode-based cells the contact that determines $\psi_B$ is created between the semiconductor and the metal, not between the semiconductor and the solution. This is true even if only a small part of the surface of a hydrogen-generating photocathode is covered by the metal catalyst [72]. The reason is that the overwhelming majority of the charge carriers reach the solution via the catalyst. The rate or current of hydrogen generation at the semiconductor-solution interface is usually negligible. Thus, catalyzed photoelectrodes that generate hydrogen are in essence classical metal-semiconductor or Schottky junctions. Their barriers are the same whether the cells are wet or dry [72, 84]. In the absence of chemical reactions between the catalyst and the semiconductor, $\psi_B$ depends on the catalyst work function $\phi_W$. In photocathodes, $\psi_B$ increases when $\phi_W$ decreases. In photoanodes, the reverse is true.

In some cases the barrier height varies at different points of the semiconductor surface. The resulting leakage lowers the efficiency, unless the surface regions where the current might leak are covered by an electrical insulator or by a layer that blocks the transport of ions to and from the surface. Thus, there are three ways to reduce leakage losses: electrical insulation of the leakage sites; blockage of reactant transport to, or product transport from, the sites (for example, by a polymer layer); and maintenance of uniformly high barriers across the semiconductor surface. The last method is implemented in hydrogen-generating p-InP photocathodes where the same $V_{oc}$ is observed when hydrogen is adsorbed on the semiconductor surface, regardless of whether it is catalyst-covered or bare [72]. In oxygen-evolving photoanodes, such as Si with Pt-silicide islands [85], leakage is probably reduced by an insulating $SiO_2$ layer covering the nonmetallized regions between the islands.

Dissolution of gases in catalysts and adsorption of gases on catalyst surfaces cause drastic changes in $\phi_W$, and therefore in $\psi_B$. Hydrogen dissolution in Pt, Pd, Rh, Re, and Ru lowers $\phi_W$, increasing $\psi_B$ and $\Delta V_{OC}$, and thereby increasing the solar-to-hydrogen conversion efficiency of photocathodes [72, 84]. These are decreased correspondingly in photoanodes. Thus, a catalyzed hydrogen-generating photocathode is most efficient when it is enveloped by hydrogen; a catalyzed oxygen- of halogen-evolving photoanode is most efficient when its products reactively eliminate traces of hydrogen from the catalyst. Values of $\psi_B$ in air and in hydrogen [84] are listed in Table 1 for 12 semiconductor catalyst contacts. While in 11 cases $\psi_B$ is drastically different in air and in hydrogen, in CdS/Pt the change is minor. The reason for the small change is that Pt reacts with CdS to form a sulfide of Pt. Thus the contact is not Pt/CdS, but a hydrogen-insensitive contact $(PtS_x\text{-}Cd\text{-}CdS)/CdS$.

In the absence of leakage and recombination, each tenfold increase in irradiance increases $\Delta V_{OC}$ by 59 mV until $\Delta V_{OC}$ approaches $\psi_B$. The photocathode p-InP/(hydrogen-saturated Pt) shows such ideal behavior [72, 86].

## IV.  EFFICIENT HYDROGEN-EVOLVING PHOTOCATHODES

From the preceding sections it is evident that solar-to-hydrogen conversion can be efficient only if several requirements are simultaneously met. These include the classical requirements of all semiconductor-based converters, such as an appropriate band gap and an adequate minority carrier diffusion length, as well as requirements unique to hydrogen-generating photoelectrodes, such as chemical passivation of corrosion and photocorrosion reactions; acceleration of the hydrogen evolution kinetics by catalyst islands that are small relative to the wavelength of the exciting sunlight; distribution of these islands at distances smaller than the minority carrier surface diffusion length; and nonleaking, high barriers at the catalyst-semiconductor contacts.

Although numerous photocathodes have been explored, only catalyzed p-InP meets all the requirements and is the only photocathode now known that converts solar energy to chemical energy stored in hydrogen gas at a Gibbs free energy efficiency of 13.3% [80]. Some of the less efficient photocathodes have, nevertheless,

TABLE 1

Barrier Heights of Catalyst Contacts to Semiconductors
in Air and in Hydrogen

| Semiconductor | Catalyst | Barrier in air (eV) | Barrier in hydrogen (eV) |
|---|---|---|---|
| n-TiO$_2$ | Pt | $\geq 0.5$ | 0 |
| | Rh | $\geq 0.6$ | 0 |
| | Ru | $\geq 0.1$ | 0 |
| n-SrTiO$_3$ | Pt | $\geq 0.6$ | $>0.1$ |
| | Rh | $\geq 1.0$ | $>0.1$ |
| | Ru | $\geq 0.7$ | $>0.2$ |
| n-CdS | Pt | 1.84 | 1.63 |
| | Rh | 1.00 | 0.75 |
| | Ru | 1.28 | 0.75 |
| p-InP | Pt | -0.72 | -0.91 |
| | Rh | -0.31 | -0.81 |
| | Ru | -0.57 | -0.85 |

interesting features. In p-LuRhO$_3$, an external catalyst is not needed [87]; in another, p-type iron (magnesium) oxide has been used [88]; in a third, based on p-Si, the catalyst has been placed in a polymer in contact with the semiconductor rather than directly on it [89].

The i/$\Delta$V characteristics of the p-InP/(hydrogen-saturated rhodium) hydrogen-evolving photocathode under sunlight at 81.7 mW/cm$^2$ are shown in Fig. 6. The maximum reduction in bias at threshold for hydrogen evolution, $\Delta V_{oc}$, is 0.64 V, and the quantum or current efficiency at $\Delta V = 0$ is $\sim 0.9$. The maximum solar-to-fuel conversion efficiency point is at $i_{max}$ = 23 mA/cm$^2$ and $\Delta V_{max}$ = 0.45 V, where the Gibbs free energy efficiency of solar-to-hydrogen conversion reaches 13.3% [80].

## V. EFFICIENT PHOTOANODES

In photoanode-based cells, hydrogen is evolved at their non-illuminated, catalyst-activated counterelectrodes. The requirements that must be met for the cells to be efficient, particularly

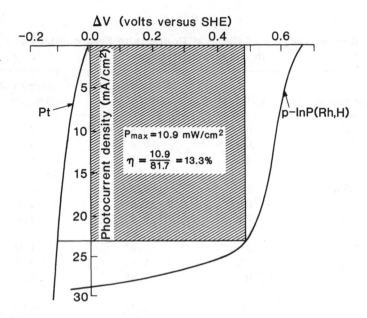

FIG. 6. Photocurrent bias reduction characteristics of a p-InP/ (hydrogen saturated rhodium) photocathode in 1 $\underline{M}$ HClO$_4$ under sunlight at 81.7 mW/cm$^2$. The solar-to-hydrogen conversion efficiency is 13.3% [80].

those relating to electron-hole recombination, leakage, and barrier heights, apply equally to photocathodes and photoanodes. The catalyst requirements are, however, relaxed as some photoanodes function adequately in the absence of a metallic catalyst on their surface. For example, the oxidation of $S^{2-}$ to $S_2^{2-}$ [22-29, 90] and that of cysteine to cystine [90, 91] proceed on "bare" n-CdS at sufficient rates. Similarly, water is oxidized to oxygen on noncatalyzed n-TiO$_2$ [12], n-SrTiO$_3$ [92-96], n-WO$_3$ [97], and n-Fe$_2$O$_3$ [98]. Nevertheless, oxygen- and chlorine-evolving n-Si photoanodes do require catalysts: Examples of catalyzed electrodes are n-Si/platinum silicide [85] and n-Si/indium tin oxide/RuO$_2$ [99].

The delicate relationship between the need for a catalyst and recombination is best seen in layered chalcogenide photoanodes, like MoS$_2$, MoSe$_2$, WS$_2$, and WSe$_2$, on which $I^-$ is oxidized to $I_2$ and and $I_3^-$, Br$^-$ to Br$_2$ and Br$_3^-$, and Cl$^-$ to Cl$_2$. Because recombination of photogenerated electron-hole pairs is much more rapid on crystal faces perpendicular to the van der Waals planes than in the van der Waals planes themselves, a high $\Delta V$ can be reached, in the absence of a catalyst, only with perfectly smooth van der Waals surfaces [63, 67]. However, if there is a high density of steps in these surfaces, then $\Delta V$ is small and is increased to values approaching those of smooth surfaces by incorporating Pt islands in the surface [100].

So far, most photoanodes of greater than 10% efficiency were made for and evaluated in electrical power producing cells [65]. Among the photoanodes that were operated in electrolytic cells, n-MoSe$_2$, n-WSe$_2$ [63, 67], and Pt-catalyzed n-Si [101] were shown to produce a sufficient $\Delta V$ for the spontaneous photoelectrolysis of HI.


## VI.   TWO PHOTOELECTRODE CELLS

Splitting the incident sunlight between a photoanode and a photocathode connected by an electrolytic solution has two potential advantages. First, the sum of the two $\Delta V$'s is obtained, and the unassisted photoelectrolysis of species that cannot be photoelectrolyzed spontaneously in single photoelectrode cells becomes possible [102-104]. Second, different segments of the solar spectrum can be used for each of the two photoelectrodes, and the solar conversion efficiency can be increased. As pointed out in the introduction, optimal efficiency is reached in two semiconductor systems when one semiconductor has a band gap of about 1.0 eV and the second of about 1.8 eV [4].

For a two-photoelectrode cell to be efficient, the photocurrents of the two electrodes must be the same. This requires precise balancing of the spectral segments of sunlight that reach each of the electrodes and necessitates a system of a complexity substantially beyond that of present photoelectrochemical cells. For this reason, no optimized photoelectrochemical cell utilizing split segments of the solar spectrum has been reported. Furthermore, although many two-photoelectrode cells have been described since the introduction of the concept [102-107], only a few have both a photoanode and a photocathode with band gaps appropriate for solar conversion. Those that do have such gaps contain either an n-MoSe$_2$ or an n-WSe$_2$ photoanode, and either a p-InP(Pt) or a p-InP(Rh) photocathode [107], and can photoelectrolyze aqueous HBr to H$_2$ and Br$_2$ and Br$_3^-$ without an external bias. Their action spectra (Fig. 7) overlap the visible and near-infrared segments of the solar spectrum. If optimized, the cells should approach the Gibbs free energy efficiency of their individual photoelectrodes ($\sim$10%).

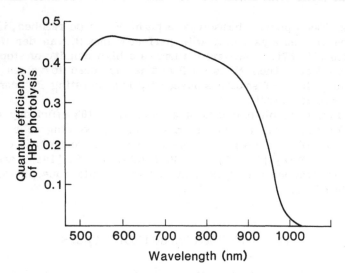

FIG. 7. Action spectrum for unassisted photoelectrolysis of aqueous HBr in the p-InP(Pt)|2 M HBr|n-WSe$_2$ cell [107].

## VII. p-n JUNCTION CELLS

The classical p-n junction silicon solar cell [108], in which all the requirements for efficiency are met, develops a sufficient photovoltage to electrolyze HI to H$_2$ and I$_3^-$ [109]. The photovoltage of a pair of n-p and p-n cells, connected in series, is adequate to electrolyze HBr to H$_2$ and Br$_2$/Br$_3^-$ [110, 111]. To accelerate the rate of hydrogen evolution at the SiO$_2$ layer of the Si-solution interface, and to generate conductive paths through the insulating SiO$_2$ layer that forms upon exposure of Si to aqueous bromine, catalysts are incorporated in the solution-exposed surfaces of both the photocathode and the photoanode. The catalyst must not form metal-semiconductor contacts with barriers that impede electron transport to the cathode catalyst, or hole transport to the anode catalyst.

The only commercial product-oriented engineering attempt in electrochemical solar cells in the United States utilizes a pair of silicon p-n and n-p junctions. These were produced in the form of glass embedded, etched, and catalyzed microspheres which electrolyzed HBr. Photoelectrochemically formed hydrogen was stored as a metal hydride. When electric power was needed, the products were allowed to recombine in a fuel cell [110, 111]. The

Gibbs free energy efficiency of the $H_2$- and $Br_2$-generating solar panels was 8.6% [89]. Similar efficiencies have been reached in laboratory experiments on HI electrolysis with catalyzed p-n junctions [109].

## VIII. PHOTOELECTROLYSIS BY SUSPENDED SEMICONDUCTOR PARTICLES

Illuminated semiconductor particles suspended in a solution constitute a collection of microcells. The surface of each particle contains two chemically and physically distinct zones: a microanode and a microcathode.

The suspensions that have been explored thus far are of n-type particles. The analysis of the photoanode requirements applies to these, though with some modifications. Because the distances traversed by photogenerated charge carriers on their way to the microelectrodes are short, recombination in bulk impurity and defect centers is less important. Surface processes, including recombination, leakage, and electrochemical kinetics, dominate the characteristics. Surface recombination velocities decrease approximately exponentially with $E_{BG}$. The three semiconductors used thus far, n-$TiO_2$ [112-116], n-$SrTiO_3$ [117, 118], and n-CdS [90-92, 119-121], have large band gaps (3.0, 3.2, and 2.4 eV, respectively). While the large band gaps reduce recombination, they also rule out the possibility of attaining high solar conversion efficiencies. In very small particles (50 Å diameter or less) the band gaps increase beyond their normal bulk value [122], further lowering the efficiency of the suspensions.

The fundamental requirement for photoelectrolysis by a microcell is

$$|\Delta V| > |\Delta G|/nF \qquad (2)$$

where $\Delta G$ is the Gibbs free energy of formation of the photoelectrolyzed species from the electrolytic reaction products. $\Delta V_{oc}$, and therefore $\Delta V$, are determined in the microcell by the difference between the barrier height at the microanode, $\psi_B{}^a$, and that at the microcathode, $\psi_B{}^c$ (Fig. 9). At very high irradiance

$$|\Delta V_{oc}| \rightarrow |\psi_B{}^a - \psi_B{}^c| \qquad (3)$$

When the anode and cathode barriers are equal in height, light cannot be converted to chemical energy (Fig. 8c). Any work gained upon transport of a hole to the microanode must be reinvested in

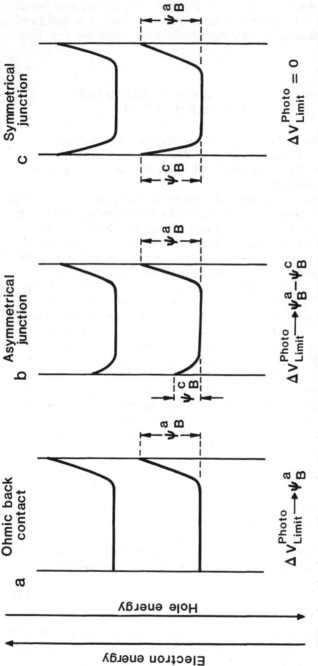

FIG. 8. Maximum bias gain in photoelectrolysis. The limit to $\Delta V_{OC}$ in a microcell consisting of a suspended semiconductor particle is the difference between the barrier height at the microcathode ($\psi_B^c$) and the barrier height at the microanode ($\psi_B^a$). (a) In an n-type semiconductor based cell with an ohmic microcathode contact, $\Delta V_{OC}$ equals the microanode barrier height. (b) In an asymmetric cell, more energy is gained in the transport of holes to the microanode than is lost in the transport of electrons to the microcathode. (c) In a symmetrical cell, all of the anode gain is lost at the cathode and there is no net reduction in bias.

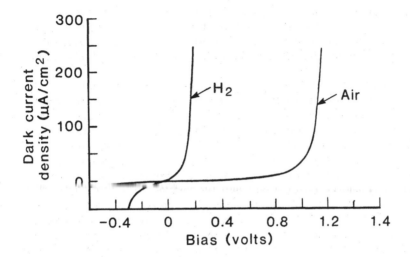

FIG. 9. Characteristics of the n-SrTiO$_3$/Rh contact in air and in hydrogen [84].

the transport of an electron, across a blocking barrier, to the microcathode. The best microcells have microcathodes that are ohmic ($\psi_B{}^c = 0$) (fig. 8a). The formation of ohmic microcontacts, a science in itself, is a formidable problem in suspended semiconductor particles, where ohmic contacts must be formed at some spots while high barrier contacts must be maintained at others. Fortunately, different microelectrode environments created by the chemical products of the microanode and microcathode reactions enhance the asymmetry [84].

The thermodynamic asymmetry requirement for the photoelectrolysis of water is

$$|\psi_B{}^a - \psi_B{}^c| \geq 1.23 \text{ eV} \tag{4}$$

When Group VIII metal-hydrogen evolution catalysts like Pt are used, part or all of the asymmetry derives from hydrogen alloying of the catalyst at microcathode sites. As discussed earlier and as seen in Table 1, hydrogen dissolution lowers the work function of the catalyst, decreasing the barrier heights of its contacts with n-type semiconductors and increasing the barrier heights with p-type semiconductors. Figure 9 shows an example of the reversible change in the diode characteristics of the Rh/n-SrTiO$_3$ Schottky diode in air and in hydrogen. While in air, the contact has a high barrier and is rectifying; it shows nearly

ohmic behavior in hydrogen. The asymmetry of the barriers in
the two environments is adequate to allow the Rh/n-SrTiO$_3$ contact to photoelectrolyze water. Note that hydrogen-saturated
catalysts form microcathodes, while O$_2$-exposed catalysts form
microanodes [84].

The likelihood for photoelectrolyzing water diminishes as the
band gap of the semiconductor particles approaches the domain
that is relevant for solar conversion, because $\psi_B$ is always smaller
than $E_{BG}$. Nevertheless, substrates that are less stable than water (Eqs. 2 and 3) might be photoelectrolyzed by particles with appropriate band gaps for solar conversion. For example, the free
energy of formation of H$_2$S is much smaller than that of water.
The asymmetry requirement can be relaxed to 0.14 eV (Eq. 3)
at a penalty of reducing the amount of stored chemical energy
from 1.23 to 0.14 eV per hydrogen atom. Indeed, photoelectrolysis of H$_2$S proceeds at high quantum yield on catalyzed n-CdS
microcrystals [121], and might also proceed efficiently on microcrystals of materials with smaller gap.

Because none of the suspensions meet as yet the essential
band gap requirement, their solar-to-fuel conversion efficiencies
remain low. The challenge is to find a system where $|\psi_B{}^a - \psi_B{}^c|$
is high enough to drive a substantially thermodynamically uphill
reaction, where $E_{BG}$ is in the 1.0-1.7 eV range, where the recombination velocity of the photogenerated electrons and holes
at the semiconductor surface is slower than their rate of reaction with the electrolyte, and where the chemical and photochemical corrosion stabilities are adequate. Until such a system is
found, efficient solar-to-hydrogen conversion will continue to
require macroscopic photoelectrodes.


## IX.  CONCLUSION

For efficient solar-to-hydrogen conversion, the photogenerated
charge carrier reaction rates must be fast relative to the rates of
recombination and the rates of leakage across the barrier. Conversion is efficient only under the following conditions: when the
rate of electron-hole recombination at photoelectrode surface regions between the catalyst islands is slowed by a controlled interfacial chemistry, when an appropriate electrocatalyst accelerates
the rate of hydrogen evolution, and when the chemistries of both
the semiconductor-catalyst and the semiconductor-solution interfaces are such that the barrier separating the photogenerated
electron and hole is either uniformly high or such that no current can flow through low-barrier regions.

To avoid light absorption losses, the catalyst must cover only a fraction of the semiconductor surface and must form islands with diameters small relative to wavelengths of the solar spectrum exciting the semiconductor. To ensure efficient collection of the photogenerated charge carriers, these islands must be spaced at distances smaller than the surface diffusion length of the minority carriers.

Unassisted ("spontaneous") but inefficient solar photoelectrolysis of water is possible with large-band-gap semiconductor photoelectrodes or with microcells based on suspended, large-band-gap, semiconductor particles. The photovoltage, $\Delta V$, and thus the very feasibility of photoelectrolysis by a suspended particle, depends on the asymmetry of its microcathode and its microanode barriers, i.e., on the difference in the work functions of the cathode- and anode-forming catalysts. This difference is greatly enhanced by, and in some cases is entirely due to, a combination of hydrogen saturation of the microcathode and of oxidative stripping of all traces of hydrogen from the microanode.

The barrier height is increased and the external voltage bias required for electrolysis is reduced in macroscopic, catalyzed photocathodes upon hydrogen saturation of the catalyst. In all of the efficient photoelectrodes, $\Delta V$ is sufficient to allow unassisted solar electrolysis of HI. By pairing a photocathode with a photoanode, or a p-n junction with an n-p junction, HBr can also be spontaneously and efficiently photoelectrolyzed with sunlight. Although efficient solar electrolysis of HCl or water requires an external bias, a substantial part of the energy consumed in the generation of either hydrogen and chlorine, or of hydrogen and oxygen, can now be derived from sunlight.

## Acknowledgments

I thank David E. Aspnes and Barry Miller for valuable discussions and, together with Edwin A. Chandross and Field H. Winslow, for critically reviewing the manuscript.

## REFERENCES

[1] S. J. Fonash, Solar Cell Device Physics, Academic, New York, 1981, p. 127.

[2] M. Almgren, Photochem. Photobiol., 27, 603 (1978).

[3] G. Porter, J. Chem. Soc., Faraday Trans. 2, 79, 473 (1983).

[4] J. C. C. Fan, B. Y. Tsaur, and B. J. Palm, Proc. 16th IEEE Photovoltaic Specialists Conf., September 1982, p. 692.

[5]   J. M. Woodall and H. J. Hovel, Appl. Phys. Lett., 30, 492 (1977).

[6]   W. H. Brattain and G. C. B. Garrett, Phys. Rev., 99, 376 (1955).

[7]   J. F. Dewald, Bell Syst. Tech. J., 39, 615 (1960).

[8]   M. Green, "Electrochemistry of the Semiconductor-Electrolyte Interface," in Modern Aspects of Electrochemistry (J. O'M. Bockris, ed.), Academic, New York, 1959, p. 343.

[9]   R. Williams, J. Chem. Phys., 32, 1505 (1960).

[10]  H. Gerischer, "Semiconductor Electrode Reactions," in Advances in Electrochemistry and Electrochemical Engineering (P. Delahay, ed.), Interscience, New York, 1961, p. 139.

[11]  Yu. V. Pleskov and V. A. Tyagai, Dokl. Akad. Nauk. SSSR, 141, 1135 (1961).

[12]  P. J. Boddy, J. Electroanal. Chem., 10, 199 (1965).

[13]  R. Memming and G. Schwandt, Angew. Chem., Int. Ed. Engl., 6, 851 (1967).

[14]  V. A. Myamlin and Yu. V. Pleskov, Electrochemistry of Semiconductors, Plenum, New York, 1967.

[15]  W. Gomes and F. Cardon, Ber. Bunsenges. Phys. Chem., 74, 431 (1970).

[16]  H. Gerischer, "Semiconductor Electrochemistry," in Physical Chemistry: An Advanced Treatise, Vol. 9A: Electrochemistry (H. Eyring, D. Henderson, and W. Jost, eds.), Academic, New York, 1970, p. 463.

[17]  P. Boddy, J. Electrochem. Soc., 115, 199 (1968).

[18]  A. Fujishima and K. Honda, Nature (London), 238, 37 (1972).

[19]  H. Gerischer, "Solar Energy Conversion," in Topics in Applied Physics, Vol. 31, Springer, New York, 1979, p. 115.

[20]  A. J. Nozik, Adv. Hydrogen Energy: Hydrogen Energy Syst., 3, 1217 (1979).

[21]  L. A. Harris and R. H. Wilson, Annu. Rev. Mat. Sci., 8, 99 (1978).

[22]  M. Tomkiewicz and H. Fay, Appl. Phys., 18, 1 (1979).

[23]  M. A. Butler and D. S. Ginley, J. Mater. Sci., 15, 1 (1980).

[24]  A. J. Bard, Science, 207, 4427 (1980).

[25]  H. Gerischer, J. Electroanal. Chem., 58, 263 (1975).

[26]  G. C. Barker, discussing paper by H. Gerischer, J. Electrochem. Soc., 113, 1182 (1966).

[27]  A. B. Ellis, S. W. Kaiser, J. M. Bolts, and M. Wrighton, J. Am. Chem. Soc., 98, 1635 (1976).

[28]  G. Hodes, J. Manassen, and D. Cahen, Nature (London),
      260, 312 (1976); 261, 403 (1976).
[29]  B. Miller and A. Heller, Ibid., 262, 680 (1976).
[30]  A. Heller (ed.), Semiconductor-Liquid Junction Solar Cells,
      The Electrochemical Society, Princeton, New Jersey, 1977.
[31]  K. C. Chang, A. Heller, B. Schwartz, S. Menezes, and
      B. Miller, "Stable Semiconductor-Liquid Junction Cell with
      9% Solar to Electrical Conversion Efficiency," Ref. 30, p.
      132; Science, 196, 1097 (1977).
[32]  H. Gerischer, "The Photoelectrochemical Cell: Principles,
      Energetics and Electrode Stability," Ref. 30, p. 1; J.
      Electroanal. Chem., 82, 133 (1977).
[33]  A. J. Bard and M. S. Wrighton, J. Electrochem. Soc.,
      124, 1706 (1977).
[34]  A. B. Ellis, J. M. Bolts, S. W. Kaiser, and M. S.
      Wrighton, J. Am. Chem. Soc., 99, 2848 (1977).
[35]  H. Tributsch, Z. Naturforsch., A32, 972 (1977); Ber.
      Bunsenges. Phys. Chem., 81, 361 (1977).
[36]  B. Miller, S. Menezes, and A. Heller, "Stability and Volt-
      ammetry of Illuminated Semiconductor Liquid Interfaces,"
      Ref. 30, p. 186.
[37]  A. Heller and B. Miller: "Photoelectrochemical Solar Cells:
      Chemistry of the Semiconductor Liquid Junction," in Inter-
      facial Photoprocesses: Energy Conversion and Synthesis
      (M. S. Wrighton, ed.), Adv. Chem. Ser. Vol. 184, Ameri-
      can Chemical Society, Washington, D.C., 1980, p. 215.
[38]  A. Heller, K. C. Chang and B. Miller, J. Am. Chem. Soc.,
      100, 684 (1978).
[39]  A. Heller, G. P. Schwartz, R. G. Vadimsky, S. Menezes,
      and B. Miller, J. Electrochem. Soc., 125, 1156 (1978).
[40]  H. Tributsch, Ibid., 125, 1086 (1978).
[41]  J. Gobrecht, H. Gerischer, and H. Tributsch, Ber.
      Bunsenges. Phys. Chem., 82, 1331 (1978); see also S.
      Menezes, F. J. DiSalvo, and B. Miller, J. Electrochem.
      Soc., 127, 1751 (1980).
[42]  J. Gobrecht, H. Tributsch, and H. Gerischer, J. Elec-
      trochem. Soc., 125, 2085 (1978).
[43]  W. Kautek, H. Gerischer, and H. Tributsch, Ber.
      Bunsenges. Phys. Chem., 83, 1000 (1979).
[44]  B. Miller, S. Menezes, and A. Heller, J. Electrochem.
      Soc., 126, 1483 (1979).
[45]  S. Menezes, A. Heller, and B. Miller, Ibid., 127, 1263
      (1980).
[46]  H. Gerischer, Faraday Discuss., 70, 137 (1980).
[47]  F. Cardon, W. P. Gomes, F. Vanden Kerchove, D. Van
      Mackelbergh, and F. Van Overmeire, Ibid., 70, 153 (1980).

[48] B. Miller, A. Heller, S. Menezes, and H. J. Lewerenz, Ibid., 70, 223 (1980).

[49] K. W. Frese Jr., M. S. Madou, and S. R. Morrison, J. Phys. Chem., 84, 3172 (1980); J. Electrochem. Soc., 128, 1527, 1939 (1981).

[50] D. Cahen, G. Hodes, J. Manassen, and R. Tenne, "Stability of Cadmium-Chalcogenide Based Photoelectrochemical Cells," in Photoeffects at Semiconductor-Electrolyte Interfaces (A. J. Nozik, ed.), Am. Chem. Soc. Symp. Ser. Vol. 146, American Chemical Society, Washington, D.C., 1981, p. 369.

[51] A. Heller, B. A. Parkinson, and B. Miller, "12% Efficient Semiconductor-Liquid Junction Solar Cell," in Proc. 13th IEEE Photovoltaic Specialists Conf., IEEE, 1978, p. 1253.

[52] B. A. Parkinson, A. Heller, and B. Miller, Appl. Phys. Lett., 33, 521 (1978).

[53] A. Heller, K. C. Chang, and B. Miller, J. Am. Chem. Soc., 100, 684 (1978).

[54] B. A. Parkinson, A. Heller, and B. Miller, J. Electrochem. Soc., 126, 954 (1979).

[55] A. Heller, H. J. Lewerenz, and B. Miller, Ber. Bunsenges. Phys. Chem., 84, 592 (1980).

[56] R. J. Nelson, J. S. Williams, H. J. Leamy, B. Miller, H. C. Casey Jr., B. A. Parkinson, and A. Heller, Appl. Phys. Lett., 36, 76 (1981).

[57] H. J. Lewerenz, A. Heller, and F. J. DiSalvo, J. Am. Chem. Soc., 102, 1877 (1980).

[58] A. Heller and B. Miller, Electrochim. Acta, 25, 29 (1980).

[59] A. Heller, "Chemical Control of Surface and Grain Boundary Recombination in Semiconductors," in Photoeffects at Semiconductor-Electrolyte Interfaces (A. J. Nozik, ed.), Am. Chem. Soc. Symp. Ser. Vol. 146, American Chemical Society, Washington, D.C., 1981, p. 57.

[60] H. J. Lewerenz, A. Heller, H. J. Leamy, and S. D. Ferris, "Carrier Recombination at Steps in Surfaces of Layered Compound Photoelectrodes," Ibid., p. 17.

[61] H. Gerischer, "The Influence of Surface Orientation and Crystal Imperfections on Photoelectrochemical Reactions at Semiconductor Electrodes," Ibid., p. 1.

[62] A. Heller, B. Miller, H. J. Lewerenz, and K. J. Bachmann, J. Am. Chem. Soc., 102, 6555 (1980).

[63] B. A. Parkinson, T. E. Furtak, D. Canfield, K. Kam, and G. Kline, Faraday Discuss., 70, 233 (1980).

[64] A. Heller, B. Miller, and F. A. Thiel, Appl. Phys. Lett., 38, 282 (1981).

[65] A. Heller, Acc. Chem. Res., 14, 154 (1981).

[66]  A. Heller, H. J. Lewerenz, and B. Miller, J. Am. Chem.
      Soc., 103, 200 (1981).
[67]  G. Kline, K. Kam, D. Canfield, and B. Parkinson, Solar
      Energy Mater., 4, 301 (1981); B. A. Parkinson, J. Chem.
      Educ., 60, 338 (1983).
[68]  A. Heller, "Hydrogen Generating Solar Cells Based on
      Platinum Group Metal Activated Photocathodes," in En-
      ergy Resources by Photochemistry and Catalysis (M.
      Grätzel, ed.), Academic, New York, 1983, p. 385.
[69]  A. Heller and R. G. Vadimsky, Phys. Rev. Lett., 46,
      1153 (1981).
[70]  A. Heller, Solar Energy, 19, 180 (1980).
[71]  A. Heller, "Hydrogen Generating Electrochemical Solar
      Cells," in Photochemical Conversion and Storage of Solar
      Energy, Part A (J. Rabani, ed.), Weizmann Science Press,
      Jerusalem, 1982, p. 63.
[72]  A. Heller, E. Aharon-Shalom, W. A. Bonner, and B.
      Miller, J. Am. Chem. Soc., 104, 6942 (1982).
[73]  H. J. Lewerenz, D. E. Aspnes, B. Miller, D. L. Malm,
      and A. Heller, Ibid., 104, 3325 (1982).
[74]  L. F. Schneemeyer and B. Miller, J. Electrochem. Soc.,
      129, 1977 (1982).
[75]  S. Menezes, B. Miller, and K. J. Bachmann, J. Vac. Sci.
      Technol., B1, 48 (1983).
[76]  D. E. Aspnes, Thin Solid Films, 89, 249 (1982).
[77]  D. A. G. Bruggeman, Ann. Phys., 24, 636, 665 (1935);
      25, 645 (1936).
[78]  A. Heller and R. G. Vadimsky, Unpublished Results.
[79]  E. Aharon-Shalom and A. Heller, J. Phys. Chem., 87,
      4913 (1983).
[80]  E. Aharon-Shalom and A. Heller, J. Electrochem. Soc.,
      129, 2865 (1982).
[81]  C. P. Kubiak, L. F. Schneemeyer, and M. S. Wrighton,
      J. Am. Chem. Soc., 102, 6896 (1980).
[82]  S. M. Ahmed and H. Gerischer, Electrochim. Acta, 24, 705
      (1979).
[83]  D. E. Aspnes and A. Heller, J. Vac. Sci. Technol., B1,
      602 (1983).
[84]  D. E. Aspnes and A. Heller, J. Phys. Chem., 87, 4919
      (1983).
[85]  F. R. F. Fan, R. Keil, and A. J. Bard, J. Am. Chem.
      Soc., 105, 220 (1983).
[86]  A. Heller, R. G. Vadimsky, W. D. Johnston Jr., K. E.
      Strege, H. J. Leamy, and B. Miller, "p-InP Photocath-
      odes: Solar to Hydrogen Conversion and Improvements
      of Polycrystalline Films by Reacting Silver with the Grain

Boundaries," in Proc. 15th IEEE Photovoltaic Specialists Conf., IEEE, 1981, p. 1722.

[87]  H. S. Jarrett, A. W. Sleight, H. H. Kung, and H. H. Gillson, Surf. Sci., 101, 205 (1980); J. Appl. Phys., 51, 3916 (1980).

[88]  C. Leygraf, M. Hendewerk, and G. A. Somorjai, J. Catal., 78, 341 (1982); J. Phys. Chem., 86, 4484 (1982); J. Solid State Chem., 48, 357 (1983).

[89]  R. N. Dominey, N. S. Lewis, J. A. Bruce, D. C. Bookbinder, and M. S. Wrighton, J. Am. Chem. Soc., 104, 467 (1982); H. D. Abruna and A. J. Bard, Ibid., 103, 6898 (1981).

[90]  J. R. Darwent and G. Porter, J. Chem. Soc., Chem. Commun., p. 145 (1981).

[91]  J. R. Darwent, J. Chem. Soc., Faraday Trans. 2, 77, 1703 (1981).

[92]  D. J. Meissner, R. Memming, and B. Kastening, Chem. Phys. Lett., 96, 34 (1983).

[93]  J. G. Mavroides, D. I. Tchernev, J. A. Kafalas, and D. F. Kolesar, Mater. Res. Bull., 10, 1023 (1975).

[94]  J. G. Mavroides, J. A. Kafalas, and D. Kolesar, Appl. Phys. Lett., 28, 241 (1976).

[95]  T. Watanabe, A. Fujishima, and K. Honda, Bull. Chem. Soc. Jpn., 49, 355 (1976).

[96]  M. S. Wrighton, A. B. Ellis, P. T. Wolczanski, D. L. Morse, H. B. Abrahamson, and D. S. Ginley, J. Am. Chem. Soc., 98, 2774 (1976).

[97]  M. A. Butler, R. D. Nasby, and R. K. Quinn, Solid-State Commun., 19, 1011 (1976).

[98]  R. K. Quinn, R. D. Nasby, and R. J. Baughman, Mater. Res. Bull., 11, 1011 (1976); K. L. Hardee and A. J. Bard, J. Electrochem. Soc., 123, 1024 (1976).

[99]  G. Hodes, L. Thompson, J. DuBow, and K. Rajeshwar, J. Am. Chem. Soc., 105, 324 (1983).

[100]  R. A. Simon, A. J. Ricco, D. J. Harrison, and M. S. Wrighton, J. Phys. Chem., 87, 4446 (1983).

[101]  Y. Nakato, A. Tsumura, and H. Tsubomura, Chem. Lett., p. 1071 (1982).

[102]  H. Yoneyama, H. Sakamoto, and H. Tamura, Electrochim. Acta, 20, 341 (1975).

[103]  A. J. Nozik, Appl. Phys. Lett., 28, 150 (1976).

[104]  K. Ohashi, J. McCann, and J. O'M. Bockris, Nature (London), 266, 610 (1977).

[105]  A. J. Nozik, Appl. Phys. Lett., 30, 567 (1977).

[106]  K. Ohashi, K. Uosaki, and J. O'M. Bockris, Int. J. Energy Res., 1, 259 (1977).

[107] C. Levy-Clement, A. Heller, W. A. Bonner, and B. A. Parkinson, J. Electrochem. Soc., 129, 1702 (1982).

[108] R. S. Ohl, U.S. Patent 2,402,662 (filed May 27, 1941, issued June 25, 1946); see also U.S. Patent 2,443,542 (September 17, 1942).

[109] H. Tsubomura, Y. Nakato, A. Tsumura, and E. Yoshihiro, Denki Kagaku Oyobi Kogyo Butsuri Kagaku, 51, 71 (1983).

[110] J. S. Kilby, J. W. Lathrop and W. A. Porter, U.S. Patents 4,021,323 (May 3, 1977), 4,100,051 (July 11, 1978), 4,136,436 (January 30, 1979).

[111] E. L. Johnson, "Recent Progress in Photovoltaic/Electrochemical Energy System Application," in Electrochemistry in Industry (U. Landau, E. Yeager, and D. Kortan, eds.), Plenum, New York, 1982, pp. 299-306; Proc. Intersoc. Energy Convers. Eng. Conf., 16, 798 (1981); I. Trachtenberg, Private Communication.

[112] A. V. Bulatov and M. L. Khidekel, Izv. Akad. Nauk SSSR, Ser. Khim., p. 1902 (1976).

[113] S. Sato and J. M. White, J. Am. Chem. Soc., 102, 7206 (1980).

[114] T. Kawai and T. Sakata, J. Chem. Soc., Chem. Commun., p. 1047 (1979); Chem. Phys. Lett., 72, 87 (1980).

[115] E. Borgarello, J. Kiwi, E. Pelizetti, M. Visca, and M. Graetzel, Nature (London), 289, 158 (1981); J. Am. Chem. Soc., 103, 6324 (1981).

[116] J. Kiwi, E. Borgarello, E. Pelizetti, M. Visca, and M. Graetzel, Angew. Chem., 92, 663 (1980).

[117] J. M. Lehn, J. P. Sauvage, and R. Ziessel, Nouv. J. Chem., 4, 623 (1980); 5, 291 (1981).

[118] J. M. Lehn, J. P. Sauvage, R. Ziessel, and L. Hilaire, Isr. J. Chem., 22, 168 (1982).

[119] E. Borgarello, K. Kalyanasundaram, M. Graetzel, and E. Pelizetti, Helv. Chim. Acta, 65, 243 (1982).

[120] K. Kalyanasundaram, E. Borgarello, and M. Graetzel, Ibid., 64, 362 (1981).

[121] K. Kalyanasundaram, E. Borgarello, D. Duonghong, and M. Graetzel, Angew. Chem., 93, 1012 (1981).

[122] R. Rosetti, S. Nakahara, and L. E. Brus, J. Chem. Phys., 79, 1086 (1983).

# The Catalyzed Photodissociation of Water

GABOR A. SOMORJAI, M. HENDEWERK, AND J. E. TURNER
Materials and Molecular Research Division
Lawrence Berkeley Laboratory;
Department of Chemistry
University of California, Berkeley
Berkeley, California,

## I. INTRODUCTION

Chemical interactions between molecules in excited electronic, vibrational, or rotational states and surfaces is a new field of catalytic science. Until recently, catalysis of chemical reactions has only been considered for molecules in their thermodynamic ground states. Most of the surface reactions to be catalyzed were exothermic or thermodynamically downhill. In carrying out endothermic reactions, the only source of energy considered has been the addition of heat. This also assured that the molecules maintained an equilibrium energy distribution throughout the reaction.

There is growing evidence from fields other than catalysis that when molecules are allowed to react from excited electronic or vibrational states, they often take different reaction paths than would be available from a ground state configuration.

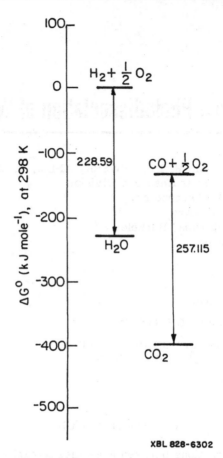

FIG. 1. Free energies for the dissociation of $H_2O$ and $CO_2$ related to the electrochemical energy scale (RHE).

Some examples of this are the deposition of materials from low energy plasmas, the formation of compounds under bombardment by high energy particles, or the application of explosions to induce bonding. Such techniques can lead to the formation of unique coatings, improved adhesion, or better electrical contact between materials.

Photoassisted thermodynamically uphill reactions such as the photodissociation of water to hydrogen and oxygen are one class of these excited state reactions. Figure 1 shows free energies for the dissociation of both $H_2O$ and $CO_2$. Both of these molecules are reactants in photosynthesis to produce hydrocarbons and oxygen. If light could be used to dissociate $H_2O$ and $CO_2$, all subsequent reactions of hydrogen with carbon dioxide or carbon monoxide with water are thermodynamically downhill. These reactions can readily be catalyzed by a large number of transition metals or transition metal compounds to produce small organic molecules. Thus, in the photodissociation of water for example, one produces oxygen, and the subsequent catalytic reactions of hydrogen with carbon dioxide might yield small organic molecules from inorganic photosynthesis even more efficiently than the chlorophyll-catalyzed process of photosynthesis. It is possible that inorganic photosynthesis was an important path in the formation of organic molecules in the prechlorophyll era of this planet.

The process of using photons to dissociate stable and abundant molecules like water and carbon dioxide could, if successful, provide for chemicals or for fuel. It is somewhat surprising how little catalytic chemistry has been published using carbon dioxide or methane as carbon sources, or using water for a source of hydrogen or oxygen.

The catalyzed photodissociation of water has been receiving increased attention since the early 70s, when the first experiments on the photoelectrochemical dissociation of water were reported [1]. When a strontium titanate ($SrTiO_3$) anode was illuminated with light of greater than bandgap energy in a basic solution using a Pt counterelectrode, hydrogen and oxygen evolution occurred at the cathode and anode, respectively, without the use of any external potential [2]. The reaction was catalytic and the electrodes showed no sign of deactivation.

The electronic state of an n-type semiconductor such as $SrTiO_3$ immersed in solution is well understood (Fig. 2). Water photodissociation may also be observed when the platinum cathode is substituted with a p-type semiconductor such as gallium phosphide. In this circumstance, both the semiconductor anode and cathode must be illuminated with light of energy greater than their respective bandgaps. While both configurations may photodissociate water, these two types of photoelectrochemical cells have one essential difference. The n-type semiconductor/metal cell requires illumination of only the semiconductor anode, while the n-type/p-type semiconductor cell requires the illumination of both electrodes simultaneously to induce water photodissociation.

The thermodynamic conditions necessary to carry out this photochemical reaction have been investigated [3] and give restrictions on the location of semiconductor band edges in solution. The conduction band edge must be anodic with respect to the $H^+/H_2$ half reaction, while the valence band edge must be cathodic with respect to the $O_2/OH^-$ redox couple. This way, electrons at the

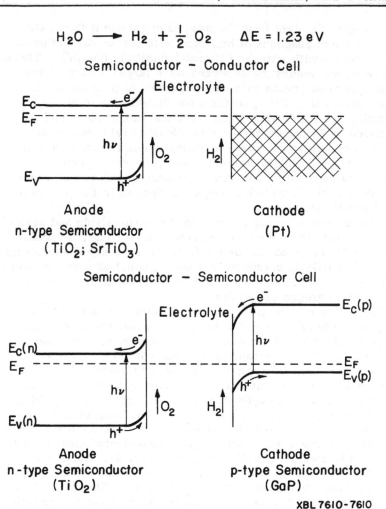

$$H_2O \longrightarrow H_2 + \frac{1}{2} O_2 \qquad \Delta E = 1.23 \, eV$$

FIG. 2.   Solid-state models of band bending for n-type and p-type semiconductor electrodes immersed in aqueous solution. Top: n-Type semiconductor electrode short circuited to a Pt electrode, Bottom: n-Type (TiO$_2$) electrode in electrical contact with a p-type (GaP) electrode.

cathode can produce hydrogen atoms, while electron transfer from OH$^-$ to hole states in the valence band will produce OH radicals, which can then be dimerized to $H_2O_2$, that splits up to produce oxygen.

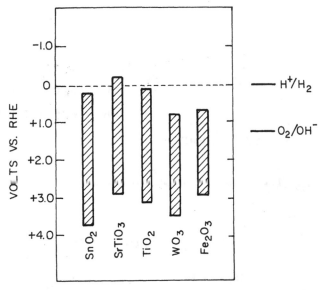

FIG. 3. Location of conduction and valence band edges of oxide semiconductors relative to the hydrogen and oxygen redox
couples.


Figure 3 shows the conduction and valence band edges for a
number of oxides. In order to qualify as a suitable electrode, a
material must satisfy three basic criteria. It must have a band-
gap larger than 1.23 eV ($\Delta G$ for the reaction $H_2O \rightarrow OH^- + H^+$).
In addition, band edges must be appropriately positioned with
respect to the redox couples as described above. Finally, the
material must be chemically stable under the reaction conditions.
As seen in Fig. 3, $SrTiO_3$ satisfies the thermodynamic conditions remarkably well, while $Fe_2O_3$ does not. In fact, $SrTiO_3$
has been found to photodissociate water with about 5% efficiency
(calculated as hydrogen energy produced divided by the total
photon energy) even in the short circuit configuration as shown
in Fig. 4. Thus when small particles or single crystal surfaces
of $SrTiO_3$ are platinized and illuminated with light of energy
greater than the bandgap of 3.1 eV, the steady-state evolution
of hydrogen and oxygen can readily be observed. The rate of
gas evolution is greater if the $SrTiO_3$ is prereduced in a hydrogen furnace and the electrolyte solution has a pH of 14 or
greater. These findings indicate that this is not only an electrochemical process, but that water photodissociation over this

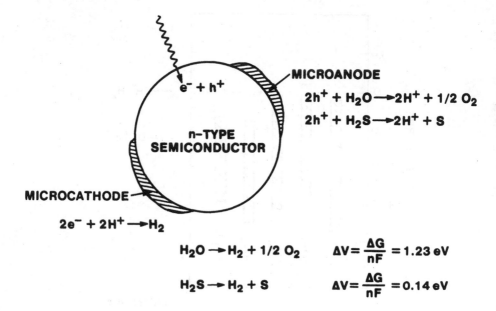

FIG. 4. Photoelectrolysis with suspensions of catalyst-coated semiconductor particles.

material involves both photoelectrochemical and photocatalytic reactions.

Carrier generation and electron transfer at the semiconductor surface are the first of several elementary steps in the photodissociation of water. Electrons and holes are usually generated near the surface in close proximity to the solution. Thus single crystals with high charge mobilities in the bulk are not necessarily needed. Alternatively, small particles or colloids may be utilized in a short circuit configuration to maximize the active surface area of the device. Indeed, colloidal platinized $TiO_2$ and $SrTiO_3$ systems have been found effective for water photodissociation [4].

In this paper we review what is known about the photoelectrochemical dissociation of water in two different configurations—the $SrTiO_3/Pt/KOH$ semiconductor metal system and the n-type/p-type iron oxide photoelectrochemical diode. Both have been shown to drive the water splitting photochemical reaction without an external potential. The $SrTiO_3$-based system requires illumination by light in the near-UV region ($h\nu \geq 3.1$ eV), while the iron oxide system can operate with light in the visible region ($h\nu \geq 2.3$

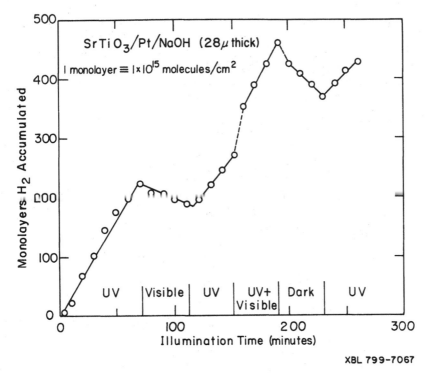

FIG. 5. Steady-state hydrogen evolution from a platinized
strontium titanium (reduced) crystal immersed in 1 $\underline{N}$ KOH.

eV). Through the discussion of these two systems for water dissociation, we will demonstrate the accomplishments and difficulties
of research in this field of inorganic photosynthesis.

## II.  THE SrTiO$_3$/Pt/KOH SYSTEM

Figure 5 shows the steady-state evolution of hydrogen from a
strontium titanate single crystal which had been prereduced in a
hydrogen furnace, platinized, and immersed in 1 $\underline{N}$ KOH solution.
The reaction appeared to continue indefinitely with no signs of
deactivation, and from the hydrogen yield this photoinduced process appeared to utilize about 5% of the incident photon energy
that is absorbed by the solid.  Upon turning off the light, the
hydrogen evolution stops.  In fact, some of the hydrogen and
and oxygen disappears due to the favorable backreaction of

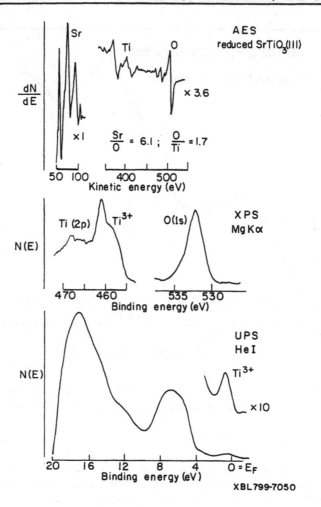

FIG. 6. Surface analyses of reduced $SrTiO_3$:Ti:O ratios determined by Auger electron spectroscopy, and the presence of $Ti^{3+}$
and $Ti^{4+}$ ions determined by x-ray photoelectron spectroscopy and
ultraviolet photoelectron spectroscopy.

recombination in the dark. When bandgap radiation illuminates the
surface again, the evolution of hydrogen recommences. Heating
the electrolyte solution increases the rate somewhat, since the activation energy for this process is about 8 kcal/mol.

A variety of surface science techniques have been used to study
the elementary steps in this photochemical surface reaction. Some
of these are shown in Fig. 6. Auger electron spectroscopy yields

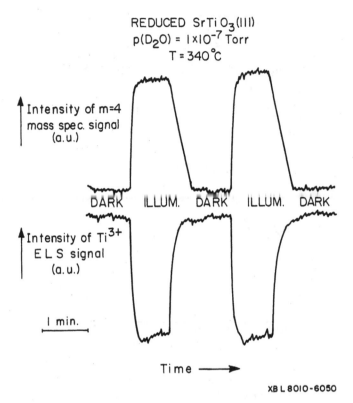

REDUCED SrTiO$_3$(III)
p(D$_2$O) = I×10$^{-7}$ Torr
T = 340°C

Intensity of m=4
mass spec. signal
(a.u.)

DARK     ILLUM.     DARK     ILLUM.     DARK

Intensity of Ti$^{3+}$
E L S signal
(a.u.)

I min.

Time ⟶

XBL 8010-6050

FIG. 7.  Simultaneous evolution of D$_2$ gas and the oxidation of
Ti$^{3+}$ and Ti$^{4+}$ after water (D$_2$O) adsorption is shown by monitoring
the D$_2$ gas by mass spectrometry and the intensity of the Ti$^{3+}$ signal with ELS.

the surface composition of the stoichiometric or hydrogen furnace
reduced strontium titanate.  Photoelectron spectroscopy reveals
the presence of Ti$^{3+}$ formal oxidation state metal ion along with
Ti$^{4+}$ ions, while ultraviolet photoelectron spectroscopy shows a
small peak near the Fermi level also associated with the presence
of a Ti$^{3+}$ ion.  It is apparent that Ti$^{3+}$ ions at the strontium titanate surface play an important role during the photodissociation
reaction.  Figure 7 shows the intensity of the Ti$^{3+}$ signal as water is adsorbed on the surface in the form of D$_2$O and the surface is illuminated.  Upon illumination, deuterium gas evolves
and the T$^{3+}$ signal diminishes, indicating that surface ions are
being oxidized to Ti$^{4+}$.  In the dark, oxygen desorption occurs

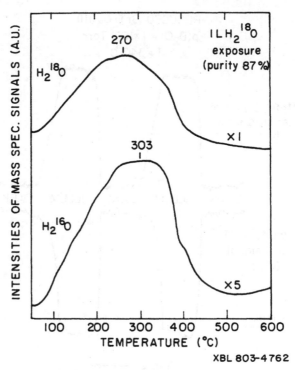

FIG. 8. Thermal desorption of water following adsorption of isotopically labeled water ($H_2{}^{18}O$) on a $SrTiO_3$ crystal.

and the surface returns to its reduced state. Upon illumination in the presence of adsorbed water, the process repeats. Thus, it appears that the photon interaction with the surface results in an oxidation/reduction reaction, involving a charge transfer for the transition metal ion from the $Ti^{3+}$ to the $Ti^{4+}$ oxidation state, which is subsequently returned to its reduced state.

Figure 8 gives direct evidence of the solid-state reaction between water and the oxide surface. When $H_2{}^{18}O$ is adsorbed on the surface and a thermal desorption experiment is carried out, part of the water desorbs as $H_2{}^{18}O$. In addition, a substantial amount of the water desorbs in the form of $H_2{}^{16}O$. This indicates an exchange of oxygen in the water with the oxygen in the oxide surface. Further information on this solid-state reaction is provided by Fig. 9, showing a thermal desorption spectrum from $SrTiO_3$ after adsorption of $D_2O$ or $D_2$ on the surface. In both

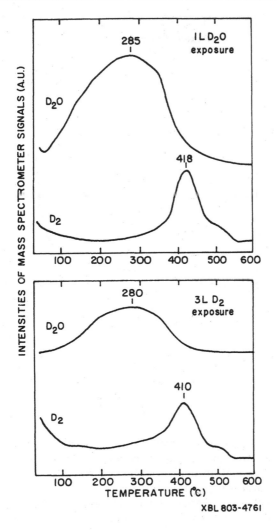

FIG. 9.  Thermal desorption from $SrTiO_3$ following adsorption of $D_2O$ or $D_2$.

cases the $D_2$ desorbs in addition to $D_2O$, indicating that the deuterium is a reducing agent which removes oxygen from the oxide lattice.  From these results, one can conclude that the photodissociation of water over $SrTiO_3$ is a solid-state surface reaction similar to the photographic process except that instead of a

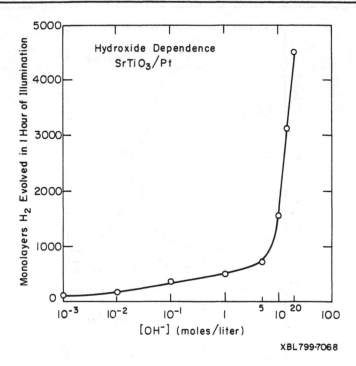

FIG. 10.  Hydrogen evolution from $SrTiO_3$/Pt.  Dependence on hydroxyl concentration in the electrolyte solution.

gross, irreversible photodecomposition of the solid itself, molecules near the surface undergo a reversible photodecomposition. The oxidation/reduction cycle of the transition metal ions on the surface is an integral part of this process.

In Fig. 10 the importance of the presence of alkali hydroxide in this process is demonstrated.  As one increases the hydroxyl ion concentration in the electrolyte solution, there is an increased rate of hydrogen evolution.  Photoelectron spectroscopy indicates that the surface is completely hydroxylated.  In Fig. 11 the top curve shows the UPS spectrum from sodium hydroxide.  The three electronic transitions are the fingerprints of the presence of $OH^-$ ions.  It is clearly seen that the strontium titanate surface in the presence of hydrogen or water shows the same fingerprint.  Thus we can infer that a strontium titanate surface which is active toward water photodissociation is completely hydroxylated.

The hydroxylation of the oxide surface does not occur readily at room temperature, and the presence of potassium hydroxide or

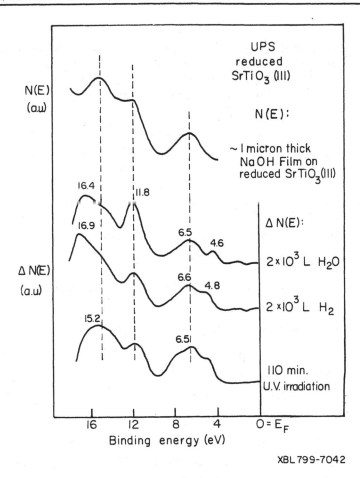

FIG. 11. The same characteristic features are seen by comparison of the UPS spectra of hydroxyl ions and $SrTiO_3$ exposed to hydrogen or water.

other alkali hydroxides seems to catalyze this process. This is the
reason that alkali hydroxide plays such an important role when the
reaction is carried out at 300 K. However, if we carry out the reaction at 450-500 K in the presence of steam, alkali hydroxides are
not needed. Under these circumstances the surface is readily hydroxylated and hydrogen evolution commences without the presence
of alkali hydroxides [5].

Next, we investigate the importance of the presence of platinum
or other metals on the surface in accelerating this reaction. Table 1
indicates that the reaction occurs faster in the presence of metals

## TABLE 1

A Summary of Hydrogen Evolution Rates from Stoichiometric
and Prereduced $SrTiO_3$ Crystals in Varied NaOH
Environments and with Metal Catalysts
Deposited on the Surface

| Crystal | Monolayers[a] $H_2$/h |
|---|---|

A. Hydrogen Production from $SrTiO_3$ Crystals Covered
by Thick (>30 μM NaOH Films Saturated
with Water Vapor

| | |
|---|---|
| Preduced, platinized | 1580 |
| Prereduced, metal-free | 100 |
| Stoichiometric, metal-free | 30 |

B. Hydrogen Production from $SrTiO_3$ Crystals
in 20 $\underline{M}$ NaOH

| | |
|---|---|
| Prereduced, platinized | 4500 |
| Stoichiometric, platinized | 120 |
| Prereduced, metal-free | 30 |
| Stoichiometric, metal-free | 50 |
| Prereduced, gold coated | 200 |

[a]1 monolayer $\equiv$ 1 X $10^{15}$ molecules/$cm^2$ illuminated surface.

on the surface. Platinum seems to be more active than gold, while
gold is more active than strontium titanate without any metal. Hydrogen atom recombination is one of the elementary reaction steps
leading to the formation of hydrogen gas after the photodissociation of water. Strontium titanate does not carry out hydrogen
atom recombination well; however, platinum is an excellent catalyst for this process. Thus, the best water-splitting system
would involve strontium titanate that carries out the photodissociation process, alkali hydroxides that keep the surface continuously hydroxylated, and metal that is needed for hydrogen atom
recombination.

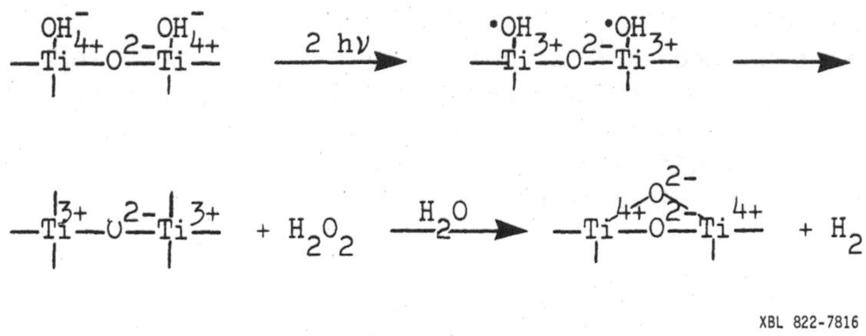

$$\underset{\substack{| \\ —Ti \\ |}}{\overset{\text{OH}^-_{4+}}{}} O^{2-} \underset{\substack{| \\ Ti \\ |}}{\overset{\text{OH}^-_{4+}}{}} \quad \xrightarrow{2\ h\nu} \quad \underset{\substack{| \\ —Ti \\ |}}{\overset{^\bullet\text{OH}_{3+}}{}} O^{2-} \underset{\substack{| \\ Ti \\ |}}{\overset{^\bullet\text{OH}_{3+}}{}} \quad \longrightarrow$$

$$\underset{\substack{| \\ —Ti \\ |}}{\overset{|3+}{}} U^{2-} \underset{\substack{| \\ Ti \\ |}}{\overset{|3+}{}} + H_2O_2 \quad \xrightarrow{H_2O} \quad \underset{\substack{| \\ —Ti \\ |}}{\overset{4+}{}} \overset{O^{2-}}{\underset{O^{2-}}{}} \underset{\substack{| \\ Ti \\ |}}{\overset{4+}{}} + H_2$$

XBL 822-7816

FIG. 12. A simple reaction mechanism for the photodissociation
of water on strontium titanate surfaces [6].

Let us summarize these important findings that indicate the
complexity of this process:

(1) Irradiation by energy $h\nu \geq 3.1$ eV is needed to induce wa-
    ter photodissociation over $SrTiO_3$.
(2) Alkali hydroxides such as KOH catalyze the hydroxylation
    of the strontium titanate surface and facilitate water dis-
    sociation.
(3) Oxygen vacancies and the presence of $Ti^{3+}$ at the surface
    are critical in the dissociation.
(4) Transition metals such as Pt at the strontium titanate sur-
    face accelerate water photodissociation by catalyzing the
    formation of hydrogen molecules from atoms.

These observations can provide information on many of the elementary reaction steps, so one can propose a simple reaction mechanism for the photodissociation of water on strontium titanate surfaces. This mechanism is shown schematically in Fig. 12 [6]. In
the dark, strontium titanate consists of titanium atoms in the $Ti^{4+}$
oxidation state. Upon illumination, a photoelectron that is generated by bandgap or higher energy radiation reduces the $Ti^{4+}$ to
$Ti^{3+}$. The electron vacancy takes the charge from the OH$^-$ ion and
converts it into a peroxide molecule that splits up oxygen. The
surface is left in a reduced state that is ready to adsorb another
molecule of water which reoxidizes the surface to $Ti^{4+}$ and produces hydrogen in the process. Then, the reaction will repeat
itself in a catalytic manner. This mechanism is consistent with
all experimental information available on the photodissociation
process over strontium titanate at present.

The difficulty with strontium titanate for the photodissociation of water is the necessity for ultraviolet radiation. Since this process works poorly with solar radiation, hydrogen generation using $SrTiO_3$ is not an economical possibility. A continued search for materials that would carry out this reaction in the solar range led us to the study of iron oxide surfaces.

## III. THE n-TYPE IRON OXIDE/p-TYPE IRON OXIDE PHOTOCHEMICAL DIODE

Undoped iron oxide is an intrinsic n-type semiconductor with a bandgap of approximately 2.3 eV. Early work by Hackerman et al. [7] demonstrated some of the desirable properties of iron oxide for water photodissociation. When $Fe_2O_3$ is used against a Pt counterelectrode, photocurrents corresponding to oxygen production are generated for an applied bias of $V_a \geq 700$ mV, RHE. It has been found in our laboratory that doping with Si reduces the magnitude of the bias that must be employed and increases the magnitude of the anodic photocurrents. This result motivated a systematic study of various dopants in $Fe_2O_3$, which showed that the introduction of Mg could yield iron oxide electrodes with p-type behavior. In preparing these p-type samples we followed a well-defined procedure that included heating the mixed oxide powders to 1400°C and then rapidly quenching them in water. The resulting material was a highly heterogeneous Mg doped iron oxide, also containing phases of magnesium ferrate ($MgFe_2O_4$) and magnesium oxide (MgO).

Individual photocurrent versus voltage characteristics for the n-type (Si doped) and p-type (Mg doped) electrodes are shown in Fig. 13. For the n-type (Si doped) electrodes, anodic photocurrents due to oxygen evolution appear for $V_a \geq 400$ mV, RHE, while the p-type (Mg doped) electrodes yielded cathodic photocurrents corresponding to $H_2$ evolution of $V_a \leq 900$ mV, RHE. Individually, these electrodes require an external bias to sustain oxygen or hydrogen evolution under illumination when employed against a platinum counterelectrode. However, when connected in a short circuit configuration as shown in Fig. 14, the p/n assembly assumes an intermediate operating bias of approximately 750 mV, RHE, so that $O_2$ and $H_2$ production will occur simultaneously without external bias.

The iron-oxide-based assembly has a photoresponse more closely matched to the solar spectrum than $SrTiO_3$. Figure 15 shows the quantum efficiency (defined as the number of chemically active electron/hole pairs produced per incident photon)

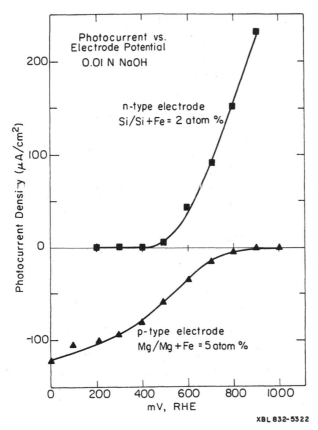

FIG. 13. Photocurrents for n-type (Si-doped) and p-type (Mg-doped) iron oxide electrodes under biased conditions.

plotted versus wavelength for three different iron oxide electrodes. For the undoped $Fe_2O_3$ electrode, the quantum efficiency was a maximum of 0.03 at a wavelength of 3800 Å. Doping with Si (Fig. 15, center) produced photocurrents about an order of magnitude larger than undoped $Fe_2O_3$, due to enhanced electronic mobility. Figure 15 (right) shows the effect of doping with Mg. In this case the photocurrents are cathodic, while features of the spectral response curves are very similar to both the undoped and Si doped electrodes.

By examining photocurrents in the near bandgap region, one can determine the bandgap. In Fig. 16 we have replotted the

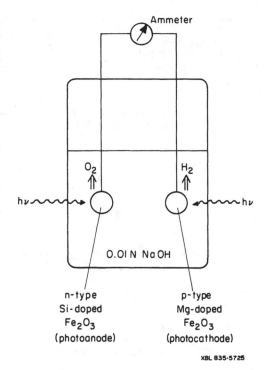

Ammeter

$O_2$

$H_2$

hν

hν

0.01 N NaOH

n-type
Si-doped
$Fe_2O_3$
(photoanode)

p-type
Mg-doped
$Fe_2O_3$
(photocathode)

XBL 835-5725

FIG. 14. Schematic illustration of p/n assembly for the photo-dissociation of water using doped iron oxide electrodes.

photocurrents from the Si doped and Mg doped electrodes as $(I_{ph}h\nu)^{1/2}$ versus photon energy $h\nu$. The linearity of these plots tells us that both n-type and p-type materials are indirect bandgap semiconductors with a bandgap of approximately 2.3 eV. These results indicate that the photoelectrochemically active component in both electrodes is most likely $Fe_2O_3$.

The locations of conduction and valence band edges with respect to redox couples in solution are of critical importance. To determine the location of these edges on the electrochemical scale, we performed Mott Schottky measurements in which the capacitance of the space charge layer was determined for the Si doped and Mg doped electrodes separately using a phase shift technique. In Fig. 17 we use the Mott Schottky relation and plot the inverse square space charge capacitance $(1/C_{sc}^2)$ versus applied potential to obtain the flatband potential from

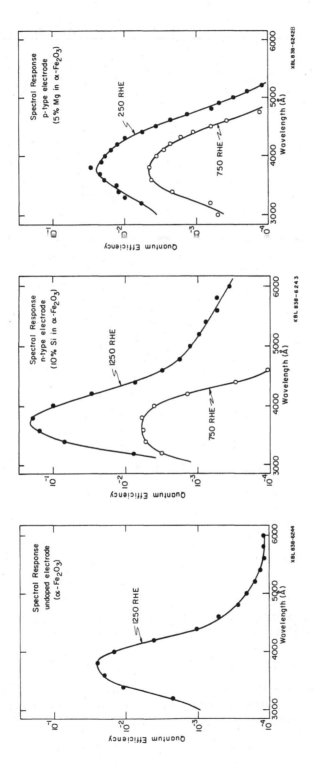

FIG. 15. Measured quantum efficiencies as a function of wavelength for doped and pure iron oxide.

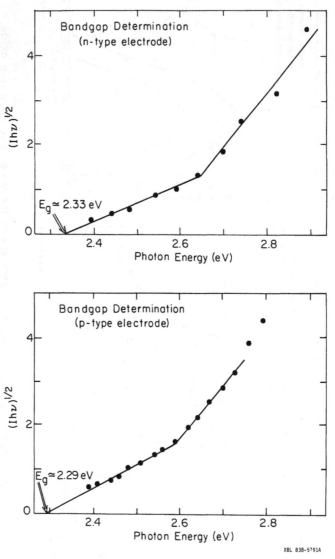

FIG. 16.   Bandgap determinations of doped iron oxide electrodes from photocurrent dependence on wavelength.

the intercept for the n-type and p-type iron oxides.  In addition, one can obtain carrier types and concentrations from the
slopes of these plots.

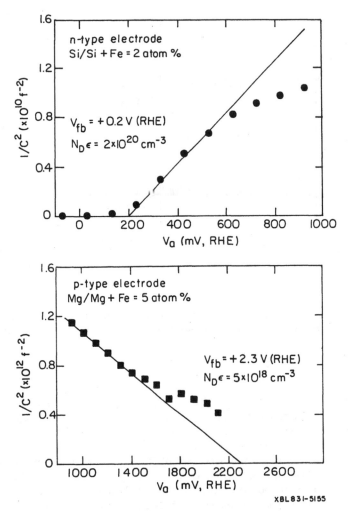

FIG. 17. Mott-Schottky plots of the inverse square of space charge capacitance versus applied potential for n-type and p-type iron oxide electrodes.

For heavily doped semiconductors such as these iron oxides, the flatband potential is nearly coincident with the band carrying the majority carriers. Thus for the n-type electrode, the measured flatband potential corresponds to the conduction band edge ($E_C$ = 200 mV, RHE); in the p-type the flatband potential gives

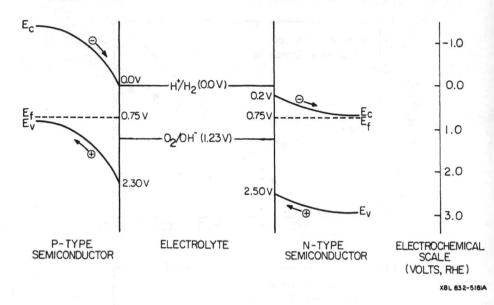

FIG. 18. Energy level diagram depicting band edge locations of doped iron oxide electrodes relative to hydrogen/oxygen redox couples.

the valence band edge ($E_V$ = 2300 mV, RHE). Information on flatband potentials can therefore be used to generate an energy level diagram as shown in Fig. 18. This shows the interfaces between the electrolyte and the n-type or p-type surfaces. We have determined band edges from Mott Schottky plots as described above and extrapolated using the measured bandgap (2.3 eV) to find the alternate band edges. The Fermi level was determined by connecting an operating assembly through a high impedance voltmeter to a reference electrode.

From this diagram, one can easily see how the p/n assembly operates. Upon illumination in solution, electron hole pairs are produced in both the p-type and n-type iron oxides. In the p-type electrode, electrons are driven through the depletion layer toward the interface where they convert hydronium ions in solution to hydrogen gas. Conversely, vacancies driven to the n-type interface convert hydroxyl species to oxygen gas. In both electrodes the majority carriers migrate away from the surface to produce the observed photocurrents. Thus hydrogen evolution occurs at the p-type electrode and oxygen evolution takes place at the n-type electrode.

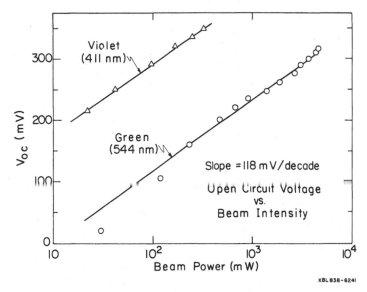

FIG. 19. Open circuit voltage generated between p-type and n-type electrodes immersed in solution as a function of illumination intensity at 411 and 544 nm.

Hydrogen generation over undoped $Fe_2O_3$ was shown to be energetically unfeasible. However, in this p/n assembly, $H_2$ evolution now occurs readily at the p-type electrode. It appears that doping the iron oxide with Mg has shifted the p-type band edges cathodically by several hundred millivolts so that hydrogen evolution is thermodynamically feasible.

To establish that the iron oxide assembly operates as a true p/n diode with a depletion layer at both interfaces, experiments were performed using a CW $Kr^+$ laser. Both p-type and n-type electrodes were illuminated with the laser at a wavelength of 411 or 544 nm. Figure 19 shows the open circuit voltage generated between the p-type and n-type electrodes in solution as a function of illumination intensity at both wavelengths. The most important result from this plot is that open circuit voltage increases 118 mV per decade increase in light intensity. For a single depletion layer, it is known that $V_{oc}$ will increase 59 mV for every tenfold increase in the intensity of illumination as the semiconductor is driven toward flatband conditions. Since we obtained twice this value in our experiments, it is clear that Schottky barriers exist at both the p-type and n-type electrode surfaces.

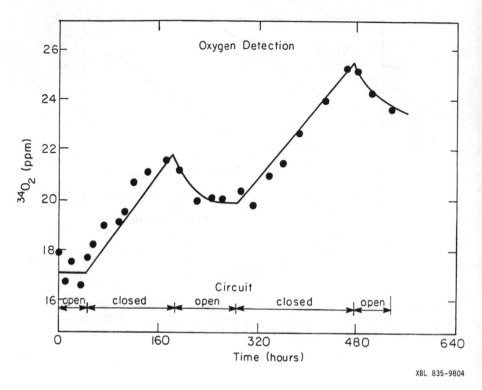

FIG. 20. Oxygen evolution from a p/n assembly with doped iron oxide electrodes.

In separate experiments, oxygen evolution has been observed from an unbiased p/n cell. An NaOH/H$_2$O solution containing isotopically labeled water (H$_2{}^{18}$O) was used as the electrolyte and the production of $^{34}$O$_2$ was monitored as a function of time with a mass spectrometer. Figure 20 clearly shows that oxygen was evolved when the p-type and n-type iron oxide electrodes were connected and illuminated in solution, but that no oxygen was produced when the electrodes were disconnected.

Another important consideration for the viability of this process for solar energy conversion is the long-term stability of these oxides in solution. Figure 21 shows the photocurrents generated by an illuminated p-type/n-type iron oxide assembly as a function of time after the first immersion in solution. Initially, photocurrents and the corresponding gas production rates were low. However, these photocurrents gradually increased

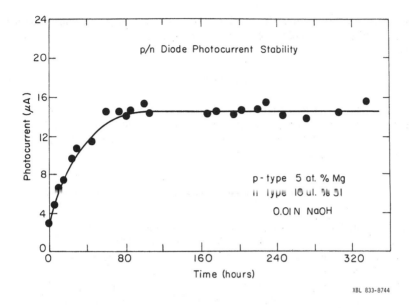

FIG. 21. Photocurrent as a function of time for a p/n assembly during the first immersion.

over an 80-h period to approximately 15 μA, where they remained for the duration of the test. The stability of the photocurrents as well as Auger analysis of the electrodes before and after testing indicate that there is no appreciable dissolution of the iron oxide electrodes.

At present, the efficiency of the iron oxide p/n assembly is low, approximately 0.1%. This efficiency is lower than strontium titanate under ultraviolet illumination, but the iron oxide assembly operates with visible light in the solar range. In addition, iron oxide is an excellent catalyst. Thus, once hydrogen is created, it might be reacted in situ with carbon dioxide or nitrogen to form organic molecules or ammonia, respectively.

It is interesting to note that the olefins which comprise over 50% of the mantle of the earth have chemical compositions very similar to the doped iron oxides we have prepared. These materials are primarily iron silicates and magnesium iron silicates, which in various compounds are known to display either n-type or p-type behavior. One can speculate that some of these minerals could form p/n diode configurations, leading to the formation of hydrogen and oxygen from water in the presence of sunlight. Subsequent chemical reactions involving hydrogen might have produced organic molecules in a thermodynamically downhill process.

## Acknowledgment

This work was supported by the Director, Office of Energy Research, Office of Basic Energy Sciences, Chemical Sciences Division of the U.S. Department of Energy under contract No. DE-AC03-76SF00098.

## REFERENCES

[1]  A. Fujishima and K. Honda, Bull. Chem. Soc. Jpn., 44, 1148 (1971); Nature (London), 238, 37 (1972).
[2]  F. T. Wagner and G. A. Somorjai, J. Am. Chem. Soc., 102, 5494 (1980).
[3]  H. Gerischer, in Solar Energy Conversion, Topics in Applied Physics, Vol. 31 (B. O. Seraphin, ed.), Springer, New York, 1979.
[4]  T. Kawai and T. Sakata, J. Chem. Soc., Chem. Commun., p. 694 (1980).
[5]  R. G. Carr and G. A. Somorjai, Nature (London), 290, 576 (1981).
[6]  H. Van Damme and W. K. Hall, J. Am. Chem. Soc., 101, 4373 (1979).
[7]  J. M. Wilhelm, K. S. Yun, L. W. Ballenger, and N. Hackerman, J. Electrochem. Soc., 126, 419 (1979).

# The Present Status of Industrial Production and Use of Photovoltaic and Photocatalytic Solar Energy Conversion Devices

F. V. WALD
Mobil Solar Energy Corporation
Waltham, Massachusetts

## I. INTRODUCTION

The recent revival of interest in the direct use of solar energy, which followed the 1973 oil crisis, has brought increased research and development activity in a number of areas. Some of these areas are discussed here, namely photocatalysis, photoelectrochemical cells, and photovoltaic devices. At the present time there is continuing research activity in photocatalysis for various purposes, including solar energy conversion, and there is also continuing research activity in photoelectrochemical methods. However, at the present time, photovoltaic devices are the only ones in limited terrestrial application for direct solar energy conversion into electricity.

407

FIG. 1. ARCO 1 MW utility connected power plant at Hesperia, California. (Courtesy of ARCO Solar, Inc., Chatsworth, California.)

Solar cells, which are the photovoltaic devices that convert so-lar energy directly into electricity, are specially constructed semi-conductor diodes. They are rugged, maintenance free, and highly modular in the milliwatt to the megawatt range, and they have a proven track record as reliable power producers for space vehicles and for remote terrestrial applications [1]. Recently, larger dem-onstration plants with several megawatts of electrical capacity have also been built in California in anticipation of using the devices for the production of utility power in favorable geographic regions (Fig. 1) [2]. Although at present the output from such plants cannot compete economically with other utility-generating equip-ment, it is expected that this situation might change, at least for certain utilities, as early as the 1990s if the pace of develop-ment in photovoltaics, which started in the 1970s, continues, and, more particularly for the United States, if present tax incentives for the use of renewable energy sources stay in place.

The technological push toward the reduction of the cost of solar cells has so far yielded a number of prospective new materials
and cell fabrication techniques which, when deployed on a large
scale, can drastically reduce the production costs for these devices.  So far, however, larger scale plants are still all based
on the use of single crystal silicon, mostly grown by the well-
known Czochralski technique and all commercially available photovoltaic devices at present are silicon based.  An increasing number of them, however, are now being fabricated on silicon sheet
manufactured by entirely new production methods which have
reached the pilot stage.

## II.  ADVANCED CRYSTALLINE SILICON
## SOLAR CELL TECHNOLOGY

In 1973, the U.S. National Science Foundation held a series of
workshops with experts from industry, government, and the academic world to lay the foundation for a national photovoltaic program [3].  At that time, single crystal silicon based solar cells
were already well established as reliable space power supplies,
and one of the workshop recommendations was that simplification and streamlining of the single crystal silicon solar panel
manufacturing sequences should have a high priority in any
photovoltaics program.  It was suggested then that such a
course could reduce photovoltaic array prices based on single
crystal silicon to perhaps $5/W in 1975 dollars.  In the past 10
years it has become clear that the photovoltaics industry has
clearly met that goal as the average selling price for photovoltaic panels world-wide was $7.50/W in 1983 [4].  And some
quotes for very large orders (megawatt scale) to be delivered
during 1983 and 1984 were made at about $5/W (in 1983 dollars)
recently.
    So far, the cost reductions which have been achieved have
been mainly due to the simplification of solar cell manufacturing
sequences used in space solar power supply fabrication.  These
in turn are based on processes derived from the general semiconductor industry and its methods of preparing devices from
single crystalline silicon material.  They are then further simplified to meet certain particular objectives of terrestrial photovoltaics without any essential change in the materials base.  The
prototype process used by the largest manufacturer at present
is described in Ref. 5, and it is stated that in the limit of its
evolution and at larger volumes (perhaps 5-10 MW/year), its
direct manufacturing cost might bottom-out around $3/W.  The
technical details on which these cost reductions are based and

the degree of automation of these processes, however, are very closely guarded secrets in an evermore competitive industry. Nevertheless, there are some glimpses on how manufacturing costs have reached the present low levels.

It is clear that the larger scale of operations has had an influence, as it has allowed more efficient manufacturing processes [6], particularly in terms of operator utilization. With the larger scale, however, came an incentive to develop methods which would allow more continuous processing of the single crystal base material, as well as reductions in the cost of materials involved in the process itself. Thus, it was quite uniformly realized that the space age multilayer vacuum metallization based on noble metals was an intolerable cost burden from the materials as well as the process standpoint. It was, therefore, very soon replaced by screen printing or plating methods which appeared to be more suitable for upscaling [5, 7, 8]. Several other developments, related for instance to junction isolation, edge etching, the diffusion process, and to antireflection coating by chemical vapor deposition, have also had an impact on costs. The latter two can now be carried out continuously in open furnaces on moving belts [6].

The general aim of such efforts is to reduce the number of processing steps, decrease the consumption of chemicals, and render the whole process fully automatable [9]. There appears to be widespread agreement on the general direction these efforts should take. When inspecting the literature, it also becomes quite clear that at this time most of the development programs in the industry are proceeding along accepted and non-controversial lines. However, as stated before, the actual process steps that each company is now using, the process capacities and yields, and the degree of automation, both in cell processing and panel, are all held strictly proprietary.

The ultimate industrial success of crystalline silicon-based photovoltaic systems very much depends on the development of simplified and fully automated cell processing and panel assembly plants, as about one-half to two-thirds of the total module cost is due to these two parts of the manufacturing process [6]. Thus, in every company active in the field, large resources are allocated to sometimes very sophisticated process and manufacturing development efforts in the cell and panel area.

However, it is thought by most that the true challenge in crystalline silicon photovoltaics lies in providing a new materials base of much lower cost, such that it can support photovoltaic panel technology at levels of less than $1/W [10], but there is yet no agreement on the materials base best suited to this purpose.

The generally recognized fundamental obstacles are (1) that the single crystal production technology requires very high purity starting silicon, (2) that single crystal preparation itself is quite costly, and (3) that the necessary slicing leads to great materials waste as well as high inherent process costs [11].

The first approach then was a suggestion of simply improving all the processes used; that is, to produce hyperpure semiconductor grade silicon by a better low-cost method, to significantly advance the state-of-the-art of single crystal growth to lower those costs, and to develop new and far more effective slicing techniques. At the same time, an inherent advantage in efficiency for the single crystal was also postulated [10]. It is now widely agreed that in total this approach is unlikely to lead to sufficiently low cost to enter the utility business in even the most favorable geographic areas, and thus there appears to be no large participant in the photovoltaic industry which is still pursuing this course as a long-term option.

In fact, as mentioned in Ref. 5, those most intimately involved with the large-scale single crystal-based production of solar cells believe that the manufacturing costs might bottom-out around $3/W. As that would seem to require selling prices around $4-$6/W, at least, for reasonable capital recovery, there is now a widespread consensus that new and lower cost base materials must be found if photovoltaics is ever to impact in a significant way the world's electric power picture. From these considerations have evolved a number of approaches to the silicon materials problem which try to improve on the weaknesses of the single crystal technology.

That first led to an approach which, simply put, is based on the reasoning that the most significant drawback of the single crystal technology lies in the cost of wasting highly pure and perfect single crystal material during slicing. It was shown, however, that much less pure and less perfect material also possesses reasonable properties in a solar cell [12, 13]. Since such silicon might be prepared quite inexpensively by casting, it was argued that the cost of the material lost during slicing would not be significant.

This view has acquired a large following in the industry at the present time, and several companies are preparing and evaluating cast silicon material of this general type. In fact, the original proponents of this view, Solarex, Inc., in the United States, and Wacker Chemitronic/Telefunken, in Germany, are already in larger pilot production, and are selling commercially panels based on their semicrystalline materials [6, 14, 15]. It can also now be shown that indeed the original predictions are correct; i.e., cells based on such materials can possess solar

conversion efficiencies very close to single crystal-based solar
cells when one considers only larger production runs and not the
properties of single laboratory-prepared cells. Solarex has recently reported efficiencies for larger batches of SEMIX cells of
100 cm$^2$ area to be between 10 and 12.5% [16]. Telefunken reported average efficiencies of 9-11% from a solar cell production
line based on Wacker's SILSO [17]. Japan Solar Energy Corporation, using a low-cost screen printing process to prepare the solar cells, reports efficiencies between 11 and 12% for cells based
on commercially available polycrystalline material [18].

However, considering the total literature on the subject, it appears to this reviewer that the reproducibility of such results is
not as good as that achieved with equivalent batches of single
crystal-based cells. It is thus likely that panels based on these
materials will continue to be somewhat lower in efficiency when
compared to single crystal-based panels of a comparable packing
density of cells in a panel. It is thus a question for the future
whether the larger scatter in efficiency in these imperfect cells
as compared to single crystal cells is inherent in the nature of
the material or is a result of the smaller experience base with
the polycrystalline material at present.

In addition, in the developmental area, although it largely has
been demonstrated for the polycrystalline materials that the low
labor and capital costs, which were shown to be necessary for
the achievement of truly low manufacturing costs, can be achieved,
there is still a serious challenge which lies in reducing the costs
of the consumables; i.e., the crucibles and heaters, during the
casting processes and also reducing the costs of the slicing, which
includes a reduction in the cost of the slicing materials itself, such
as abrasives, blades, or wires.

Yet a different approach to the provision of a low-cost starting
silicon for solar cell fabrication is based on the preparation of a
sheet or ribbon of silicon which obtains the required thickness
of 100-200 μm directly during crystal preparation [3]. The advantages inherent in such an approach are obvious as any waste
of silicon is totally avoided. Also, early cost calculations readily
demonstrated that reasonable sets of operating parameters existed
for these methods from which to predict that very low costs could
be achieved [19]. A significant research and development into
shaping such ribbons then ensued, much of that still ongoing
[11, 20]. However, at the present time, only Mobil Solar Energy
Corporation is in full pilot production with its EFG method and
sells panels based on the pilot production output. An up-to-date
summary of that effort may be found in Ref. 21.

Since in the EFG method a carbon capillary is used, it was
widely held that the ensuing high carbon concentration would be

detrimental to achieving high efficiency solar cells from the material. In addition, it is still believed that low-cost, less pure
starting material cannot be effectively used in such a method [14].
Nevertheless, production efficiency in large batches of 10-13% is
common for this material now [22], and it may be demonstrated
that the method can be quite impurity tolerant, as large concentrations of iron, aluminum, and molybdenum can be introduced
along with the inevitable carbon and oxygen concentrations while
cell efficiencies remain at quite acceptable levels.

Thus, the pilot experience with this material rather parallels
that of the cast polycrystalline silicon discussed before, and the
larger variations in efficiency from batch to batch along with the
higher sensitivity to solar cell processing conditions are also found
here [23, 24].

In sum, then, the solid-state properties which determine the
solar cell efficiency available from various forms of defective crystalline silicon do not seem to be significantly different. The top
efficiencies achieved with these materials do not seem significantly
lower than those possible with single crystal silicon. However,
there is more variability in these materials, and solar cell processing conditions exert a large influence on the final efficiency
achieved. To understand and control the latter, while also generally increasing current efficiency levels, is a challenging goal
for the materials researchers in the industry.

As mentioned before, in the developmental area it has largely
been demonstrated during pilot operations with various polycrystalline silicon-based solar cell devices that the low labor and capital costs which were shown to be necessary for the achievement
of truly low manufacturing costs can likely be achieved. The
technological challenge at present would seem to lie in reducing
the cost of consumables; that is, crucibles and heaters, and in
the case of the cast ingot methods also including abrasive material, slicing blades or wires, and the like.

In addition, at the present time, these methods are all used on
a scale of production which is significantly smaller than that for
the single crystal solar cell. Therefore, it remains to be demonstrated whether on a large scale all these processes can be operated at the high yields which are absolutely necessary to obtain the extremely low manufacturing costs postulated.


## III. THIN FILM SOLAR CELLS

Aside from the early, selenium-based photocells, the first "semiconductor age" thin film cell was that composed of copper-sulfide
and cadmium sulfide. It was discovered almost contemporaneously

to the first power producing silicon cells in the middle 1950s.  It
quickly became apparent that its manufacture on a large scale
might be significantly cheaper than that of silicon cells, since
CdS could be evaporated continuously and the $Cu_2S$ formed by
a simple chemical dip into $CuCl_2$ solution.  The economics and
some earlier results which pre-date the period before 1973, when
serious cost reduction efforts for terrestrial solar cells started,
are presented in a 1974 review [25].  The development of the
CdS cell since then is, however, somewhat symptomatic of the
difficulties which stand in the way of making low-cost thin film
cells a reality.

In order of importance the most serious problems discovered
were the following:

a.  Cell instabilities; i.e., significant reductions in efficiency
    over time occurred, with or without light impinging con-
    tinuously on the cell.
b.  Unexplained efficiency shortfalls.  Often only <25% of
    theoretical predictions were achieved vs an accepted 50%+
    for crystalline silicon devices.
c.  Routine, continuous manufacture was not demonstrated.
d.  Materials savings were not realized; i.e., truly thin films
    (of the order of 5 μm or so) were shown not to perform
    properly, their thickness could not be optimized, and be-
    cause selective deposition on the substrate only was not
    possible, large losses of material occurred during evap-
    oration.
e.  Objections against the toxicity of some of the basic materi-
    als or questions about their availability were raised.  Both
    of these, it has been suggested, might preclude truly large-
    scale deployment.

There is a still-ongoing significant research and development
effort to address these shortcomings without compromising the
important aspects of the low-cost manufacturing process, and an
up-to-date summary of that is presented in Ref. 26.

The most significant development which occurred in this effort
was the discovery that the replacement of the $Cu_2S$ with $CuInSe_2$
could successfully address objections (a) and (b) [27].  However,
the original low-cost process for the $Cu_2S$-CdS cell could not be
maintained, and the Boeing Corporation is now involved in a development program to determine whether a low-cost manufacturing
process can be evolved.  Also there seems to be enough confidence
in the basic $Cu_2S$-CdS cell that recently pilot production efforts on
an improved version have in fact started again in Germany [28].

However, in the United States all pilot operations on this system
are now shut down.

A number of alternative thin film cells, based on semiconduc-
tive compounds with a gap near the optimum value of 1.5 eV, have
also been investigated. The most successful of these, when not
considering concentrator applications, seem to be those based on
CdTe [29]. At least one company has announced that it is now
beginning pilot production with that system [30].

The latest and most exciting entry in the thin film arena undoubtedly is the cell based on hydrogenated amorphous silicon,
introduced by Carlson and Wronski [31] in 1976. This material
differs from crystalline silicon because momentum conservation
selection rules do not apply for disordered materials. Thus, a
"pseudo-direct" energy gap arises [32] and a high absorption
coefficient for light is thus obtained. That in turn reduces the
basic materials requirement to a few microns of thickness, which
can readily be deposited by continuous "thin film" processes.

Although the scientific basis for the properties of such a substance is rather unexplored, practical progress with solar cells
based on it has been quite rapid, particularly in Japan [33],
and until recently it seemed that this material would answer
all objections ((a)-(e)) in the above list. However, it has now
been discovered that amorphous silicon-based devices do de-
grade in operation [34, 35]. That puts into sharper focus the
lack of basic information on the solid-state properties of such a
material, as it seems difficult to fundamentally rectify a degradation problem when its basic nature is unclear.

However, recently a better phenomenological understanding
of how the degradation affects the device has been achieved.
The model suggests that unequal rates of extraction of holes
and electrons from the active high field region of the p-i-n
junctions lead to polarization and eventual field collapse [36].
It is proposed, therefore, that the problem may be ameliorated
by impinging the light such that the slower carrier is more effectively extracted from the front of the device. Also, graded
boron doping of the high field layers of these devices has recently been suggested to improve the overall carrier collection
efficiency [37]. It appears that this, as it increases the collection field across the active layer of the devices, might also be
found to counteract the degradation.

Nevertheless, for these devices to reach the projected very
low manufacturing cost levels, it will also be necessary to utilize
more effectively the high cost capital equipment involved in the
deposition processes. This requires that the deposition rates
for the amorphous silicon layers are significantly increased

without any attendant degradation in electrical properties. But
very little is known about the connections between deposition
method and rate on one hand, and the electrical as well as optical properties of the material produced on the other [38].

For the time being, therefore, the selling price for amorphous
silicon devices, even those produced in continuous production
lines, is not projected to fall significantly below that for crystalline silicon devices.

The discovery that disordered materials can become the basis
for thin film solar cells has a far-reaching importance and may
indeed provide a very, very, low-cost, long-range future for
the photovoltaics field. It is thus no wonder that research in
this area is quite intense. It is also not surprising that a number of companies are trying to obtain a significant early foothold
in the field by actually establishing larger pilot operations, the
product of which is sold into suitable, but as yet mainly low-
power applications [39, 40], such as hand-held calculators,
watches, and other consumer products.

## IV. APPLICATIONS

The advantages of photovoltaic devices in field applications
may be succinctly stated as three:

(a) Totally modular.
(b) Largely maintenance free and without fuel requirement.
(c) Friendly to the environment.

The high modularity of the devices directly results in their
preferred application to consumer products which are based on
solid-state semiconductor devices and which therefore have extremely low power requirements, such as hand-held calculators,
solar-powered watches, toys, and the like, in which the photovoltaic cell either replaces, or much improves the life of, the expensive batteries which are otherwise required. At the moment
this area constitutes a major market for amorphous silicon devices which can be readily manufactured in sizes particularly
suitable for these kinds of applications.

However, after the 1955 publication of the Bell Telephone Laboratories report that described silicon solar cells with efficiencies
of 6%, the ink was hardly dry when engineers discovered that the
solar cell's high degree of modularity and its reliable maintenance-
free as well as fuel-free nature made it the ideal power source for
remote applications in communications. Soon, solar cell power

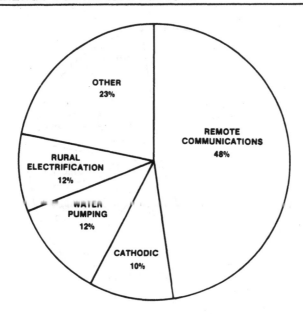

FIG. 2. Distribution of world markets for photovoltaics, 1981. (Courtesy of R. Hammond, Mobil Solar Energy Corporation, Waltham, Massachusetts, and Strategies Unlimited, Mountain View, California.)

supplies were installed experimentally on a rural telephone line in Americus, Georgia [1]. Since then, the largest market for photovoltaics has been in remote communications (Fig. 2, Ref. 41), and this is expected to continue well into the future.

Other markets have already developed and are expected to increase as the price of photovoltaic systems continues its downward trend (Fig. 3). In fact, current costs of solar cell systems are such that they can compete with diesel engines producing up to 3 kW of continuous power (Fig. 4, Ref. 41).

Future solar cell systems, therefore, may be particularly attractive for deployment on a large scale in developing countries, where photovoltaic panels are coming into increasing use for a variety of village power needs [42]. In many cases, such as water pumping for drinking and irrigation, the question here is not so much that the photovoltaic power source is in competition with another power source. Oftentimes, there is no other power source, and the electricity provided is "literally priceless," as it has a life-sustaining impact.

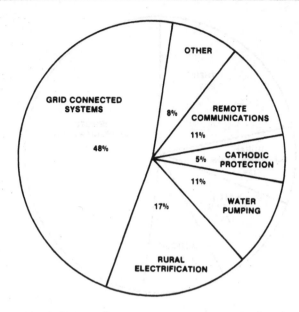

FIG. 3. Projected distribution of world markets for photovoltaics in the year 2000. (Courtesy of R. Hammond, Mobil Solar
Energy Corporation, Waltham, Massachusetts, and Strategies Unlimited, Mountain View, California.)

However, the economic problem still exists, because the poorer
countries which are most in need of these devices cannot finance
the acquisition of large numbers of them unless prices for such
systems are drastically reduced.

It is also quite essential that photovoltaic panels be extremely
reliable in these applications. There are furthermore infrastruc-
tural problems, both during the installation of the panel and after-
wards, when they relate more to the working systems, such as
pumps, which do have to be maintained either on the site or by
a regular replacement schedule. It is apparent, then, that systems design for such applications should emphasize simplicity and
ruggedness and the suppliers of these systems must be willing to
provide strong educational, training, and systems support.

It is also obvious, then, to think of extending such applications further toward higher power levels as the highly modular
nature of photovoltaic devices does make possible a gradual build-
up of power generating capability simply by adding units, almost
watt by watt. Also, the older units do not become obsolete when

Basis: $1.50/Gal Delivered Diesel Fuel Cost
>5 km from Grid

| | $/kWhr | | |
| | Diesel/ | PV[1] | |
| Case | Electric | $8/W | $15/W |
| 1 kW PV/.3 kW Diesel | $5.50 | $0.90 | $1.50 |
| 3 kW PV/1 kW Diesel | $1.80 | $0.90 | $1.50 |

[1]Diesel and PV costs/kWhr include all balance of system costs

FIG. 4. Cost comparison between electricity production costs using small diesel engines and photovoltaic systems. (Courtesy of R. Hammond, Mobil Solar Energy Corporation, Waltham, Massachusetts, and Strategies Unlimited, Mountain View, California.)

newer ones are installed next to them, clearly an economic plus for a user who wants to build up more capacity to satisfy larger and more varied demands for electric power. In the limit, one could thus imagine an eventual solar electric grid to be built up from a number of small, distributed power plants spread over many villages. In this way, photovoltaics could have a very significant impact on developing countries. It could make small amounts of power available practically anywhere, which would allow these countries to circumvent to a large extent the prohibitively expensive process of investing in very large conventional power plants and then extending an electric grid to all areas.

To cite some examples of these current markets in more remote areas, Fig. 5 shows a cathodic protection system now operating on the Arabian Peninsula, and Fig. 6, which is a good example for the type of increasing application which we have just discussed, shows the power supply on a remote Navajo Indian dwelling in Arizona. Only a little further out are desalination systems in which the photovoltaic device drives high-pressure pumps in a reverse osmosis unit to provide drinking water. Mobil Solar pioneered such systems with its first 8 kW installation at Jeddah, Saudi Arabia, next to the Red Sea in 1981 [43], Fig. 7. A number of systems of this type are now being installed around the world by Mobil Solar and several other suppliers.

FIG. 5. Abu Dhabi, U.A.E.: A 12-V, 400-W (peak) photovol-
taic-powered cathodic protection system, supplying an impressed
current to ARAMCO's Well Head #32. (Courtesy of Mobil Solar
Energy Corporation, Waltham, Massachusetts.)

At the present time there are also in operation in the United
States a number of systems in which the photovoltaic power supply is located on the roof of a dwelling [44]. Many consider this
an excellent solution to expanding electrical capacity. The power
is generated where it is used, and transmission losses or extra
land use are totally avoided. The problem, however, is to decide what to do when there is no sunshine. One house (Fig. 8)
which Mobil Solar recently equipped with a photovoltaic roof
avoids the problem by hooking up to the local electric utility.
Excess power can be sold through the utility when there is
good sunshine and needed power can be bought at night or on
rainy days. However, it is clear that there are certain limitations on overall penetration into an electric network by such distributed power plants, which fundamentally aim at using the
overall grid as a storage/standby system.
    One can then consider either central photovoltaic power plants,
perhaps of the general type which is presented in Fig. 1, which
are integrated into a large electrical grid in which several conventional and unconventional means of generating power are
inter-tied [2]. Otherwise, for residential systems one must
consider standby electricity and/or storage on site. Such

FIG. 6. New Mexico: A photovoltaic array, rated at 12 V, 320 W (peak), supplying electricity to the home of a Navajo Indian family. (Courtesy of Mobil Solar Energy Corporation, Waltham, Massachusetts.)

FIG. 7. Reverse osmosis desalination unit with an 8-kW photovoltaic power supply installed at Jeddah, Saudi Arabia. (Courtesy of Mobil Solar Energy Corporation, Massachusetts.)

FIG. 8. Four-kilowatt photovoltaic power supply installed on a privately owned total solar house in southeastern Massachusetts. This is a utility inter-tied system. (Courtesy of Mobil Solar Energy Corporation, Waltham, Massachusetts.)

systems also have already been built. One of them is shown in Fig. 9. It shows a commercial building on Block Island which is equipped with battery storage. At this site also, the competing local utility sells power at 30¢/kWh, along with significant connection charges for new customers, which makes photovoltaics nearly economical, even at current solar panel prices.

## V. CONCLUSIONS AND OUTLOOK

The current embryonic photovoltaic industry is entirely based on silicon with the largest proportion of the capacity still resting on single crystal silicon-based panels.

However, increasingly, panels based on newer forms of crystalline silicon, such as cast material and sheet directly grown from the melt, are becoming available for commercial sale. In various field tests, panels based on these materials have performed well.

FIG. 9. Commercial building on Block Island, Rhode Island. This 1 kW system is not utility connected. Battery storage is used. The building houses a small custom printing shop and a dealership of solar devices. Both are supplied with electricity exclusively from the photovoltaic system. (Courtesy of Mobil Solar Energy Corporation, Waltham, Massachusetts.)

In the processing of crystalline silicon material into solar cells, much progress has been made in terms of streamlining the original processes, automating them, and simplifying many of the steps, most notably those involved with the application of the metallic contacts and the antireflection coating. In the preparation of the contacts, screen printing and plating methods are almost exclusively in use, and all antireflection coating today is being done by simplified chemical vapor deposition processes.

In addition, one can observe that devices and systems which have now been for several years in field tests around the globe have performed well, and it may be concluded at this time that crystalline silicon-based systems fulfill all the technological requirements which are necessary to establish them in a variety of applications. Also, the costs have been reduced to a level of approximately $5-$8/W, which has increased the range of remote power applications considerably.

Nevertheless, much further progress toward the goal of producing these panels at costs compatible with selling prices of around \$1-\$2/W must be made, and such progress will undoubtedly still require significant fundamental advances in both the materials base as well as in automated and simplified fabrication techniques.

These cost levels are necessary to introduce photovoltaic devices into power applications where they can replace more conventional means of electric generation in favorable geographic areas, in particular at this time, oil-based generation systems.

It has widely been hoped that the introduction of thin film cells would allow more rapid progress toward low-cost objectives. At the present time that seems not to be the case.

However, there is increasing interest in the use of thin film cells based on hydrogenated amorphous silicon. Even though the scientific basis for the electrical transport in this class of substances is very poorly explored, much progress has been made with devices based on such a material, to the extent that at the present time it seems clear that the most generally useful thin film system of the future is likely to be based on amorphous silicon.

The question, however, of the competition between amorphous silicon and various advanced forms of crystalline silicon, particularly those directly produced in sheet form, must still be considered open at this time.


## REFERENCES

[1]  G. Raisbeck, "The Solar Battery," Sci. Am., p. 102-110 (December 1955).

[2]  J. Johnson, Photovoltaics Int., pp. 6-7, 30 (April/May 1983).

[3]  U.S. National Science Foundation, Document #NSF-RA-N-74-013, Photovoltaic Conversion of Solar Energy for Terrestrial Applications, Available from National Technical Information Service, U.S. Department of Commerce, Springfield, Virginia, 22151.

[4]  R. Hammond, Mobil Solar Energy Corporation, Private Communication.

[5]  V. K. Kapur, J. E. Avery, and C. F. Gay, in Materials and New Processing Technologies for Photovoltaics 1981 (J. A. Amick, ed.), The Electrochemical Society, Pennington, New Jersey, pp. 140-146.

[6]  J. Lindmayer, in Third European Photovoltaic Solar Energy Conference (W. Palz, ed.), Reidel, Dordrecht, 1981, pp. 178-185.

[7]   M. G. Coleman, R. A. Pryor, and T. G. Sparks, in 14th
      IEEE Photovoltaic Specialists Conference, Institute of Electri-
      cal and Electronics Engineers, New York, 1980, pp. 793-799.
[8]   R. C. Peterson and A. Muleo, in Ref. 6, pp. 684-690.
[9]   J. Ponon, H. Lanonay, P. Aubril, G. David, and P. Loubly,
      in 4th European Photovoltaic Solar Energy Conference
      (W. Bloss, ed.), Reidel, Dordrecht, 1982, pp. 399-403.
[10]  L. M. Magid, in Ref. 6, p. 160.
[11]  P. D. Maycock and E. D. Stirewalt, Photovoltaics, Brick
      House Publishing, Andover, Massachusetts, 1981.
[12]  J. Lindmayer in 12th IEEE Photovoltaic Specialists Confer-
      ence, Institute of Electrical and Electronics Engineers,
      New York, 1976, pp. 82-85.
[13]  H. Fischer and W. Pschunder, IEEE Trans. Electron De-
      vices, ED24, 438-441 (1977).
[14]  J. Dietl, D. Helmreich, and E. Sirtl, in Crystals, Growth
      Properties and Applications, Vol. 5 (J. Grabmaier, ed.),
      Springer Verlag, Berlin, 1981, pp. 43-108.
[15]  D. Helmreich, J. Phys. (Paris), Colloq. C1, Suppl. Vol.
      43-10, C1-289 to C1-305 (1982).
[16]  Z. Putney, T. Rosenfield, and C. Wrigley, in Ref. 9, pp.
      990-993.
[17]  K. Roy and W. Pschunder, in Ref. 6, pp. 263-269.
[18]  Japan Solar Energy Corporation, Kyoto, Japan, Private
      Communication.
[19]  K. V. Ravi and A. I. Mlavsky, in Heliotechnique and De-
      velopment, Vol. 1, Development Analysis Associates,
      Cambridge, Massachusetts, 1976, pp. 563-587.
[20]  T. F. Ciszek, in Ref. 5, pp. 70-88.
[21]  F. V. Wald, in Ref. 14, pp. 147-198.
[22]  F. V. Wald and K. V. Ravi, Toward Industrial Production
      of Silicon Sheet for Solar Cells, Presented at Fifth European
      Photovoltaic Specialists Conference, Athens, October 1983,
      To Be Published in 1984.
[23]  B. Mackintosh, J. P. Kalejs, C. T. Ho, and F. V. Wald,
      in Ref. 6, pp. 553-557.
[24]  K. V. Ravi, R. Gonsiorawski, A. R. Chaudhuri, C. V.
      H. N. Rao, J. I. Hanoka, B. R. Bathey, and C. T. Ho,
      in 15th IEEE Photovoltaic Specialists Conference, Institute
      of Electrical and Electronics Engineers, New York, 1981,
      pp. 928-933.
[25]  R. J. Mytton, Sol. Energy, 16, 33-44 (1974).
[26]  K. W. Boer, J. Cryst. Growth, 59, 111-120 (1982).
[27]  R. Michelson and W. S. Chen, in Ref. 24, pp. 800-804.
[28]  H. Huschka, B. Schurich, and J. Wörner, in Ref. 9, pp.
      399-403.

[29]  M. Rodot, Rev. Phys. Appl., 12, 411-416 (1976).

[30]  Ametek Inc., Announcement at the U.S. Solar Energy Indus-
      tries Association Meeting, Atlanta, Georgia, November 8,
      1982.

[31]  D. E. Carlson and C. R. Wronski, Appl. Phys. Lett., 28,
      671-673 (1976).

[32]  E. A. Davis, Endeavour, 30, 55-61 (1971).

[33]  Y. Hamakawa, Sol. Energy Mater., 8, 101-121 (1982).

[34]  D. E. Carlson, Ibid., 8, 129-140 (1982).

[35]  J. I. Pankove, Ibid., 8, 141-151 (1982).

[36]  G. Mück, M. Simon, G. Müller, and G. Winterling, Fifth
      European Photovoltaic Solar Energy Conference, Athens,
      Greece, October 1983, To Be Published 1984.

[37]  P. Sichanugrist, M. Kumada, M. Konagai, K. Takahashi,
      and K. Komori, J. Appl. Phys., 54, 6705-6707 (1983).

[38]  A. Matsuda, T. Kaga, H. Tanaka, L. Malhotra, and K.
      Tanaka, Jpn. Appl. Phys., 22, L115-L117 (1983).

[39]  Y. Kuwano, M. Ohnishi, T. Fukatsu and S. Tsuda, in
      Materials and New Processing Technologies for Photovoltaics
      1982 (J. P. Dismukes, E. Sirtl, P. Rai-Choudhury, and
      L. P. Hunt, eds.), The Electrochemical Society, Penning-
      ton, New Jersey, pp. 89-105.

[40]  Y. Uchida, H. Sakai, N. Furushu, M. Nishiura, and H.
      Haruki, J. Electrochem. Soc., 130, 712-716 (1983).

[41]  Courtesy of Strategies Unlimited, Mountain View, Califor-
      nia.

[42]  See, for instance, Ref. 9, pp. 30-151.

[43]  J. Crutcher, Mod. Power Syst., 2, 28-32 (1982).

[44]  S. Strong, Photovoltaics Int., pp. 8-11 (August/September
      1983).

# Index